Optimization of Trustworthy Biomolecular Quantitative Analysis Using Cyber-Physical Microfluidic Platforms

Optimization of Trustworthy Biomolecular Quantitative Analysis Using Cyber-Physical Microfluidic Platforms

Mohamed Ibrahim
Krishnendu Chakrabarty

CRC Press
Taylor & Francis Group
Boca Raton London New York

CRC Press is an imprint of the
Taylor & Francis Group, an **informa** business

First edition published 2020
by CRC Press
6000 Broken Sound Parkway NW, Suite 300, Boca Raton, FL 33487-2742

and by CRC Press
2 Park Square, Milton Park, Abingdon, Oxon, OX14 4RN

© 2020 Taylor & Francis Group, LLC

CRC Press is an imprint of Taylor & Francis Group, LLC

Library of Congress Control Number: 2020938233

ISBN: 978-0-367-22352-6 (hbk)
ISBN: 978-0-367-51291-0 (pbk)
ISBN: 978-1-003-05318-7 (ebk)

Visit the Taylor & Francis Web site at
`ttp://www.taylorandfrancis.com

` the CRC Press Web site at
`/`/www.crcpress.com

*This book is dedicated to the memory
of my father, Saber R. Ibrahim.
— Mohamed S. Ibrahim*

Contents

Foreword   xi

Preface   xiii

1 Introduction   **1**
    1.1   Overview of Digital Microfluidics   4
    1.2   Overview of Continuous-Flow Microfluidics   10
    1.3   Design Automation and Optimization of
          Microfluidic Biochips   12
    1.4   Cyber-Physical Adaptation for Quantitative Analysis   15
    1.5   Security Assessment of Biomolecular Quantitative Analysis .   17
    1.6   Proposed Research Methodology   18
    1.7   Book Outline   26

I   Real-Time Execution of Multi-Sample
Biomolecular Analysis   **31**

2 Synthesis for Multiple Sample Pathways: Gene-Expression
Analysis   **33**
    2.1   Benchtop Protocol for Gene-Expression Analysis   34
    2.2   Digital Microfluidics for Gene-Expression Analysis   39
    2.3   Spatial Reconfiguration   43
    2.4   Shared-Resource Allocation   46
    2.5   Firmware for Quantitative Analysis   49
    2.6   Simulation Results   51
    2.7   Chapter Summary   57

3 Synthesis of Protocols with Temporal Constraints: Epigenetic
Analysis   **59**
    3.1   Miniaturization of Epigenetic-Regulation Analysis   60
    3.2   System Model   64
    3.3   Task Assignment and Scheduling   68
    3.4   Simulation Results and Experimental Demonstration   73
    3.5   Chapter Summary   79

4 **A Microfluidics-Driven Cloud Service: Genomic Association Studies**   **81**
  4.1 Background   82
  4.2 Biological Pathway of Gene Expression and Omic Data . . .   84
  4.3 Case Study: Integrative Multi-Omic Investigation of Breast Cancer   86
  4.4 The Proposed Framework: BioCyBig   90
  4.5 BioCyBig Application Stack   94
  4.6 Design of Microfluidics for Genomic Association Studies . . .   100
  4.7 Distributed-System Interfacing and Integration   105
  4.8 Chapter Summary   107

II High-Throughput Single-Cell Analysis   **109**

5 **Synthesis of Protocols with Indexed Samples: Single-Cell Analysis**   **111**
  5.1 Hybrid Platform and Single-Cell Analysis   113
  5.2 Mapping to Algorithmic Models   120
  5.3 Co-Synthesis Methodology   123
  5.4 Valve-Based Synthesizer   126
  5.5 Protocol Modeling Using Markov Chains   130
  5.6 Simulation Results   135
  5.7 Chapter Summary   141

6 **Timing-Driven Synthesis with Pin Constraints: Single-Cell Screening**   **143**
  6.1 Preliminaries   145
  6.2 Multiplexed Control and Delay   150
  6.3 Sortex: Synthesis Solution   160
  6.4 Experimental Results   168
  6.5 Chapter Summary   176

III Parameter-Space Exploration and Error Recovery   **177**

7 **Synthesis for Parameter-Space Exploration: Synthetic Biocircuits**   **179**
  7.1 Background   180
  7.2 PSE Based on MEDA Biochips   183
  7.3 Sampling of Concentration Factor Space   185
  7.4 Synthesis Methodology   188
  7.5 High-Level Synthesis   191
  7.6 Physical-Level Synthesis   193
  7.7 Simulation Results   202
  7.8 Chapter Summary   207

8 Fault-Tolerant Realization of Biomolecular Assays **209**

  8.1 Physical Defects and Prior Error-Recovery Solutions 209

  8.2 Adaptation of the $\mathcal{C}^5$ Architecture to Error Recovery 211

  8.3 System Design 213

  8.4 Dictionary-Based Error Recovery 216

  8.5 Experiment Results and Demonstration 219

  8.6 Chapter Summary 221

IV Security Vulnerabilities and Countermeasures **223**

9 Security Vulnerabilities of Quantitative-Analysis Frameworks **225**

  9.1 Threats Assessment of DMFBs 226

  9.2 Manipulation Attacks on Glucose-Test Results 230

  9.3 Attacks in the Presence of Cyber-Physical Integration 238

  9.4 DNA-Forgery Attacks on DNA Preparation 240

  9.5 Chapter Summary 248

10 Security Countermeasures of Quantitative-Analysis Frameworks **249**

  10.1 Microfluidic Encryption 250

  10.2 Aging Reinforces DMFB Security 256

  10.3 Encryption Security Analysis and Simulation Results 257

  10.4 DNA Barcoding as a Biochemical-Level Defense Mechanism 262

  10.5 Benchtop Demonstration of DNA Barcoding 265

  10.6 Chapter Summary 273

11 Conclusion and Future Outlook **275**

  11.1 Book Summary 275

  11.2 Future Research Directions 278

Appendix A Proof of Theorem 5.1: A Fully Connected Routing Crossbar **283**

Appendix B Modeling a Fully Connected Routing Crossbar **287**

Appendix C Proof of Lemma 6.1: Derivation of Control Delay Vector Ψ **291**

Appendix D Proof of Theorem 6.1: Derivation of Control Latency α_i **297**

Appendix E Proof of Lemma 7.1: Properties
of Aliquot-Generation Trees   **303**
   E.1 Overlapping-Subproblems Property   304
   E.2 Optimal-Substructure Property   305

Appendix F Proof of Theorem 7.1: Recursion in
Aliquot-Generation Trees   **307**

Bibliography   **313**

Index   **341**

Foreword

Lab-on-a-chip systems (LoCs), often referred to as microfluidic biochips, are revolutionizing many biochemical analysis procedures such as clinical diagnostics and DNA sequencing. Biochips offer the advantages of miniaturization, simple instrumentation, dynamic reconfigurability, and ease of integration with other technologies. The rapid evolution of biochip technologies opens up opportunities for new biological or chemical science that can be directly facilitated by microfluidics control. For example, automated LoCs have been commercialized to enable automated syndromic testing that can be performed at the patient's bedside.

Despite advances in research, the adoption of biochips in molecular biology has been slow. Recent studies suggest that state-of-the-art design techniques of microfluidics have two major drawbacks: (1) current LoCs were only optimized as auxiliary components and are only suitable for sample-limited analyses; therefore, they cannot cope with the requirements of contemporary molecular biology applications; (2) the integrity of these automated LoCs and their biochemical operations is still an open question, since no protection schemes were developed against adversarial contamination or result-manipulation attacks.

To address these challenges, this book presents a new design flow that is based on the realistic modeling of contemporary molecular biology protocols. It also presents a microfluidic security solution that provides a high level of confidence in the integrity of such protocols. The results presented in this book open up new research directions by bridging the gap between microfluidic systems and molecular biology protocols, and enabling a new level of synergy between molecular biology, computing, and electrical engineering.

Preface

Considerable effort has been devoted in recent years to the design and implementation of microfluidic platforms for biomolecular quantitative analysis. However, today's platforms suffer from two major limitations: (1) they were optimized for sample-limited analyses, thus they are inadequate for practical quantitative analysis and the processing of multiple samples through independent pathways; (2) the integrity of these platforms and their biochemical operations is still an open question, since no protection schemes were developed against adversarial contamination or result-manipulation risks.

Design optimization techniques for microfluidics have been studied in recent years, but they overlook the myriad complexities of biomolecular protocols and are yet to make an impact in microbiology research. The realization of microfluidic platforms for real-life quantitative analysis requires: (1) a new optimization flow that is based on the realistic modeling of biomolecular protocols, and (2) a microfluidic security flow that provides a high-level of confidence in the integrity of miniaturized quantitative analysis.

Motivated by the above needs, this book is focused on optimized and trustworthy transfer of benchtop biomolecular analysis, particularly epigenetic studies, to programmable and cyber-physical microfluidic biochips. The book first presents a set of optimization mechanisms that leverages cyber-physical integration to enable real-time execution of multi-sample biomolecular analysis. The proposed methods include a resource-allocation scheme that responds to decisions about the protocol flow, an interactive firmware that collects and analyzes sensor data, and a spatio-temporal reconfiguration technique that aims to enhance the reliability of the microfluidic system. An envisioned design for an Internet-of-Things (IoT)-based microfluidics-driven service is also presented to cope with the complexity of coordinated biomolecular research.

Next, this book advances single-cell protocols by presenting optimized microfluidic methods for high-throughput cell differentiation. The proposed methods target pin-constrained design of reconfigurable microfluidic systems and real-time synthesis of a pool of heterogeneous cells through the complete flow of single-cell analysis. A performance model related to single-cell screening is also presented based on computational fluid-dynamics simulations.

With the increasing complexity of microbiology research, optimized protocol preparation and fault-tolerant execution have become critical requirements in today's biomolecular frameworks. This book presents a design method for reagent preparation for parameter-space exploration. Trade-offs between reagent usage and protocol efficiency are investigated. Moreover, an integrated design for automated error recovery in cyber-physical biochips is demonstrated using a fabricated chip.

In order to ensure high confidence in the outcome of biomolecular experiments, appropriate security mechanisms must be applied to the microfluidic design flow. This book provides an assessment of potential security threats that are unique to biomolecular analysis. Security countermeasures are also proposed at different stages of the biomolecular information flow to secure the execution of a quantitative-analysis framework. Related benchtop studies are also reported.

In summary, the book tackles important problems related to key stages of the biomolecular workflow. The results emerging from this book provide the first set of optimization and security methodologies for the realization of biomolecular protocols using microfluidic biochips.

1

Introduction

A microfluidic system is an engineered fluidic device that controls the flow of analytes, thereby enabling a variety of useful applications. According to recent studies, the fields that are best set to benefit from the microfluidics technology, also known as lab-on-chip technology, include genetic analysis, DNA amplification, clinical chemistry, cell-based assays, single-cell analysis, proteomics, point-of-care (PoC) diagnostics, drug discovery, and nanomaterial synthesis [50, 129]. The growth in such fields has significantly amplified the impact of microfluidic-based PoC technology, whose market value is forecast to grow from $4 billion in 2017 to $13.2 billion by the end of 2023 [291]. The rapid evolution of lab-on-chip technologies opens up opportunities for new biological or chemical science that can be directly facilitated by microfluidic control [37].

The early generation of microfluidic devices, referred to as continuous-flow microfluidic biochips (CMFBs) [265, 271], consist of micro-fabricated channels, micropumps, and microvalves which are permanently etched in a silicon or a glass substrate. Initially, CMFBs have relied on simple topologies and only a few channels; however, commercial microfluidic devices nowadays require large-scale networks of channels so that they can be used in implementing complex assays, such as organ-on-a-chip assays [219], or be exploited for real-world applications. For example, a product from Fluidigm [3], a biotechnology company that produces CMFBs, can perform a series of gene analyses steps, including enrichment for a target DNA sequence, sample barcoding, for multiplexed sequencing, and preparation of the sequencing library. While early designs of continuous-flow biochips are highly efficient and robust in the miniaturization and high-throughput execution of biochemical assays, the structure and the functionality of such devices are tightly coupled. As a result, each biochip is typically applicable to a narrow class of applications, and the reconfigurability of this early technology is quite limited [134].

Recent research has focused on making traditional CMFB designs more flexible by using valve-based reconfigurable components. For example, Silva *et al.* [232] introduced a programmable flow-based primitive, called a transposer, that can be scaled to create a field-programmable flow-based fabric. Also, a software-programmable microchannel structure has been introduced in [71], which allows a software program to control numerous valves by using microfluidic multiplexer circuitry. Similarly, in [134], Kim *et al.* presented a design of field-programmable flow-based devices, called programmable

1

microfluidic platforms (PMPs). In these devices, fluid can be flexibly manipulated using a 2-D array of valves that can act as both flow-switching elements and reaction chambers. A variety of biochemical applications such as single-cell screening and high-throughput cell barcoding can efficiently be implemented using these reconfigurable devices. These reconfigurable devices have motivated researchers to develop design-automation and testing solutions, thus leveraging the flexibility and programmability provided by these devices [276, 152, 113, 119].

Despite the advances in field-programmable CMFBs, they are still limited in their ability to support multiple sample pathways and on-demand execution of biochemical reactions; such characteristics are inherent in typical quantitative-analysis applications such as quantitative polymerase chain reaction (qPCR) [153]. In contrast, a digital microfluidic biochip (DMFB) is a reconfigurable lab-on-a-chip technology that is well-suited for cyber-physical integration, thus allowing biochemical reactions to be carried out on-demand based on a feedback sensory signal [83, 157]. DMFBs have therefore achieved remarkable success in enabling miniaturized analysis systems for several contemporary biochemical applications. As an indicator of commercial success, Illumina, a market leader in DNA sequencing, has recently transitioned DMFBs to the marketplace for sample preparation [121]. This technology has also been deployed by GenMark for infectious disease testing [80] and by Baebies for the detection of lysosomal enzymes in newborns [6]. These significant milestones highlight the emergence of DMFBs for commercial exploitation and their potential for immunoassays for point-of-care diagnosis [234, 239], DNA sequencing [31], environmental monitoring [297], and synthetic biology [230]. The above advances have been a major driver for bridging the gap between the microfluidic technology and quantitative biomolecular analysis, even though DMFBs are not as effective as CMFBs for up-stream analysis and interfacing to the, external world especially in single-cell applications [113, 229].

Digital-microfluidics technology allows the discretization of continuous biochemical fluids into nanoliter droplets, which can be manipulated using a patterned array of electrodes based on a phenomenon called *electrowetting-on-dielectric* [68]. Therefore, fluidic operations such as dispensing, mixing, incubation, heating, and electroporation can be executed on-demand under software control in a cost-effective manner [52]. Advantages of DMFBs are further extended by introducing a special architecture that supports real-time detection and allows mixing of various reagents with considerably high resolution. This architecture, known as the micro-electrode-dot-array (MEDA) architecture [274], contains smaller electrodes (micro-electrodes), which can be dynamically grouped to form a single micro-component (e.g., mixer or diluter) that can perform on-chip biochemical operations. Unlike conventional DMFBs, MEDA biochips naturally offer fine-grained mixing capabilities that can be leveraged for producing complex biomolecular mixtures.

As the applications of molecular biology grow, greater demands are placed on the research community to recognize the following realities about microfluidic systems [50, 112]:

- Different microfluidic tools are not equal enablers in all scenarios. In other words, it is imperative to identify the right applications for each class of microfluidic biochips, which can employ appropriate microfluidic control to improve key performance factors. For example, pressure-driven continuous-flow microfluidics can be leveraged to enable ultra-high-throughput single-cell screening (up-stream analysis) [224, 35, 170, 119], whereas digital microfluidics can be the basis for optimizing the sensitivity of sample preparation or improving the precision of quantitative analysis (down-stream analysis) [228, 212]. Moreover, microfluidic devices can be viewed as modular components that can be integrated on board to develop a complete analytical toolkit, thus enabling advanced flows for molecular biology [113, 229].

- A major goal in recent advances of microfluidics is to enable the miniaturization of complex biomolecular assays. This goal, however, requires hundreds to thousands of molecular samples to be processed on a single platform [218]. This requirement can be realized only if appropriate design-automation (or "synthesis") and error-recovery methods are developed. New microbiology-driven synthesis methods are especially required to cope with advances in microfluidics and microbiology technologies. These software tools need to use insights from real benchtop experiments or to interact with a model that is generated based on computational fluid dynamics (CFD) simulations; therefore, these tools can help predict the performance of a microfluidic application with high accuracy [225, 119]. Similarly, developing schemes for dynamic error recovery is equally important to verify the correctness of on-chip fluidic interactions during assay execution [108].

- Microfluidics-driven biomolecular analysis can offer remarkable benefits, especially for PoC diagnostics and mission-critical applications such as forensic DNA analysis. Several microfluidic commercial developers, including U.S.-based IntegenX [4], ANDE [1], and Lockheed Martin [5], have already started to roll out prototypes aiming to replace traditional benchtop procedures for biomolecular analysis, and it is anticipated that design automation and cyber-physical integration will play a significant role in advancing this technology. However, the pressure to drive down costs will lead to cheap untrusted devices and a multitude of unanticipated privacy violations if preventative measures are not taken. Sensitive information in a microfluidic device can include data collected after processing of the fluids and personally identifying metadata. Irresponsible handling of patient data has led to the breakup of companies in the past [55], and current device makers would do well to learn from those mistakes. Other issues include trust in the sensor readings themselves; the rise and fall of Theranos, and the invalidation of two years worth of test results set a poor precedent for microfluidics diagnostics [127]. The nascent nature of microfluidics in biomolecular quantitative analysis presents an opportunity to incorporate security and trust in such critical applications before it becomes too late to rescue this rising industry from security threats.

In this chapter, we introduce basic concepts and background related to microfluidic technologies, provide motivation for the book topic, and present an outline of the book. Section 1.1 presents an overview of DMFBs and MEDA biochips, and it also describes the working principle of these devices. Section 1.2 presents an overview of CMFBs with a focus on reconfigurable CMFBs that are used for biomolecular analysis. A review of previous design-automation and optimization techniques for different microfluidic technologies along with advances in cyber-physical integration is presented in Section 1.3. For the miniaturization of biomolecular analysis protocols, we present our vision for a multi-layered architecture for cyber-physical microfluidic biochips in Section 1.4. A brief discussion on security assessment of biomolecular quantitative analysis is presented in Section 1.5. In Section 1.6, the workflow of biomolecular quantitative analysis is highlighted, and a discussion on design-automation and security challenges are introduced. Finally, an outline of this book is provided in Section 1.7.

1.1   Overview of Digital Microfluidics

A typical DMFB consists of a two-dimensional electrode array, on-chip reservoirs, optical sensors, and heaters, as shown in Figure 1.1(a). A cell in a DMFB consists of two parallel plates; see Figure 1.1(b). The electrode surface is coated with a thin layer of an insulator such as Paralyene [234]. Both plates are also coated with a thin film to provide a hydrophobic platform that is necessary for smooth electrowetting-based droplet actuation [202]. The gap between the top and bottom plates is usually filled with silicon oil which acts as a filler medium, reducing surface contamination. When an electric field $\mathcal{V}_e$ is applied between the parallel plates of a DMFB, the interfacial surface energies are modulated and an electrical double layer is created, which in turn alters the apparent contact angle $\theta_e(\mathcal{V}_e)$ of a conductive liquid droplet that is in contact with the hydrophobic surface (Figure 1.1(b)). The change in the contact angle, in turn, influences the wetting behavior of the droplet. This phenomenon is known as electrowetting-on-dielectric (EWOD), and it can be modeled using the Lippmann-Young equation [202]:

$$cos\ \theta_e(\mathcal{V}_e) = cos\ \theta_e(0) + \frac{\epsilon_0 \epsilon_r \mathcal{V}_e^2}{2d_e \cdot \gamma_{LG}} \qquad (1.1)$$

where γ_{LG} is the liquid-gas interfacial tension, ϵ_0 is the permittivity of vacuum, ϵ_r is the permittivity of the bottom insulator, and d_e is its thickness.

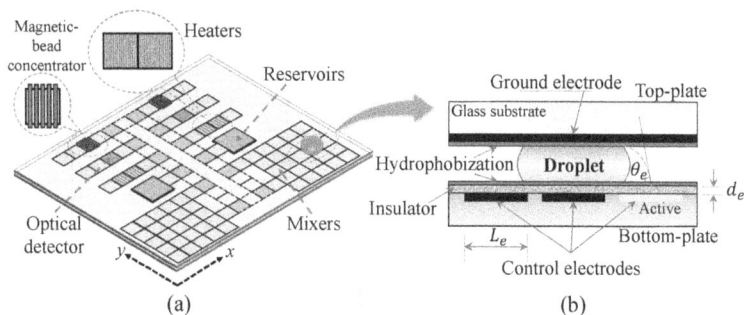

FIGURE 1.1
Schematic view of a DMFB: (a) A DMFB with a 2D array of electrodes and (b) A side-view of the DMFB.

Using the EWOD mechanism, picoliter droplets can be electrically transported and manipulated through various fluidic operations such as mixing, splitting, heating, and incubation [68].

Since discrete droplets are manipulated independently on the biochip, DMFBs are especially suitable for microbiologists and biomedical engineers to seamlessly process several steps of biomolecular reactions [228] without the need for a bulky surrounding instrumentation or a complex microfluidic-channel structure [166]. Digital-microfluidics technology has proven its ability to handle complex biological protocols, from manipulation of cell suspensions [226], or cell harvesting [67] to parallel cell-based screening [30] or highly multiplexed enzymatic reaction [236]. The flexibility provided by this technology for integrated biomolecular analysis has been supported by advances in integrated sensing technologies [53], magnetic bead manipulation [46], on-chip thermal-cycling procedures [99], and DMFB-enabled electroporation techniques [163]. These advances lay the groundwork for enabling complex bench-top biomolecular procedures on programmable and cyber-physical DMFBs.

Sensing Systems for Biomolecular Analysis

Sensing systems on DMFBs can be used to detect a biological event or quantify the existence of a constituent biochemical analyte during an analytical procedure. Examples of biochemical analytes are toxins in food, pollutants in air/water, and intracellular biomolecules. Researchers have developed numerous integrated sensing techniques that are used in concert with DMFBs; these techniques can be functionally classified into three main categories [165]: (1) optical detection (e.g., absorbance, fluorescence, chemiluminescence, and surface plasmon reasonance detection); (2) mass spectrometry; (3) electrochemical detection. Fluorescence detection is the dominant sensing method for biomolecular analysis—it is widely used because the fluorescent labeling techniques are well-established and they provide quantification results with

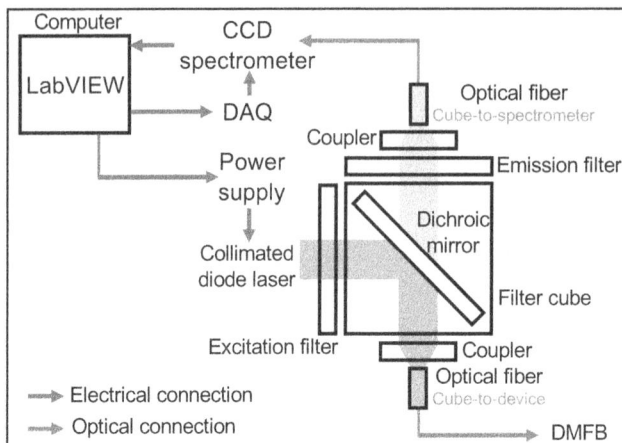

FIGURE 1.2
A schematic view of custom-built fluorescence sensor.

superior selectivity and sensitivity despite small sample volumes [138, 234]. We briefly describe fluorescence detection below.

Fluorescence detection: Contemporary biomolecular research relies extensively on analyzing amplified DNA signals in correspondance with biological events such as chromatin packing alteration [279]. To develop a single-chip solution that incorporates the purification and detection of DNA signals, a fluorescence sensor along with a fluorescent dye injected into the sample is used to analyze the fluorescence generated by the on-chip quantitative polymerase chain reaction (qPCR) protocol [104, 212]. In [99], Hsu constructed a custom-built fluorescence sensor that is adequate for qPCR experiments, and it consists of five major components (Figure 1.2):

1. A 4.5 mW 532 nm collimated diode laser (light source), whose power supply is controlled by a software program.

2. A filter cube that is installed with three filters (excitation filter, dichroic mirror, and long-pass emission filter); the filters are used to provide the appropriate wavelength of light to excite the fluorescent molecules in the sample and also to filter the light reflected from the sample.

3. Two optical fibers (cube-to-device and cube-to-spectrometer) to focus the laser beam.

4. A CCD spectrometer that captures the reflected fluorescence signal.

5. A software program (LabVIEW) that controls the setup.

A significant advantage of fluorescence sensors is that they can be used to visualize multiple fluorescent molecules by simply adjusting the excitation and emission filters to detect the fluorophores in a sample—this feature enables multiplex genomic analysis even with multiple samples.

Lensless sensing: Although optical microscopy provides high-resolution imaging of biomolecular samples, it relies on high-magnification and high-numerical aperture objective lenses. A new trend has recently emerged to design lensless microscopy that can provide high-resolution images without the use of any focusing lenses, offering cost-effectiveness and portability features [57, 193]. Such technqiues can be easily used to detect and monitor cells, and they can be widely utilized in the early phases of quantitative-analysis workflow.

Magnetic-Bead Manipulation for Particle Detection

In recent years, digital microfluidic (DMF) methods have been employed for separating molecules of interest from a mixture of biological samples—this utility has been widely used for the miniaturization of immunoassays [234, 45] and nucleic-acid purification protocols [99]. For this purpose, a droplet containing antibody-coated magnetic beads of sizes ranging from micrometres to nanometres is mixed with the biological mixture, and a magnetic field is applied to segregate the molecules of interest from the mixture [235]; see Figure 1.3. Early methods of magnetic-bead manipulation relied on permanent magnets and a mechanical system to control the segregation process [184, 143]; thus limiting their potential in developing integrated microfluidic devices. In addition, although permanent magnets can have strong field gradients and therefore they are capable of trapping large quantities of beads, they are less successful in separating magnetic beads with small magnetic moment differences.

To overcome the above drawbacks, on-chip magnetic-bead manipulation has recently been demonstrated using an integrated current-wire digital microfluidic device [46]. The significance of this approach is that different kinds of magnetic beads can be separated within a droplet, since magnetic beads (with different sizes) may react differently to alterations of current within the wires [45]. This capability allows selective multiplexed quantitative analysis on chip such as multiplex chromatin immuneprecipitation (ChIP). In this setting, we can conduct an experiment with multiple tagged cell samples; these samples could be identical or may be cells exposed to different (drug/epigenetic) conditions if needed. A specific cell sample would need to be bound to a selected antibody attached to magnetic beads of a certain size. This process may be streamlined by allowing a specific sample (e.g., green-/yellow-tagged sample) to be bound with a pre-specified antibody, based on the "detected" sample tag. For example, we can control the system such that if tag "A" is detected, then use antibodies attached to magnetic beads of sizes "S_1, S_2" that will be separated individually. Therefore, the use of integrated wires for on-chip magnetic-bead manipulation can be harnessed to achieve controlled multiplicity [46].

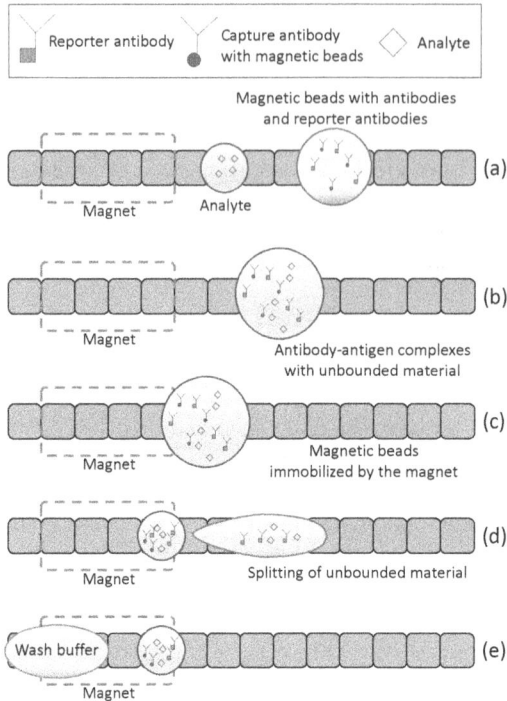

FIGURE 1.3
Implementation of immunoassay on a DMFB using magnetic beads: (a) dispensing of reagents, (b) incubation, (c) immobilization of magnetic beads, (d) removal of supernatant and washing, (e) adding fresh wash buffer.

On-Chip Heaters

Precise cycling of temperatures is a critical step for implementing qPCR-based biomolecular analysis. As a result, the design of on-chip thermocyclers has received a great deal of attention in the literature [104, 234, 212, 187, 99]. The goal here is to minimize the thermal mass and thus the PCR time via system integration. In [104], Hua *et al.* in presented a spatial qPCR design by making use of the capability to shuttle droplets between fixed heating zones to achieve cycling—this scheme has been also adopted in [234]. An alternative temperature-cycling approach has been presented by Rival *et al.* in [212], Norian *et al.* in [187], and Hsu in [99] elucidating a temporal qPCR design in which the droplets are kept stationary on thermocyclers' sites while temperature variations are precisely controlled over specific time intervals. Despite its complexity, the latter approach has become increasingly valuable as researchers get better control of thermocycling during the course of the

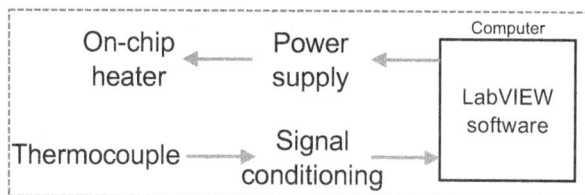

FIGURE 1.4
A closed-loop temperature control of an on-chip heater.

experiment. Furthermore, this approach provides the option of using the integrated thermocyclers for other purposes, such as thermal cell lysis.

The design of a temperature-varying thermocycler traditionally relies on a resistive heater such as a high-resistance polysilicon resistor, a temperature control system, and a temperature sensor (e.g., a thermocoupler); see Figure 1.4. The temperature control system is responsible for modulating the on-chip heater to perform thermocycling, whereas a temperature sensor can be used to monitor the thermal cycling thus determine the progression of the PCR reaction in correspondance with fluorescence sensors (explained in Section 1.1) [187]. The temperature control system relies on a feedback from the temperature sensor to conduct two functions. First, the system regulates the on-chip heater to carry out thermocycling using a PID algorithm. Second, when cooled to a preset temperature, it will trigger the fluorescence sensing system to perform spectrum acquisition. The temperature sensor, in turn, can be permanently mounted on the chip substrate [187] or manually aligned and attached to the bottom of the microfluidic biochip [99].

MEDA Biochips for Biomolecular Mixture Preparation

The MEDA architecture is a digital microfluidic technology, which consists of an array of identical basic microfluidic units called *microelectrode cells* (MCs); see Figure 1.5(a-b) [274]. Each MC consists of a microelectrode and a control/sensing circuit. Using this configuration, MEDA biochips can employ the the concept of a sea-of-micro-electrodes, where microelectrodes can be dynamically grouped to form a single micro-component (e.g., mixer or diluter) that can perform on-chip biochemical operations. A prototype of MEDA biochips has been demonstrated in [145].

By using a conventional DMFB, a biomolecular mixture that comprises several reagents can be systematically prepared using a sequence of (1 : 1) mixing-splitting operations only. In this (1 : 1) model, an intermediate droplet is generated by mixing two reagent droplets of equal volume, and a large droplet can be split only into two small droplets of equal volume. Clearly, this model has limitations if the requirement is to produce a mixture that has a

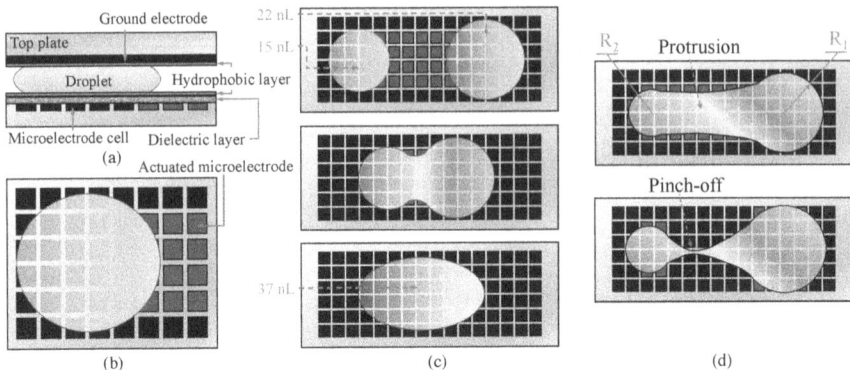

FIGURE 1.5
Droplet actuation in MEDA biochip: (a) side view; (c) top view; (c) MEDA-enabled (m, n) mixing model; (d) MEDA-enabled (m, n) aliquoting/splitting model.

large number of constituent reagents with different volumetric ratios. MEDA biochips, on the other hand, utilizes the dynamic grouping of microelectrodes to extend the above model to a general $(m : n)$ mixing-splitting model [148]; thus enabling droplets with different volume to be mixed or split with high resolution. As shown in Figure 1.5(c), MEDA enables two reagent droplets of volume 15 and 22 nL to be mixed to generate a 37 nL mixture with volumetric ratios 40% and 60%, respectively. Also, Figure 1.5(d) shows a generalized MEDA-enabled splitting (or aliquoting) operation. Such a powerful capability can be exploited to produce biomolecular mixtures with high precision.

1.2   Overview of Continuous-Flow Microfluidics

A typical CMFB device consists of channels and valves, which can control fluid flow using deformable polydimethylsiloxane (PDMS) membranes. CMFB components can be constructed and assembled using one of two microfabrication technologies: monolithic membrane valves [87] and multi-layer soft lithography [63]. Using monolithic membrane valves, a microfluidic device can be produced by assembling two rigid glass wafers and bonding them using a sandwiched PDMS membrane, as shown in Fig. 1.6(a). Note that fluid channels are etched in the glass wafers. When the chip is assembled and bonded, the PDMS membrane remains rigid, except at the valves that control the fluid flow [134].

On the other hand, to use multi-layer soft lithography to fabricate CMFB devices, a soft polymer such as PDMS is cast onto a multi-layer mold, which in

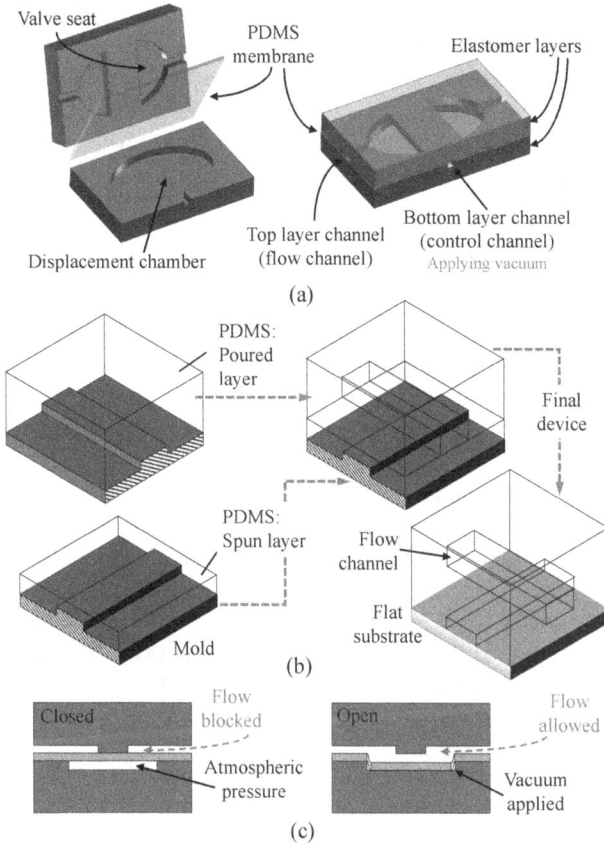

FIGURE 1.6

A schematic of flow-based components: (a) the components of a biochip fabricated using monolithic membrane valves; (b) the components of a biochip fabricated using soft lithography; (c) the operation of a normally closed valve (2-D cross-sectional view) when it is kept closed (left) and when it is forced to open (right).

turn contains engraved patterns. Fig. 1.6(b) shows an example of a CMFB that is fabricated using the soft lithography technology. Nowadays, the vast majority of CMFBs are produced using multi-layer soft lithography and PDMS substrate, as opposed to monolithic membrane valves and glass wafers. The reason for such a preference is that PDMS is biocompatible, transparent, gas-permeable, water-impermeable, fairly inexpensive, and rapidly prototyped with high precision using simple procedures, e.g., 3D printing [93].

Typically, CMFBs are fabricated considering that biochemical fluids are carried at the top layer (flow layer) and the other layer (control layer) provides the vacuum to deflect the PDMS membrane. Considering a normally closed valve (Fig. 1.6(c)), fluid flow can be permitted once the membrane is deflected "outside" the flow channel. Such a control mechanism can be used to control the movement of a continuum of liquid or a collection of droplets enclosed in an oil medium, referred to as a two-phase system [88]. A major benefit of this mechanism is its capability to perform single-cell isolation and sorting in a high-throughput manner, making CMFBs superior in single-cell screening applications. Given the nature of fluid control in CMFBs, these devices can also be referred to as flow-based microfluidic biochips.

Field-programmable flow-based devices can play a key role in the automation of chemical and biochemical analysis. Not only do they allow programmable execution of biomolecular applications, but they can also be integrated with detection platforms, thereby improving both performance and portability. In [134], a pneumatically actuated PMP was used to perform genetic analysis, allowing quantitative reactions to be performed inside the valve chambers and an integrated fluorescence sensor to examine gene-expression levels. This technology can be further advanced by using multiple fluidic and pneumatic layers with parallel pumping mechanisms (analogous to multi-core processors), thus enabling high-throughput parallel analyses [134].

Despite the obvious benefits of field-programmable flow-based devices, the underlying pneumatic control still needs appropriate design-automation tools to enable on-demand execution of biomolecular reactions. Furthermore, reducing expensive off-chip control is still a major challenge that has yet to be addressed in a systematic manner.

1.3 Design Automation and Optimization of Microfluidic Biochips

Motivated by the increasing complexity of biochemistry-on-chip, research on design automation for both digital microfluidic and flow-based microfluidic devices has received much attention over the past decade [118, 41, 203, 100]. The goals of such design-automation methods are twofold: (1) non-expert users of microfluidic platforms (e.g., a local clinician in a rural area [49, 185]) should be able to utilize only a high-level specification of a protocol, e.g., a sequencing graph (Figure 1.7(a)), to program a biochip; (2) using information about a microfluidic device such as resources of a DMFB (Figure 1.7(c)) and pre-specified operation constraints (Figure 1.7(b)), a well-established synthesis framework must convert the abstract representation of a protocol into an executable code, e.g., actuation sequences in DMFBs. The synthesis outcome can be used to timely control the actuators, e.g., electrodes of a DMFB; thus moving droplets to execute the associated protocol, as shown in Figure 1.7(e).

Early research on design automation for both technologies focused on scheduling, resource mapping, droplet routing, and sharing of control pins

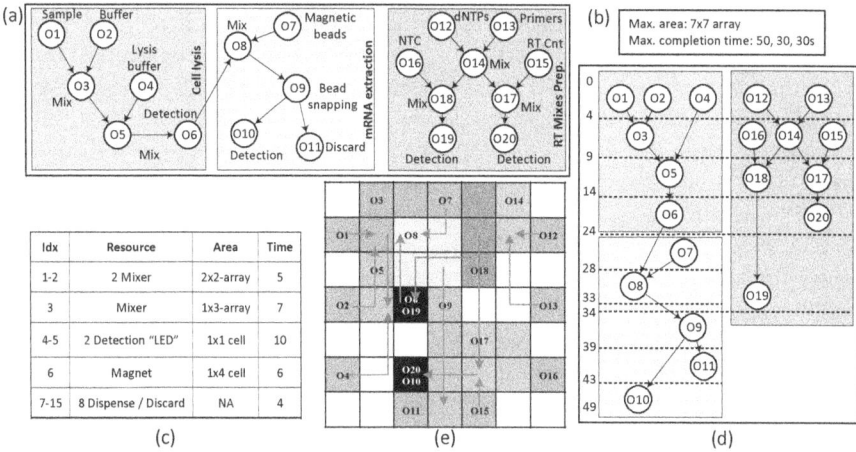

FIGURE 1.7
Synthesis flow for a DMFB that processes mRNA samples: (a) sequencing graph, (b) design specification, (c) library of DMFB modules, (d) scheduled sequencing graph, and (e) droplets control using actuation sequences.

(or pressure sources) [248, 245, 247, 246, 249, 285, 268, 171, 172, 173, 290, 288, 286, 292, 293, 51, 85, 86, 149, 101]. Algorithms have been also developed for on-chip solution preparation [175, 214, 137] and droplet routing while avoiding cross-contamination [295, 174, 151, 105]. However, these methods are limited to single sample pathways and they are inadequate for single-cell analysis. Moreover, a major drawback with these methods is that they cannot handle uncertainties in the order of fluidic steps in bioassay execution—these characteristics are prevailing in biomolecular quantitative-analysis protocols. Also, the interplay between hardware and software in the biochip platform (cyber-physical system design) or the real-time coordination between different microfluidic solutions was not considered in these early methods.

Based on the existing hardware platform of biochips and emerging on-chip sensing techniques, the concept of a cyber-physical microfluidic biochip has been mainly introduced to DMFBs [159]. In such a biochip (Fig. 1.8(a)), the control software (cyber space) and the hardware platform (physical space) are tightly coupled. Such a design consists of two main functional components: (1) the feedback sensor connectivity that ensures real-time detection of the sample droplets; (2) intelligent control software (online synthesis) that captures the detection outcome and provides an appropriate action; see Figure 1.8(b). Design and optimization techniques for cyber-physical DMFBs have thus far considered only error recovery [159, 14, 122, 102, 109, 201] and termination control of a biochemical procedure such as PCR [157].

However, these designs do not support real-time decision making for protocols designated for multiple samples, and they do not exploit the potential of

DMFBs for quantitative analysis-based biochemical reactions. Gao *et al.* [79] proposed a modular design with built-in electronic control to enhance the reliability and robustness of DMFBs. However, this design can only be used for droplet volume measurement and position control. With this computationally expensive method, the control of more than a few droplets transported concurrently on an array becomes impractical. In [103], an experimental demonstration of hardware-based real-time error recovery in integrated cyber-physical DMFBs has been shown. In this work, a capacitive sensor, signal conditioning circuit along with a hardware control machine have been used to address reliable bioassay execution in a practical setting. Yet this framework lacks support for real-time quantitative measurement.

Therefore, a major limitation of all prior work on design automation and cyber-physical integration is that they are limited to simple droplet manipulation on a chip; however, in order to make DMFBs useful to a microbiologist, we need a new design paradigm to demonstrate that these chips can be used for actual biomolecular protocols from microbiology [110].

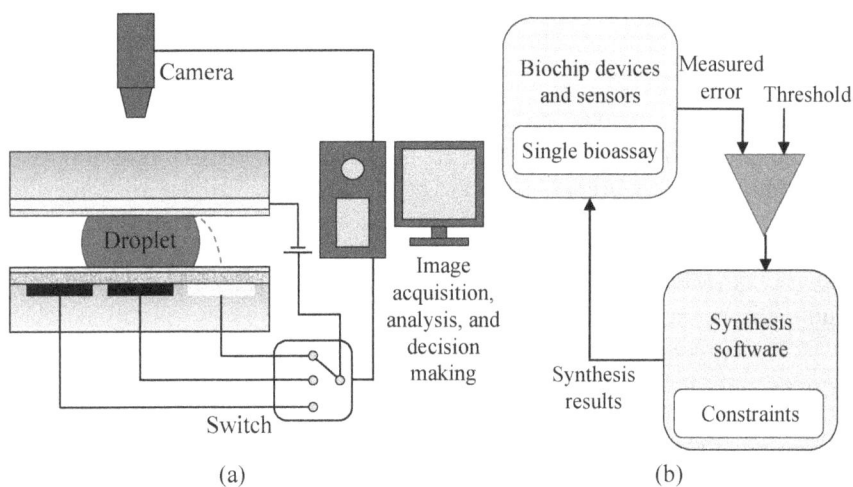

FIGURE 1.8
Cyber-physical integration of DMFBs for error recovery: (a) Feedback system configuration (not drawn to scale); (b) System block diagram.

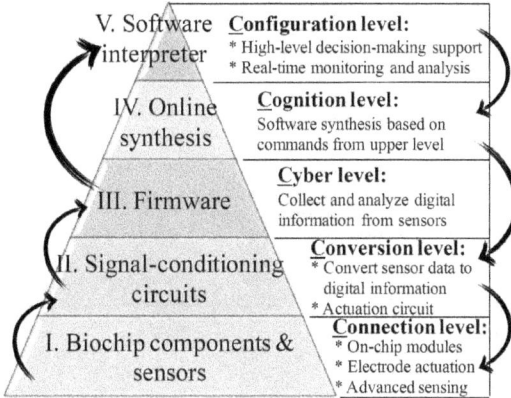

FIGURE 1.9
The proposed unified 5-layer ($\mathcal{C}^5$) architecture for cyber-physical DMFBs.

1.4 Cyber-Physical Adaptation for Quantitative Analysis

Our premise is that significant rethinking in system design is needed to ensure cyber-physical system (CPS) adaptation for quantitative analysis on-chip. Figure 1.9 illustrates the proposed CPS-inspired 5-layer $\mathcal{C}^5$ (based on the C for each level) architecture [110]. Today's design methods incorporate online biochemistry-on-chip synthesis, which provides reconfiguration capability to recover from operational errors (Level IV in Figure 1.9). However, there is still a gap between the physical space and online synthesis, which impedes the use of DMFBs for quantitative analysis due to the lack of autonomous data analysis and intelligent decision-making. There is a need to explore algorithmic innovations that fill the gap between the control and monitoring of the physical space on one side, and the cyber space (i.e., online biochemistry-on-chip synthesis) on the other side. Coupling the firmware with online synthesis will potentially open new opportunities for dynamic synthesis. For example, the need for short time-to-result might require the prioritizing and selection of samples for detection. These ideas can also be extended to prioritize the bioassays selected for synthesis in a multi-assay setting. Such coupling between the firmware and online synthesis is key in type-driven single-cell analysis [95], where the selection of a quantitative protocol depends on the cell type.

   An important objective of our research is to integrate these levels to enable the seamless on-chip execution of complex biochemical protocols. As shown in Figure 1.9, each level is expected to play a distinct role. With the integration of sensors at the hardware level (Level I), there is a need to provide analog signal acquisition and digital signal processing capabilities to transform the received

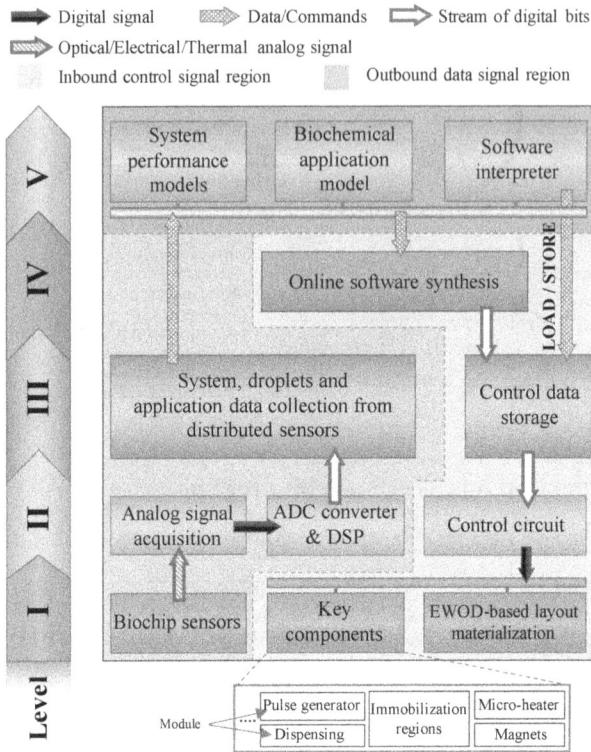

➡ Digital signal   ▧▷ Data/Commands   ⟹ Stream of digital bits

▧▷ Optical/Electrical/Thermal analog signal

    Inbound control signal region   Outbound data signal region

FIGURE 1.10

The interactions among the various levels of the system infrastructure.

signals into readable data. This can be achieved via a signal conditioning level (Level II). Previous designs of cyber-physical DMFBs have attempted to integrate this level with the system infrastructure, but only for the limited purpose of error recovery [159].

The uppermost level (Level V) will be the system coordinator. It will be responsible for adapting the flow of protocol execution based on the decisions conveyed from the firmware (Level III). Adaptation is therefore supported by an application model, which keeps track of application progress, and a performance model that keeps track of the alterations in a target performance metric such as chip utilization, degradation level, or droplet evaporation. Figure 1.10 illustrates the anticipated interactions among the various levels of the system infrastructure. We explain the adaptation of the $\mathcal{C}^5$ architecture for gene-expression analysis, single-cell analysis, and dynamic error recovery in Chapter 2, Chapter 5, and Chapter 8, respectively.

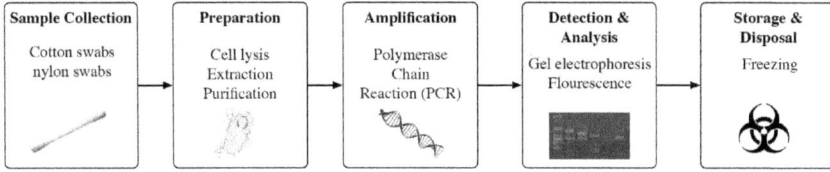

Sample Collection	Preparation	Amplification	Detection & Analysis	Storage & Disposal
Cotton swabs nylon swabs	Cell lysis Extraction Purification	Polymerase Chain Reaction (PCR)	Gel electrophoresis Flourescence	Freezing

FIGURE 1.11
The DNA processing flow.

1.5 Security Assessment of Biomolecular Quantitative Analysis

With an increasing interest in biochip-based automation of quantitative analysis especially for mission-critical applications, a malicious adversary can seek an opportunity to tamper with the results of a DNA quantitative analysis. Tampering can result in the complete destruction of a sample, or modification of the biochemical process such that it produces a misleading result. Motivations are varied and may include the obstruction of justice, activism, or sabotage of a medical-services corporation.

In real-life scenarios, the adversary can be present in any stage of the analysis process—evidence exists for integrity compromises at every phase of the DNA processing flow (Figure 1.11). The capabilities of an attacker will vary depending on where the attack takes place. An attacker at the sample collection phase does not require much technical expertise; it is sufficient to substitute another person's DNA sample at an opportune time. An attacker who wishes to tamper at one of the processing phases may require a sophisticated knowledge of how the processing equipment is constructed and its security vulnerabilities.

Hence, there is a broad set of attacks against the DNA processing flow, classified based on where an attack can take place within the flow of information. Note that this analysis applies in more traditional analysis settings, where samples are transported to laboratories, as well as in emerging lab-on-a-chip platforms where processing can be completed within hours.

Biochemical Level: Attacks on the integrity of DNA processing can occur even before DNA analysis begins. DNA processing flows currently assume that the samples under investigation were collected in a trustworthy manner. That is, the samples do not have contamination or were not exposed to environmental conditions that would lead to fouling of the samples. In a more adversarial setting, trust in the collections requires that the samples were not deliberately replaced by a malicious actor. This assumption has proven to be invalid in several occurrences [74, 198].

Bioassay Level: Once the DNA samples are collected, they must be processed and analyzed for comparison with reference samples or a DNA database. The instructions for carrying out the processing are specified with a bioassay that instructs when samples are mixed, heated, or split with other reagents to obtain a desired result, such as in the polymerase chain reaction (PCR) used for DNA amplification. Tampering of the bioassay can lead to the loss of precious samples that may be impossible to re-collect, or can lead to misleading results that are inconclusive.

System Level: The support systems that surround the DNA processing are potential attack surfaces that must be evaluated. As shown in Section 1.3, microfluidic biochips are often supported by computer-based controllers and sensor feedback. The results obtained from sensors must be transmitted electronically, which could potentially be tampered with. Furthermore, the storage and disposal of samples leaves open questions about how secure the entire flow of information is.

1.6   Proposed Research Methodology

Quantitative-analysis protocols rely on the estimation of the dose-response relation curve for biomolecular compounds (e.g., protein, gene, cell, tissue, and organ), wherein the response of a biomolecular compound is studied with changing dose. Using the dose-response curve, we can estimate the dose or con-

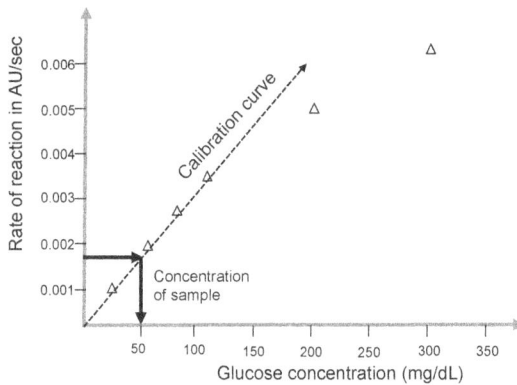

FIGURE 1.12
Glucose-calibration curve: The X-axis represents the different concentrations formed by these dilutions (in mg/dL) and the Y-axis represents the rate of reaction quantified by the change in absorbance degree reported as AU/sec (absorbance unit per second).

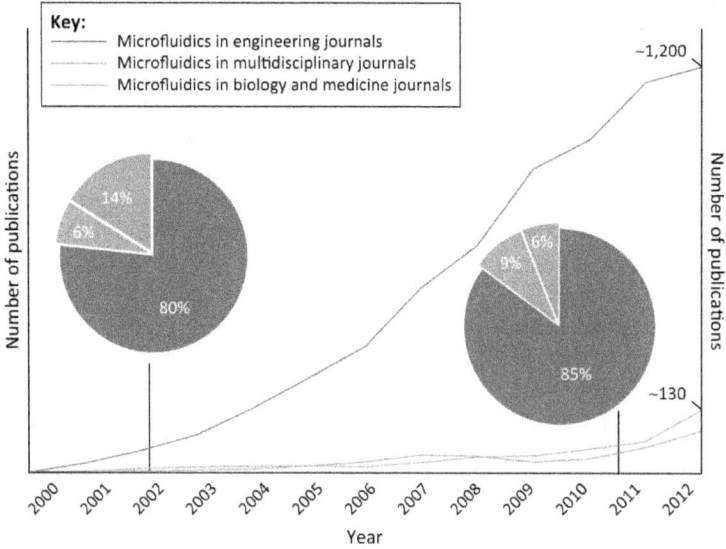

FIGURE 1.13
Microfluidics publications in engineering, multidisciplinary, and biology and medicine journals from 2000 to 2012.

centration of a substance, such as dye fluorescence intensity, associated with a specific biological response, such as gene expression. Figure 1.12 shows an example of a dose-response relation curve for investigating blood-glucose concentration using a colorimetric assay; the change of dye color is associated with glucose concentration and it is detected using an on-chip absorbance measurement system such as LED-photodiode [239]. In this context, the dose-response relation curve is known as glucose-calibration curve. This curve therefore helps in interpolating the concentration of a glucose sample under test. As shown in Figure 1.12, the reaction rate of the sample is a point on the Y-axis and the corresponding point on the X-axis is the sample concentration. Similarly, the dose-response relation is also applied through qPCR to investigate gene expression, producing what is known as qPCR standard curve [116].

As briefly shown in Sections 1.1 and 1.2, several engineering solutions have been developed for chip-scale implementation of biomolecular steps [221]. Although these designs are completely sensible from an engineering perspective, they may not be appropriate for real-life scientific applications—these methods oversimplify the dynamics of large-scale biomolecular protocols, leading to a large gap between academic engineers and biologists [111]. Figure 1.13 shows the growth of microfluidics publications in engineering journals compared with biology and medicine journals, and the gap between them.

Similar to the gap between biochip engineering and translational research, there is also a gap between what microfluidics can offer for DNA-processing applications and the perception of this technology by the end users. For example, in the DNA forensics application, a survey performed by Walsh [273] concluded that current research for improving DNA forensic analysis has been only directed at technological advancement, and research that establishes trust in DNA forensic science has been limited. As briefly shown in Section 1.5, lack of trust in biochip-based DNA processing flows is upheld due to potential vulnerabilities in current systems. Hence, filling the trust gap is necessary to introduce the microfluidic technology as a true enabler for a wide scale of DNA-processing applications.

As a result, the dissemination of microfluidic technology in translational research and real-life applications needs to come from an alliance between engineering and microbiology parties. For this purpose, this book addresses challenges associated with design optimization and security threats; these challenges arise from mapping realistic implementation of quantitative-analysis protocols to the chip-scale. The proposed research is led by a set of benchtop experiments from the epigenetics domain.

Workflow of Biomolecular Quantitative Analysis

The proposed research is an effort to bridge the above gaps—that is to streamline design methodologies related to the miniaturization of quantitative-analysis protocols. For this purpose and motivated by our benchtop studies, explained later in Chapters 2, 9, and 10, our first goal is to pinpoint the main characteristics of these protocols, whereby these characteristics can be incorporated into the algorithms underlying the design-automation and security flows.

An experimental framework for quantitative-analysis studies must ultimately be designed with considering the following key objectives:

1. From a technical perspective, a protocol framework needs to be optimized to utilize the lowest amount of samples to obtain answers to questions in biology with the highest possible precision. A clear example of this challenge is the realization of ChIP-seq protocol, wherein reducing the number of cells is a major goal to make ChIP-seq widely adopted [281]. However, sample reduction in turn entails a significant decrease in the precision of reported DNA-protein bindings (i.e., sequencing coverage) [233]. Another example is the realization of parameter-space exploration in synthetic biology, where a number of DNA mixtures are prepared and used to examine the gene-regulatory space of a biological circuit [94]. Reducing the number of samples clearly lowers the system cost and increases the chance of utilizing the framework for larger circuits. Nevertheless, sample reduction may lead to imprecise identification of a gene-expression profile. As a result, co-optimization of framework cost and application performance is essential.

2. From a biology perspective, an experimental framework aims at the *identification* of biomolecules—also known as up-stream analysis—that exhibit

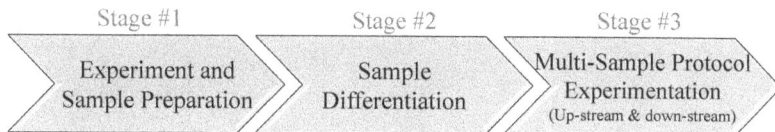

Stage #1	Stage #2	Stage #3
Experiment and Sample Preparation	Sample Differentiation	Multi-Sample Protocol Experimentation (Up-stream & down-stream)

FIGURE 1.14

The proposed workflow of biomolecular quantitative analysis.

specific or abnormal biological behavior (e.g., fluorescence-based gene expression/suppression due to enzymatic reaction) or the *causative exploration* of biomolecules—also known as down-stream analysis—that contribute to the *in vivo* interactions leading to such biological behavior (e.g., the impact of protein binding to DNA at specific genomic regions). Such causative exploration therefore requires a method for indexing/barcoding samples at the end of up-stream analysis to account for cellular heterogeneity [300].

As a result, the workflow of biomolecular analysis adopted in this book consists of three main stages: experiment and sample preparation, sample labeling and differentiation, and multi-sample experimentation, as shown in Figure 1.14. The first stage, i.e., biomolecular sample preparation, is carried out seeking co-optimization of experiment goals (e.g., amount of input samples, resources, and application precision) [233]. A decision at this stage impacts the following protocol procedures. The second stage, i.e., sample labeling and differentiation, is focused on developing scalable methods for differentiation of heterogeneous samples within down-stream analysis [136, 113, 119]. The third stage, i.e., multi-sample protocol experimentation, represents the implementation of protocol procedures using multiple sample pathways in order to obtain quantitative results; this stage applies to both up-stream and down-stream analyses [107, 115, 116]. Our goal is to address design-automation and trust challenges associated with these stages, considering all levels of information flow described in Section 1.5: biochemical level, bioassay level, and system level.

Workflow-Related Design-Automation Challenges

We highlight the main design-automation challenges associated with the above stages (see Figure 1.15(a)):

1. **Co-optimization of protocol parameters for precise results:** Biomolecular protocols are employed to capture precise biochemical interactions at the molecule scale. For example, in synthetic biology, protocols are implemented to build complex networks of molecular components and gene-regulatory parameters that can perform new biological functions. Optimal configuration of these parameters requires producing numerous DNA mixtures in order to modulate the parameter space of the biological circuit in a precise

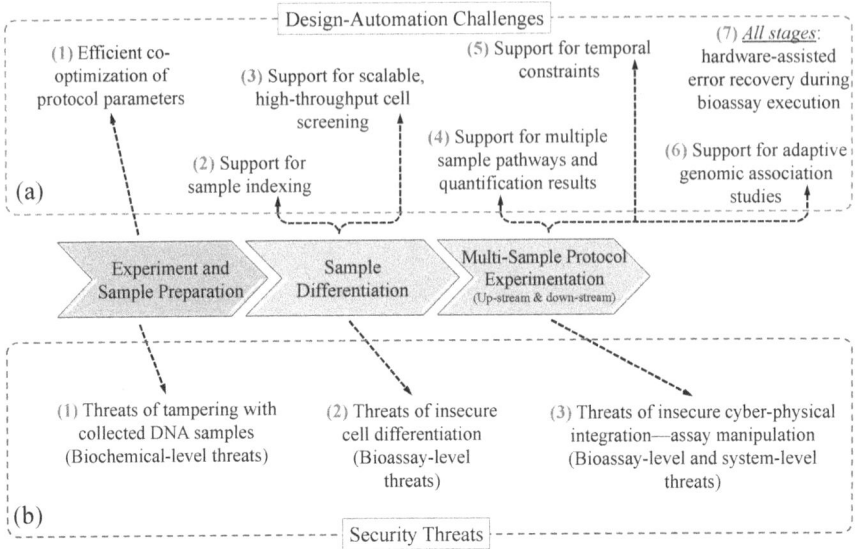

FIGURE 1.15
Research challenges associated with the proposed biomolecular workflow: (a) design-automation challenges; (b) potential security threats.

manner [94, 106]. This process however can be cost-prohibitive, especially for a design with a large number of parameters. Therefore, a co-optimization design methodology is needed to investigate the correlation between the cost of DNA sample production and protocol precision.

2. **Support for indexing/labeling samples (cells):** Previous microfluidic solutions have been developed to classify cell types (using cell sorting) or perform biochemical analysis at the cell-level on pre-isolated types of cells [95]. With the emergence of point-of-care microfluidic platforms for single-cell analysis, new techniques are needed to efficiently classify cells and conduct biochemical experiments on multiple cell types concurrently. System integration and design automation are major challenges in this context. To enable such an integrated system, it is necessary to develop a microfluidic platform and a design-automation framework that combine the following steps of single-cell experimental flow: (1) cell encapsulation and differentiation; (2) sample-indexing (barcoding) to keep track of the samples identity; (3) type-driven biochemical analysis [113, 114].

3. **High-throughput and scalable screening of heterogeneous samples:** Single-cell screening is a necessary step for sample differentiation (Stage #2 in Figure 1.14), and many microfluidic solutions for single-cell screening/sorting have been developed [35]. However, such solutions need significant improvements so that a fully integrated microfluidic workflow for scalable,

high-throughput single-cell screening becomes a reality. There are three major drawbacks in these solutions: (1) limited sorting capacity: most of the current microfluidic solutions are well-suited for separating only two types of samples via a T-junction configuration [19], even with the use of different-sized magnetic beads for trying to scale the sorting scheme; (2) unreliable sorting: using digital microfluidics for cell soring improves the throughput capacity, however, it relies on a repeated sequence of splitting to separate single cells—the splitting operation on DMFBs is highly variable and may lead to unreliable screening; (3) lack of cell barcoding: screened samples need to be barcoded (indexed) to be prepared for down-stream analysis—such an integrated capability has not been realized with today's screening methods. Hence, it is required to develop a microfluidic system and a design-automation framework to support scalable, high-throughput single-cell screening [119, 120].

4. **Independent sample pathways, online decision making, and quantification of results:** Quantitative results are reported numerically and are compared against an accompanying reference interval for interpretation. For this purpose, multiple samples are treated through independent sample pathways. Subsequently, the distributed detection results of the sample pathways are collected and fused to report a final quantitative result. Therefore, to support multiple sample pathways, it is required to extend the traditional sequencing graph model to represent the associated bioassay protocols for all samples. Note that the procedures associated with these protocols can either be identical for all samples or be unique to each sample type according to the application specifications. This modeling technique is similar to the multifunctional design scheme presented in [286], which combines a set of multiplexed bioassay protocols in the same platform. Our solution, however, executes the bioassay protocols concurrently and collects detection results from all these protocols (using firmware) to reach a meaningul conclusion [116, 115].

Another important feature of quantitative protocols, especially epigenetic protocols, is the inherent uncertainty about the order of basic fluidic steps for each sample. In other words, our recent benchtop experience with multiple sample pathways highlights the need for incorporating decision-making and adaptation capability into our framework. For example, nucleic-acid isolation during the execution of qPCR-based gene-expression analysis must be followed with a decision on whether the quality of the isolated NA allows subsequent cDNA synthesis or requires further steps to get rid of the residual protein and debris—such "if-then-else" reconfiguration can be performed either automatically using cyberphysical integration or using human-in-the-loop if necessary.

5. **Support for temporal constraints:** Temporal constraints may arise during protocol execution due to physical phenomena such as droplet evaporation or deadlines imposed by the target chemistry, e.g., degradation of samples and reagents [123]. Therefore, we need to provide a systematic approach that handles such constraints during the course of an experiment in order to ensure robust and reproducible quantitative results. An example of an upper-bound

temporal constraint can be seen in ChIP protocol, in which the protein is initially cross-linked to DNA. It has been shown that an excessively long cross-linking time results in the majority of chromatin being resistant to shearing, thus degrading chromatin shearing process [33]. Therefore, an upper-bound temporal constraint on cross-linking reactions must be specified; this temporal constraint is often dependent on the sample type [107].

6. **Support for scalable and adaptive genomic-association studies:** One of the most important application areas in quantitative analysis is cancer-genomics, which relies on an interplay between single-cell biology and omics technologies (i.e., DNA sequences, RNA expression levels, proteomics, and other epigenetic markers) and its impact on disease development and evolution (i.e., association studies) [82]. Single-cell biology utilizes several microfluidic and computational techniques, with an ultimate goal of constructing a unified and genome-wide catalog of genetic, epigenetic, and transcriptomic elements [34, 40]. Nevertheless, the application of genome-wide association studies at the single-cell level is extremely complicated and it is hindered by several technical limitations, one of which is the need for a reliable, self-adaptive, and high-throughput control scheme that readily coordinates the experiments of single-cell analysis at a large scale. A framework for such an application may require a cloud-based microfluidic service and a big data infrastructure to enable collaborative and coordinated molecular biology studies [117, 72].

7. **Hardware support for real-time error recovery during the execution of a biomolecular assay:** Due to the inherent randomness and complexity of the component interactions that are ubiquitous in biomolecular assays, it is necessary to verify the correctness of on-chip fluidic interactions during bioassay execution and recover from unexpected errors. As a result, to enable reliable on-chip implementation of biomolecular quantitative analysis, the design of a microfluidic biochip needs to take into consideration error recoverability. Several studies have recently been reported for empowering DMFBs with a control system that can enable real-time recovery from errors during bioassay execution [159, 158, 102, 14, 108]. However, in these methods, a software approach is used to accomplish dynamic reconfiguration for the biochip; that is, a desktop computer must be involved in the control system. The drawbacks of this approach are: (1) increased complexity of the cyber-physical system; (2) then need for a software-resynthesis step, which leads to increased bioassay response time when errors occur. Hence, providing hardware support for error recovery is an essential step towards integrated and reliable microfluidic platforms [103].

In Section 1.7, we specify the scope of this book that is aligned with the above design-automation challenges.

Workflow-Related Security Threats

Concurrent with the above design-automation challenges, adopting the proposed workflow for DNA-processing applications presents security threats, some of which are described below; see Figure 1.15(b).

1. **Threats of tampering with DNA preparation:** Currently, biomolecular analysis frameworks assume that the cells under study were collected and processed in a trustworthy manner and therefore allow making clinical or judicial decisions based on the genomic study. However, an attacker can fabricate blood and saliva samples containing DNA from a person other than the donor of the blood and saliva. Forged evidence can be realized using a strand of hair or drinking cup used by the target person. Forged DNA can then be extracted, amplified, and planted within the collected samples. The attacker can be either an intruder or a technician in a diagnostics or a crime lab. This attack, which is considered as a biochemical-level attack, does not require access to the analysis instrumentation. Obviously, the impact of such an attack is catastrophic; therefore, there is a need to investigate potential attack surfaces and provide suitable defense mechanisms against malicious alteration of DNA samples.

2. **Implications of insecure microfluidic cell-differentiation mechanisms:** As described earlier, a number of single-cell analysis methods have been established using microfluidic devices, which can be widely adopted in low-template DNA settings such as forensic analysis. A cell-differentiation mechanism based on DNA barcoding is utilized to preserve the identity of the genomic samples. However, if cellular samples are not properly barcoded, the integrity of a single-cell application may be compromised even before regular genomic analysis begins [257]. Hence, insecure single-cell microfluidics can have a significant adverse impact on human lives. Implications of insecure microfluidic barcoding utilities at the bioassay level need to be investigated [261].

3. **Security implications of cyber-physical integration on biomolecular protocols:** A typical DMFB platform can be leveraged in quantitative analysis, but it requires other components in addition to the electrode array to enable this process. A general-purpose processor or an integrated hardware unit generates actuation sequences that are then sent to the biochip, off-chip camera or sensors gather information related to the status of the bioassay, and actuators may be required to enable heating or magnetic-bead segregation. These components together form a cyber-physical DMFB, as described in Section 1.3.

The fact that a DMFB is often deployed as a cyber-physical system implies unique security threats [11, 10]. Physical assets as well as protocol data can be vulnerable to alteration or destruction. Reprogrammability and connectivity of such a cyber-physical system represents potential attack surfaces that can be exploited by an adversarial attacker with the aim of modifying an assay sequence or causing denial-of-service. These attacks can be launched at the bioassay level or at the cyber-physical system level [13]. Securing cyber-physical microfluidic platforms is essential, especially those used in mission-critical applications [259, 256, 12].

This book focuses on security assessment and mitigation of threats that belong to #1 and #3. More specifically, we focus on: (1) threats of manipulating assay outcomes and their impact at the bioassay level; (2) security implications of cyber-physical integration in digital microfluidic systems; (3) biochemical-level threats of DNA forgery; In Section 1.7, we specify the scope of this book that is aligned with these threats. Assessment and mitigation of threats that belong to #2 are investigated in [262].

1.7   Book Outline

This book addresses design-automation challenges and security threats associated with the biomolecular workflow. The book follows a front-back approach that is demonstrated using the block diagram in Figure 1.16. In other words, for tackling design-automation challenges, we first present solutions for the multi-sample experimentation stage (the third and front stage), then we move to the second and first (back) stages, respectively. Similarly, discussion on security vulnerabilities and countermeasures of biomolecular assays starts with assay-manipulation and insecure cyber-physical integration threats (vulnerabilities of the third stage) then we move to the threats of tampering with/falsifying collected DNA samples (vulnerabilities of the first stage). The remainder of this book is organized as follows:

Part 1 - A Framework for Real-Time Execution of Multi-Sample Biomolecular Analysis: This part consists of three chapters, which are mainly focused on the third stage of the biomolecular workflow. Chapter 2 starts the discussion on design-automation challenges associated with biomolecular assays. This chapter describes our first benchtop experimental study, which was carried out to understand gene-expression analysis and its relationship to the biochip design specification. Chapter 2 also presents an integrated framework for quantitative gene-expression analysis using DMFBs. The proposed framework includes: (i) a spatial-reconfiguration technique that incorporates resource-sharing specifications into the synthesis flow; (ii) an interactive firmware that collects and analyzes sensor data based on quantitative polymerase chain reaction; (iii) a real-time resource-allocation scheme that responds promptly to decisions about the protocol flow received from the firmware layer. This framework is combined with cyber-physical integration to develop the first design-automation framework for quantitative gene expression. Simulation results show that our adaptive framework efficiently utilizes on-chip resources to reduce time-to-result without sacrificing the chip's lifetime. The work presented in this chapter appears in [116, 115, 112, 111, 110].

Chapter 3 satisfies the temporal constraints by modeling biochip design in terms of real-time multiprocessor scheduling and utilizes a heuristic algorithm to solve this NP-hard problem. The presented framework performs fluidic task assignment, scheduling, and dynamic decision-making for quantitative epigenetics. Simulation results show that the proposed algorithm is computationally efficient and it generates effective solutions for multiple sample pathways, while taking into consideration the temporal constraint imposed by droplet evaporation. Experimental results using an embedded micro-controller as a testbed are also presented. The work presented in this chapter has been published in [107, 112, 111].

Chapter 4 starts the discussion on scalable single-cell analysis. This chapter presents a research vision to design a large-scale CPS experimental framework to enable collaborative and coordinated single-cell molecular biology studies. This framework will be based on the integration of CPS with microfluidic biochips and cloud computing. It has the potential to drastically advance personalized medicine through knowledge fusion among many research groups, and synchronization of research planning. This framework therefore leads to a better understanding of diseases such as cancer, and helps researchers in

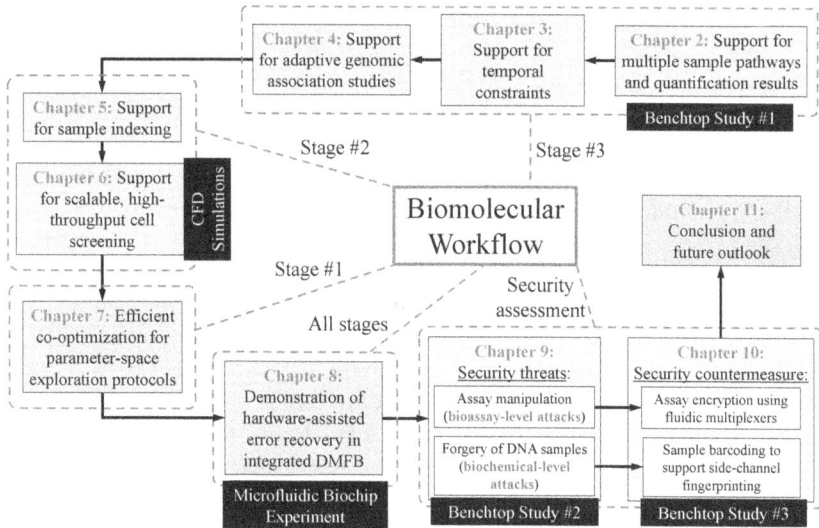

FIGURE 1.16
A block diagram of the book outline.

identifying effective treatments. A case study from cancer research is discussed to explain the significance of our framework in promoting coordinated genomic studies. The work presented in this chapter appears in [117, 72].

Part 2 - A Framework for High-Throughput Single-Cell Analysis: This part consists of two chapters, which focus on single-cell analysis and support for sample differentiation, i.e., the second stage of the biomolecular workflow. Chapter 5 presents a hybrid microfluidic platform that enables complete single-cell analysis on a heterogeneous pool of cells. This architecture is combined with an associated design-automation and optimization framework, referred to as Co-Synthesis (CoSyn), which facilitates scalable sample indexing/barcoding. The proposed framework employs real-time resource allocation to coordinate the progression of concurrent cell analysis. Simulation results show that CoSyn efficiently utilizes platform resources and outperforms baseline techniques. The work presented in this chapter appears in [113, 114].

Chapter 6 presents a pin-constrained design methodology for high-throughput single-cell screening using reconfigurable flow-based microfluidics. The proposed design is analyzed using computational fluid dynamics simulations, mapped to an RC-distributed model, and combined with inter-valve connectivity information to construct a high-level synthesis framework, referred to as Sortex. Simulation results show that Sortex significantly reduces the number of control pins and fulfills the timing requirements of single-cell screening. The work presented in this chapter appears in [119, 120].

Part 3 - A Framework for Biomolecular Parameter-Space Exploration and Hardware-Assisted Error Recovery: This part consists of two chapters, which address the remaining design-automation challenges—more specifically, it presents design methods for protocol parameter-space co-optimization (the first stage of the biomolecular workflow) and real-time error recovery for biomolecular quantitative analysis. Chapter 7 introduces a design-automation flow for protocol parameter-space co-optimization, which is especially useful for enabling biomolecular analysis using synthetic biocircuits. A fundamental challenge in building reliable biocircuits has been the lack of cost-effective design methodologies that enable optimal configuration of gene-regulatory parameters. Such a configuration can be obtained only using a systematic microfluidic assay, named biocircuit-regulatory scanning (BRS), that produces numerous (and representative) reagent mixtures to modulate the parameter space of a biocircuit. The design flow proposed in this chapter addresses the above challenge using: (1) a statistical method that computes volumetric ratios of reagents in each mixture droplet; (2) an Integer Linear Programming (ILP)-based synthesis method that implements a BRS assay on a MEDA biochip; and (3) an iterative decision-making utility based on regression analysis to control the accuracy of the parameter-space. Simulation results show that the proposed flow efficiently utilizes reagent fluids and enhances the prediction accuracy of a parameter space. The work presented in this chapter appears in [106].

Chapter 8 introduces error recoverability to DMFB implementation of quantitative-analysis protocols. This chapter describes an experimental demonstration of the first practical and fully integrated cyber-physical error-recovery system that can be implemented in real time on a field-programmable gate array (FPGA). The hardware-assisted solution, which can be applied to all stages of the biomolecular workflow, is based on an error dictionary containing the error-recovery plans for various anticipated errors. The dictionary is computed and stored in FPGA memory before the start of the biochemical experiment. Errors in droplet operations on the digital microfluidic platform are detected using capacitive sensors, the test outcome is interpreted by control hardware, and corresponding error-recovery plans are triggered in real-time. Experimental results are reported for a fabricated silicon device, and links to videos are provided for the first-ever experimental demonstration of real-time error recovery in cyber-physical DMFBs using a hardware-implemented dictionary. The work in this chapter appears in [103, 111, 110, 108, 109, 10].

Part 4 - Security Vulnerabilities and Countermeasures of Quantitative-Analysis Frameworks: This part consists of two chapters, which address security threats associated with biomolecular quantitative analysis. Chapter 9 provides an assessment of the following security vulnerabilities: (1) bioassay-result manipulation, which can be triggered by bioassay-level attacks; (2) insecure cyber-physical integration in microfluidic biochips, which can form a backdoor for system-level attacks; (3) forgery of biomolecular samples, which is imposed by biochemical-level attacks. First, result-manipulation attacks on a DMFB are identified, where these attacks can maliciously alter the assay outcomes. Two practical result-manipulation attacks are shown on a DMFB platform that performs enzymatic glucose reactions on serum. Second, we identify vulnerabilities of the current design flow for cyber-physical microfluidic systems that can represent an obvious attack surface. We show that a cyber-physical system design that was intended for error recovery can be abused by a malicious attacker to leak sensitive information or launch sensor-spoofing attacks. Third, a benchtop experimental study is presented to demonstrate three different attacks on DNA preparation for gene-expression analysis: positive denial-of-service attack, negative denial-of-service attack, and sample switching/forgery attack. The proposed study explains the catastrophic impact of each attack on the outcome of the DNA-identification process. A part of the work presented in this chapter appears in [11, 13].

Chapter 10 extends the security analysis started in Chapter 9 by presenting defense mechanisms that secure the execution of biomolecular assays. First, to prevent attacks against DMFB-enabled bioassays, an encryption method based on digital microfluidic multiplexers is introduced, where an assay is encrypted with a secret-key pattern of fluidic operations. With this scheme, only an authorized user of the DMFB can obtain the correct assay outcome. Simulation results show that for practical assays, e.g., protein dilution, an 8-bit secret key is sufficient for overcoming threats to DMFBs. Second, to address DNA-forgery threats, a benchtop experimental study is presented to demonstrate a biochemical-level countermeasure solution, where

DNA samples are barcoded using unique DNA biomarkers. The barcoding scheme can also provide effective side-channel fingerprinting mechanisms that can secure quantitative-analysis frameworks against a wide range of threats, e.g., piracy threats. Although this countermeasure is demonstrated using a benchtop setting, it can be mapped into a microfluidic platform using the framework explained in Chapter 5. A part of the work presented in this chapter appears in [12].

Finally, Chapter 11 summarizes the contributions of the book and identifies directions for future work.

Part I

Real-Time Execution of Multi-Sample Biomolecular Analysis

2

Synthesis for Multiple Sample Pathways: Gene-Expression Analysis

In this chapter, we introduce an integrated platform and design-automation solution for quantitative analysis, e.g., the study of gene expression in molecular biology [112, 116, 115]. We first describe our benchtop experiment that was implemented to pinpoint design-automation challenges. We next present a method for physical-aware resource allocation for multiple sample pathways. The main contributions of this chapter are as follows:

- We first discuss our benchtop experimental study for gene-expression analysis along with the resulting outcomes. This experimental approach is used to guide biochip synthesis for the underlying protocol.

- We present a spatial-reconfiguration technique that incorporates resource-sharing constraints into the synthesis flow.

- We describe and evaluate a physical-aware resource-allocation framework that enables tunable resource allocation among bioassays.

- Motivated by the $\mathcal{C}^5$ architecture in Section 1.4, we bridge the gap between cyber and physical spaces in DMFBs for quantitative analysis by introducing a software library for an interactive firmware layer. This layer collects and analyzes data from on-chip sensors. Thus, it communicates flow decisions to the resource-allocation scheme. We also describe a software utility that enables the firmware to provide quantitative results in existence of imprecise amplification data.

Note that even though we consider the gene-expression analysis protocol in this chapter due to its widespread use, the proposed framework is amenable to alternative methods for quantitative analysis that seek the quantification of other sample properties such as concentration (in glucose testing) [238]. In other words, it is not limited to the measurement of relative abundance of a gene expression as considered in this chapter. The proposed solution can be used to solve the synthesis problem for down-stream protocols, which use samples that were processed and differentiated at an earlier stage; see Figure 1.14. The integration of this framework with a sample-differentiation utility is introduced later in Section 5.

The rest of the chapter is organized as follows. An introduction to gene-expression analysis and the implemented benchtop experiment are presented in Section 2.1. Quantitative protocol miniaturization using DMFBs and the associated layered-software support are described in Section 2.2. Section 2.3 explains the requirement of spatial reconfiguration for the efficient realization of protocols. Next, the shared resource-allocation algorithm is presented in Section 2.4. Section 2.5 describes the software architecture of the interactive firmware layer. Finally, results of our experimental evaluation are presented in Section 2.6 and conclusions are drawn in Section 2.7.

2.1   Benchtop Protocol for Gene-Expression Analysis

The objective of the experiment was to study the transcriptional profile of a green fluorescent protein (GFP) reporter gene under epigenetic control. In this experiment, three types of *S. pombe* strains were analyzed: (i) control (GFP not under epigenetic control) strains $(\mathcal{S}_c)$; (ii) experimental (GFP under epigenetic control) strains $(\mathcal{S}_e)$; (iii) wild-type strains used as a reference to improve the outcome efficiency $(\mathcal{S}_r)$. Note that these strains were run manually and independently in a set of reactions in which the investigated gene is GFP. Ultimately, the need to improve the efficiency of the experiment and target combinations of multiple genes impose practical limitations on the benchtop approach. To reduce the experiment time and to minimize the process cost, it is necessary to design a digital-microfluidic platform that can run tens of droplets through independent sample pathways under automated control.

Figure 2.1 depicts a flowchart of the protocol for quantitative gene-expression analysis that we have studied using a benchtop setup. The expression level of *S. pombe* strains were analyzed by qPCR following cell lysis, mRNA isolation and purification, and cDNA synthesis. The details of the experiment are described below:

(1) The samples were first placed in culture medium and cells were grown overnight under controlled condition. Cell culture is a critical step that is needed to ensure to that cells remain alive after they are isolated from their natural environment; i.e., a tissue [168].

(2) In order to evaluate cell concentration and assess whether a cell environment is uncontaminated, the cells of each sample were observed under a Phase Contrast Microscope. Contaminated cells or samples with improper cell concentration must be discarded and the cell-culture process is repeated.

(3) Cell lysis was performed for each sample to release intracellular contents, e.g., protein, nucleic acid (DNA and RNA), and cell debris. Glass beads were used for cell lysis, since glass beads can mechanically disrupt cell walls.

(4) To get rid of protein and cell debris, nucleic acids (NAs) were isolated and precipitated with the aid of 100% ethanol. The NAs were then resuspended in ultra-purified DEPC water.

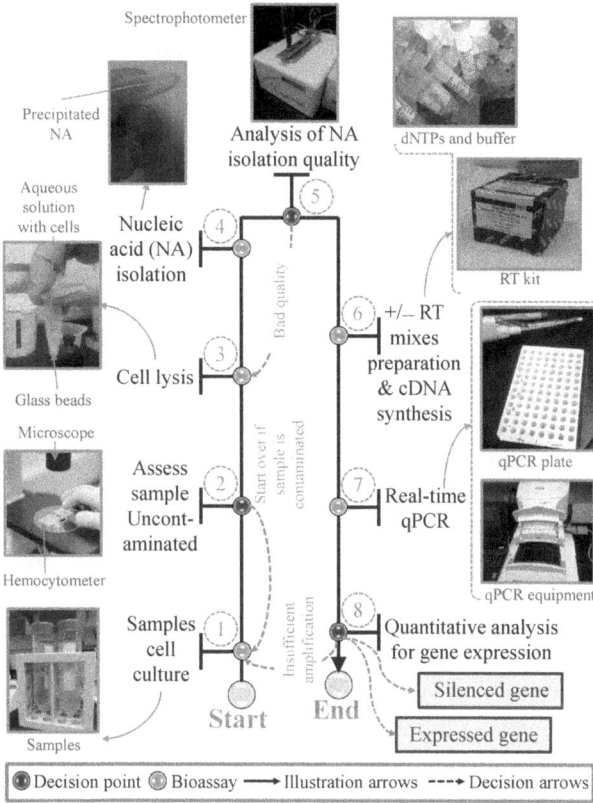

FIGURE 2.1
A protocol flowchart for quantitative analysis of gene expression using a benchtop setup.

(5) The purity of the isolated NA for each sample was assessed by spectrophotometry. A spectrophotometer was used to evaluate the light absorbance for all samples—that is an indication to the isolation quality [223]. Note that a sample with poorly isolated NA (i.e, NA co-exists with protein) has to be discarded and we have to start over with a new sample of the same type.

(6) We proceeded to this step after successful isolation of NA. Using "QuantiTect Reverse Transcription (RT)" kit (from Qiagen, Inc.), a DNA-wipe-out (DNAse) was added to each sample to eliminate all the DNA, leaving the mRNA in the solution. Recall that NAs consist of DNA and RNA molecules. Next, positive and negative RT mixes (constructed from buffers and dNTPs) were prepared to assist in generating the complementary DNA (cDNA) for these particular strains. Negative mixes (also known as negative or no-template controls) serve as a general control for extraneous nucleic acid contamination [38].

(**7**) Finally, DNA amplification through qPCR was carried out using the following steps: (i) Each sample was iteratively 10-fold diluted to create multiple copies; each being with a unique DNA concentration (serial dilution). (ii) qPCR reactions were performed in duplicate with GFP-specific primers (positive and negative RT mixes) in the presence of SYBR green using a BioRad iCycler (BioRad, Inc).

Note that the expression level of GFP in the *S. pombe* samples needs to be quantified with respect to a reference level. This was achieved by qPCR analysis using gene-specific primers to the constitutively expressed β-actin gene, which is constitutively expressed constantly under any experimental conditions [196]. Therefore, Steps (1-6) were concurrently conducted twice using the same types of samples; we used a GFP-complementary primer in the first run and a β-actin-complementary primer in the second run. The dilutes of both runs were included in the qPCR analysis.

Furthermore, note that in the described protocol, intermediate decision points have been used to control the protocol flow for every sample. To illustrate the fundamental role of these decision points, we present the outcome of NA-isolation assessment (Step 5 in Figure 2.1) for two different samples $\mathcal{S}_A$ and $\mathcal{S}_B$. According to [223], the use of spectrophotometer enables us to distinguish NA from protein, which ideally have absorbance maxima at 260 and 280 nm, respectively. Typically, the ratio of absorbances at these wavelengths (referred to as 260/280 ratio) has been used as a measure of purity in both NA and protein extractions. A resulting ratio in the range 1.7 to 2.0 indicates a purely isolated NA.

As shown in Figure 2.2, the spectrum of sample $\mathcal{S}_A$ as well as the 260/280 ratio indicate that NA was properly isolated and extracted in this sample; thus we proceeded to the following step. On the other hand, the spectrum of $\mathcal{S}_B$ and its associated ratio show that NA is contaminated with protein; i.e., it was poorly isolated. In this case, $\mathcal{S}_B$ cannot be used since it leads to inaccurate quantification results. Therefore, sample $\mathcal{S}_B$ was discarded and a new sample belonging to the same type was resuspended for the reaction.

It is therefore evident that the incorporation of sample-dependent decision points is crucial for accurate quantitative analysis.

Protocol Efficiency and Quantitative Analysis:

Determination of the amplification efficiency is a critical step in a quantitative-analysis protocol. It gives a measure of whether the DNA amplification is approximately the same for all samples—that is a necessary condition for successful gene-expression analysis. qPCR provides the following terminologies for the assessment of DNA amplification [196].

- **Sample amplification plot:** It is a sigmoid-like plot (Figure 2.3) which shows the PCR cycle number on the x-axis, whereas the fluorescence from the amplification (which is proportional to the amount of the amplified

260/280: 1.68 ~ 1.7

260/280: 1.36 ~ 1.4

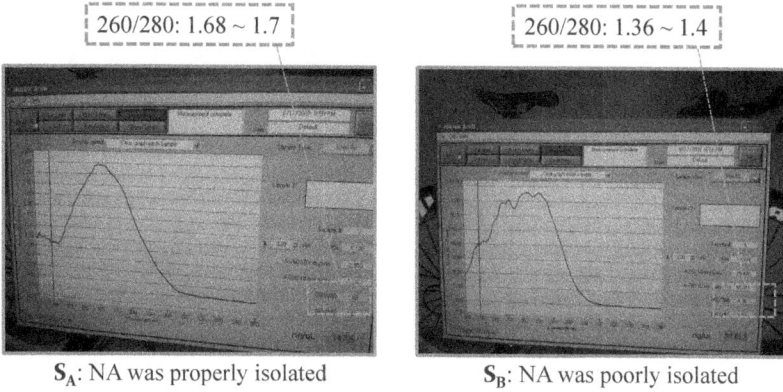

$\mathbf{S_A}$: NA was properly isolated $\mathbf{S_B}$: NA was poorly isolated

FIGURE 2.2
Assessment of NA isolation quality for two samples $\mathcal{S}_A$ and $\mathcal{S}_B$.

product in the sample) is shown on the y-axis. An important aspect of this plot is the threshold cycle $\mathcal{C}_T$, which signifies the cycle at which sufficient amplified product starts to accumulate to yield a detectable fluorescence signal. $\mathcal{C}_T$ for a sample is calculated based on a pre-specified threshold. Figure 2.3 shows amplification plots based on our benchtop experiments.

- **Standard curve:** A powerful way to determine whether qPCR implementation is optimized is to run serial dilution on every sample that has purified NA. The diluted samples are then amplified and used to generate a standard curve (Figure 2.3). Construction of a standard curve is carried out by plotting the log of the sample's dilution factor (10-fold dilution factor in our experiment) against the $\mathcal{C}_T$ value obtained during the amplification of each dilution.

Next, the amplification efficiency E^* is calculated based on the slope of the standard curve using the following formula [26]:

$$E^* = (10^{-\frac{1}{slope}} - 1) \times 100\%$$

where an ideal reaction implies that $E^* = (2 - 1) \times 100\% = 100\%$; i.e., there is a 2-fold increase in the number of DNA copies with every cycle. Normally, an amplification efficiency in the range 95–105% is acceptable.

Once amplification efficiencies are assessed to be nearly 100%, we quantify the gene-expression of GFP gene (target gene) relative to β-actin gene based on $\mathcal{S}_e$ and $\mathcal{S}_c$. The $2^{-\Delta\Delta\mathcal{C}_T}$ (Livak) method is used for quantification as follows [153]:

- **Step 1:** Normalize the $\mathcal{C}_T$ of GFP to that of β-actin for both types of samples:

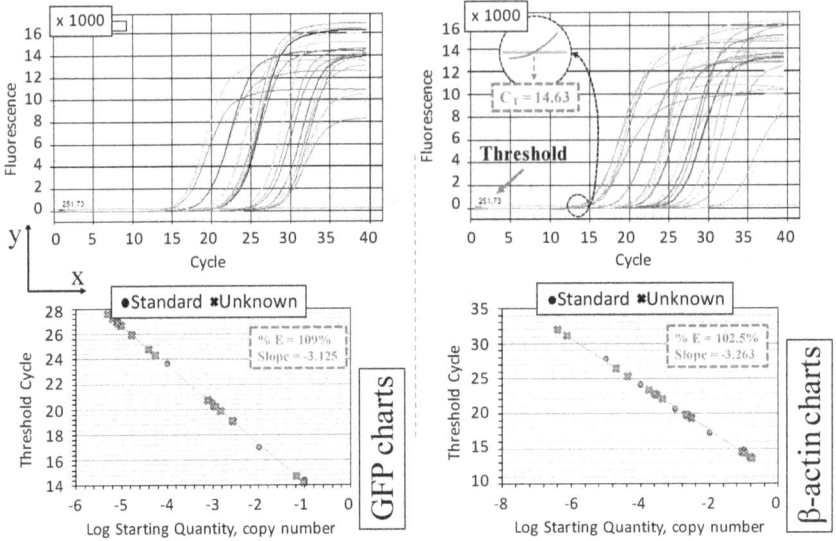

FIGURE 2.3
Generated amplification and standard curves for samples which target GFP and β-actin (housekeeping) genes. Results are obtained using dilutes of $\mathcal{S}_c$, $\mathcal{S}_e$, and $\mathcal{S}_r$ and recorded using a BioRad iCycler.

$$\Delta\mathcal{C}_{T(\mathcal{S}_e)} = \mathcal{C}_{T(\text{GFP},\mathcal{S}_e)} - \mathcal{C}_{T(\beta\text{-actin},\mathcal{S}_e)}$$

$$\Delta\mathcal{C}_{T(\mathcal{S}_c)} = \mathcal{C}_{T(\text{GFP},\mathcal{S}_c)} - \mathcal{C}_{T(\beta\text{-actin},\mathcal{S}_c)}$$

- **Step 2:** Normalize the $\Delta\mathcal{C}_T$ of $\mathcal{S}_e$ to that of $\mathcal{S}_c$:

$$\Delta\Delta\mathcal{C}_T = \Delta\mathcal{C}_{T(\mathcal{S}_e)} - \Delta\mathcal{C}_{T(\mathcal{S}_c)}$$

- **Step 3:** Finally, calculate the expression ratio as follows:

$$2^{-\Delta\Delta\mathcal{C}_T} = \text{Normalized expression ratio}$$

The result obtained in this manner is the fold increase (or decrease) of the expression of the target gene normalized to the expression of a housekeeping gene. Hence, a target gene is said to be expressed if the fold increase is above a certain limit. This global result can then be used to guide the next steps for studying epigenetic regulation.

2.2 Digital Microfluidics for Gene-Expression Analysis

In this section, we present the proposed mapping of gene-expression analysis to DMFBs.

Protocol Miniaturization

Our benchtop experience with multiple sample pathways highlights the need for incorporating decision-making and adaptation capability, hence we have developed the enhanced and miniaturized protocol shown in Figure 2.4. On-chip operation begins with the dispensing of sample droplets containing cultured cells. The cells are then lysed in order to obtain intracellular materials (DNAs, RNAs, proteins, etc.). Using magnetic beads, enzymes, and a washing step, mRNA can be isolated and then reverse-transcribed into the corresponding complementary DNA (cDNA) with primers and other reverse-transcription reagents. Next, the resulting cDNA samples are subjected to thermal cycling via qPCR to amplify the target gene.

Similar to the benchtop approach, miniaturization of gene-expression analysis requires the execution of this protocol on two sample droplets, one of which is used to quantify the amplification of the gene-under-investigation, whereas the other droplet is used to quantify the amplification of a reference gene, also known as a housekeeping gene [196]. Since the primers are gene-specific, the two droplets are chemically treated using different types of primers. However, based on the outcomes of the intermediate decision points, the two droplets can be utilized in an unpredictable manner by different bioassays; this unpredictability makes microfluidic control difficult. Furthermore, with the randomness exhibited in molecular interactions, effective gene-expression analysis requires that the experiment be conducted on at least three replicates. The expression level is first calculated for each replicate, then averaged across the three replicates.

Hence, designing an autonomous digital-microfluidic system for gene-expression analysis requires the concurrent manipulation of independent samples. Utilizing the decision points shown in Figure 2.4(b), we incorporate sample-dependent decision-making capability into the cyber-physical system. Furthermore, the specification of the protocol efficiency and the level of gene expression are included on-chip in the feedback system. Illustration of the firmware design that assists in evaluating the protocol efficiency and gene-expression level on DMFBs is presented in Section 2.5.

There are major advantages of using microfluidic platforms compared to benchtop (conventional) setup for gene-expression analysis. For example, it has been reported in [155] that PCR execution using a DMFB is three times faster than using a conventional setting, and it also offers an inexpensive miniaturized form of the PCR assay. In addition, microfluidic platforms are

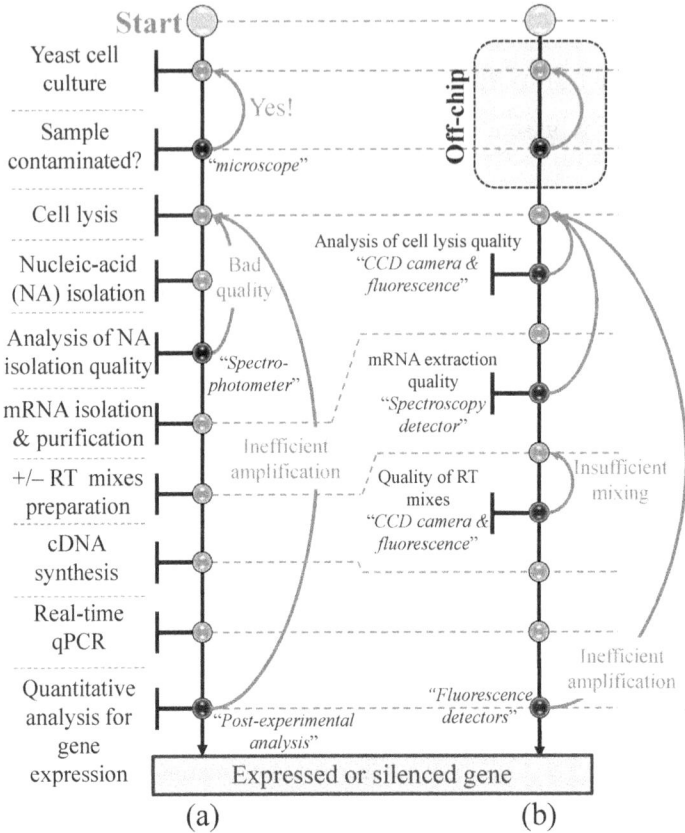

FIGURE 2.4
Mapping of the quantitative protocol for gene-expression analysis from (a) a benchtop approach to (b) DMFBs.

versatile, portable, and amenable to POC use, and therefore it can be used in the hospital or outpatient setting. Such characteristics and others motivate the use of microfluidics for microbiology applications such as gene-expression analysis.

Layered-Software Support and Protocol Model

As discussed in Section 1.3, the cyber-physical configuration shown in Figure 1.8 is suitable only for error recovery; such a configuration does not avail DMFBs to support quantitative protocols that carry out analysis based on multiple sample pathways. To fulfill the aforementioned features, the $\mathcal{C}^5$ methodology (presented in Section 1.4) assists in constructing the components of cyber-physical DMFBs to support such quantitative protocols. As shown in

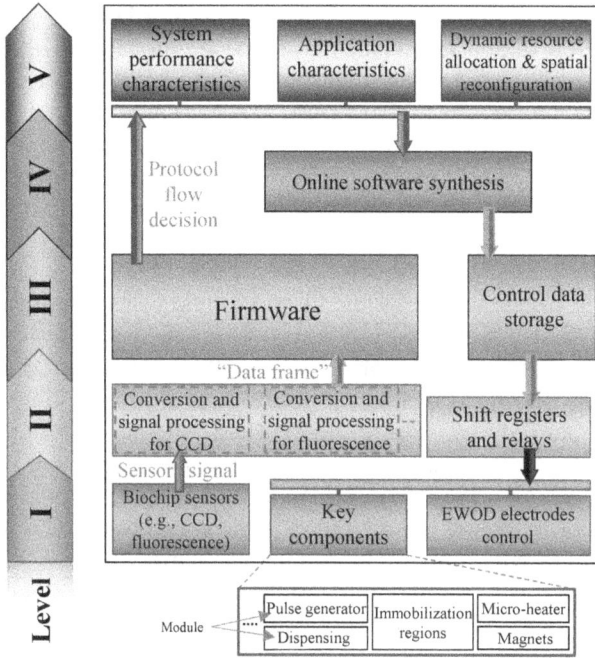

FIGURE 2.5
$\mathcal{C}^5$-compliant architecture design of cyber-physical DMFBs for gene-expression analysis.

Figure 2.5, Level I of the adapted $\mathcal{C}^5$ architecture includes a variety of on-chip sensors and monitoring systems; each designed for a specific functionality. For example, analyzing cell-lysis quality (e.g., assess that samples are not contaminated with bacteria and the cell concentration of the starting samples is acceptable) is accomplished using a CCD camera-based monitoring system, whereas DNA amplification signals are captured by an array of on-chip fluorescence sensors. The Conversion Level (Level II), in turn, is decomposed into partitions; with each partition corresponding to data acquisition, conversion, and encoding from a single sensor. The encoded signal, encapsulated within a "data frame" containing the sensor metadata, is fed into the Cyber Level (Level III) for data processing and analysis.

The Cyber Level (or generally referred to as the firmware layer) is especially important for two reasons: (1) it can be deployed on a microcontroller board and/or a desktop computer, depending on the complexity level of data-analysis task. For example, the data frame generated from a CCD camera after completing cell lysis can be processed on a microcontroller board, whereas a desktop computer is preferable for computing gene expression levels after thermal cycling. (2) it is responsible for communicating analysis decisions into the Configuration Level (Level V) so that the procedure flow of a sample pathway can be adjusted. Meanwhile, with a resource-limited DMFB, the deployment of dynamic resource allocation and spatial reconfiguration at Level V is essential and it is motivated by the following realities about biochemistry protocols [115]: (1) the benchtop characteristics of a contemporary microbiology application (i.e., quantitative analysis), where multiple sample pathways are manipulated concurrently and investigated independently (application characteristics; see Figure 2.5); (2) the need for a robust design of a DMFB, since a non-robust designs may lead to system failure and inefficient quantification (system performance characteristics; see Figure 2.5). While the application characteristics are taken care of by adopting dynamic resource allocation on DMFBs, system constraints are enforced through a spatial-reconfiguration approach. Utilizing both mechanisms in a combined manner ensures a robust miniaturization of a microbiology application on a resource-limited DMFB.

A number of hardware/software co-synthesis techniques for embedded systems capture control dependencies in the task specifications by using conditional process graphs [284, 204, 283]. Similarly, in DMFBs, in order to solve the synthesis problem for the new paradigm, we represent the protocol as a control flow graph (CFG) $G_c = (V_c, E_c)$, in which every node $v_{ci} \in V_c$ (referred to here as a *supernode*) signifies a bioassay, e.g., cell lysis, and a directed edge $e_c(v_{c1}, v_{c2}) \in E_c$ represents a potential transition (decision-making) path from v_{c1} to v_{c2}. The synthesis framework synthesizes the supernodes based on the decisions made during protocol execution.

We also model the fluid-handling operations corresponding to a supernode v_{ci} using a directed sequencing graph $G_s = (V_s, E_s)$, where the set of nodes V_s represents the types and timing of the operations, and the set of edges E_s shows interdependencies among these operations. Figure 2.6 illustrates the representation of the protocol used for quantitative analysis of gene expression. Due to the mapping of many bioassays onto this platform, the design of the underlying DMFB must include on-chip devices such as heaters, magnets, and sensors that are required to complete the execution of the protocol.

Existing biochip synthesis methods must be extended to handle dynamic adaptation for multiple sample pathways and the inherent uncertainties associated with them. Dynamic resource allocation and spatial reconfiguration in DMFBs are motivated by the following realities about biochemistry protocols: (1) the benchtop characteristics of a contemporary microbiology application (i.e., quantitative analysis), where multiple sample pathways are manipulated concurrently and investigated independently; (2) the need for a robust design of a DMFB, since a non-robust designs may lead to system failure and inefficient quantification. While we take care of the application characteristics by adopting dynamic resource allocation on DMFBs, the design constraints are enforced through a spatial-reconfiguration approach. Utilizing both mechanisms in a combined manner ensures a robust miniaturization of a microbiology application on a resource-limited DMFB.

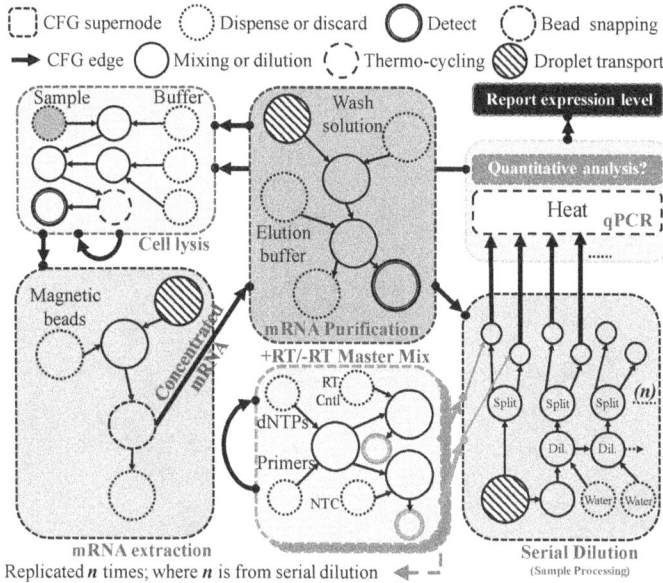

FIGURE 2.6
A CFG representation of the gene-expression analysis protocol for a single sample pathway.

2.3 Spatial Reconfiguration

Resource-Sharing Schemes

Gene-expression analysis using DMFBs relies on the concurrent manipulation of a collection of droplets. Since there is uncertainty about the order of execution of bioassays, and the allocation of on-chip devices among the samples is not known *a priori*, there is a need for a reconfiguration technique that can map specifications of the bioassays to the space of chip resources; we refer to this as *spatial reconfiguration*. We consider three levels of spatial reconfiguration:

Non-reconfigurable scheme ($\mathcal{NON}$): This scheme is adopted by current prototypes; on-chip devices are allocated *a priori* [212, 187]. Inter-bioassay communication is achieved by passing droplets between the dedicated areas.

Restricted resource sharing ($\mathcal{RR}$): The restriction here is in terms of the reconfigurability of the shared devices among bioassays. For instance, a heating device can be shared between qPCR and cell lysis for the purpose of thermal manipulation or even sample processing, but it cannot be used by other bioassays. Note that the electrodes located within the heating regions are unique in terms of their role since they control the droplets that are subjected

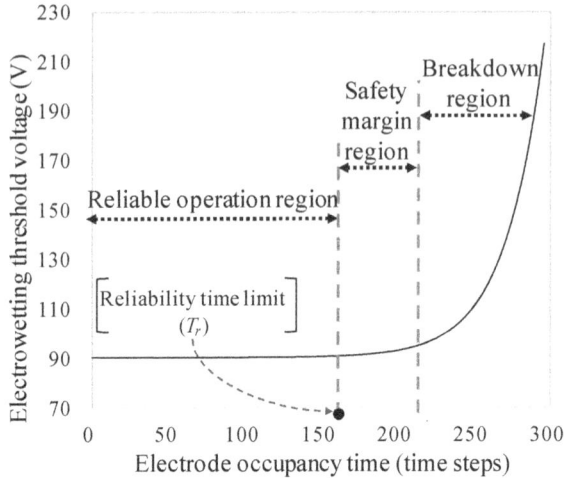

FIGURE 2.7
Electrode degradation model in terms of electrowetting threshold voltage.

to thermal manipulation. Therefore, excessive usage of these electrodes leads to degradation [187], which significantly impacts biochip lifetime. The same concern applies to other on-chip devices such as sensors and electrodes that are in proximity to magnets used for bead-based assays.

Unrestricted resource sharing ($\mathcal{NR}$): In this scheme, no restriction is imposed on resource sharing. Therefore, heating modules cannot only be used for thermal manipulation or sample processing in cell lysis or qPCR, but they can also be utilized by sample processing operations in all bioassays.

Motivation for Degradation-Aware Resource Sharing

For a reliable process, mapping the specifications of the bioassays to the space of chip resources must take into account the degradation caused by a bioassay. In [187], Norian *et al.* provided a degradation model for the electrodes. An electrode's lifetime can be divided into three regions, *reliable operation*, *safety margin*, and *breakdown*; see Figure 2.7. In the reliable operation region, the threshold voltage needed for actuation is constant. The threshold voltage increases linearly in the safety margin region. In the breakdown region, a significant increase in the electrowetting voltage is needed in order to transport a droplet. This increase in voltage, however, quickly leads to dielectric breakdown [195].

The above analysis motivates the need to pinpoint the pros and cons associated with each reconfiguration level. Therefore, we provide an example of mapping a simplified quantitative protocol, which consists of: (a) cell lysis, (b) mRNA isolation and purification, (c) sample processing and DNA

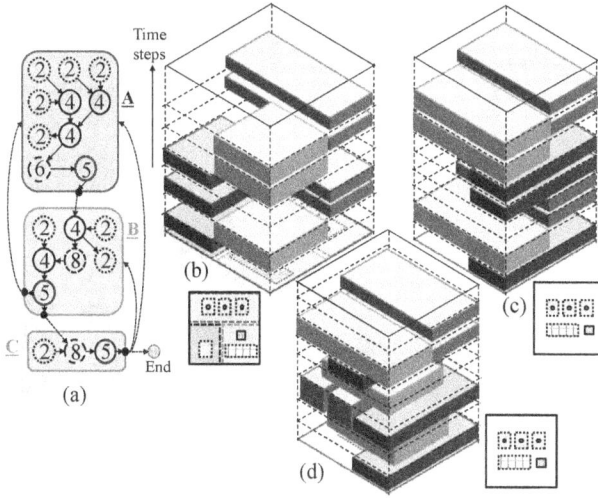

FIGURE 2.8
Illustration of spatial-reconfiguration schemes based on a quantitative protocol described in (a): (b) Non-reconfigurable scheme; (c) Restricted resource-
sharing scheme; (d) Unrestricted resource-sharing scheme.

amplification, as shown in Figure 2.8(a). Similar to our benchtop experiment
(Section 2.1), three samples $\mathcal{S}_1, \mathcal{S}_2$, and $\mathcal{S}_3$ are concurrently subjected to biochemical reactions in which the three experimental pathways are as follows: $\mathcal{S}_1$
traverses the pathway **A-B-C**, $\mathcal{S}_2$ traverses the pathway **A-B-A-B-C**, and $\mathcal{S}_3$
traverses the pathway **A-B-C-A-B-C**. Figure 2.8(b)-(d) shows the mapping
of bioassays into the space of resources for $\mathcal{NON}$, $\mathcal{RR}$, and $\mathcal{NR}$, respectively.

Recall that the electrodes located underneath the heating region are
unique; thus it is imperative to consider the impact of spatial reconfiguration on their lifetime. In Figure 2.9, we evaluate the reconfiguration levels
based on our example with respect to the following criteria: (i) total completion time, (ii) chip size, (iii) number of required heaters, (iv) worst-case heater
occupancy time, which indicates the worst-case degradation level. Based on
this example, it is obvious that the non-reconfigurable scheme is not scalable
as it uses larger chip area and more heaters to complete the protocol even
though both the completion time and the occupancy time are low. Also, it
is apparent that unrestricted resource sharing achieves minimum completion
time, but it reduces chip lifetime. On the other hand, restricted resource sharing decelerates degradation of the chip, but the completion time is higher.
Therefore, there is a need for a resource-allocation scheme that combines the
best of both worlds; i.e., lower completion time and less degradation of the
chip. We refer to this scheme as *degradation-aware resource allocation* and
describe it in the next section.

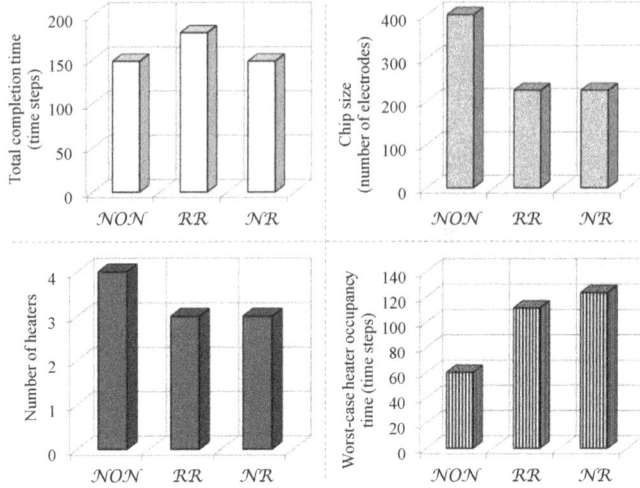

FIGURE 2.9
Comparison of spatial-reconfiguration schemes: $\mathcal{NON}$ refers to the non-reconfigurable scheme; $\mathcal{RR}$ refers to restricted resource sharing; $\mathcal{NR}$ refers to unrestricted resource sharing.

2.4   Shared-Resource Allocation

This section formulates the resource-allocation problem and describes the proposed solution.

Problem Formulation

The notation used in this chapter is listed in Table 2.1. The system described here is composed of three types of modules: (1) *non-reconfigurable* modules (input and output ports), (2) *sample-processing* modules (mixers), and (3) *reconfigurable* modules (heaters, detectors, and magnet regions). Unlike the non-reconfigurable and samples processing modules, the reconfigurable modules are shared among the protocol's bioassays, and access control is managed by the resource allocator. In addition, we make the following important observations:

- Each bioassay $ba_i \in \mathcal{B}$ is mapped to a local space of non-reconfigurable and sample processing modules.

- A shared module $sr_i \in \mathcal{R}_{sh}$ is essential for a bioassay ba_i if and only if the absence of this module leads to a failure in execution of this bioassay. An example is the heater resource for a thermal-cycling bioassay.

TABLE 2.1
Notation used in Chapter 2.

$\mathcal{B}$	The complete set of the protocol bioassays
ba_i	A bioassay ba_i
$\mathcal{R}$	The complete set of the chip modules
$\mathcal{R}_{sh}$	The set of shared, reconfigurable modules
sr_i	The set of shared resources that are essential for a bioassay ba_i
gr_j	The set of shared resources that are granted to a bioassay ba_j
DG_i^j	The degradation caused by a bioassay ba_i on the shared resources gr_j
CT_i^j	The completion time for a bioassay ba_i when the resources gr_j are used

- A shared module $gr_i \in \mathcal{R}_{sh}$ that is granted to a bioassay ba_i *may not* be essential for the execution of the bioassay, *i.e.*, $sr_i \subset gr_i$. Typically, this module can be used for sample processing.

Our problem formulation is as follows:
Inputs: **(1)** The protocol CFG $G_c = (V_c, E_c)$, where $V_c = \{v_{c1}, v_{c2}, ..., v_{cm}\}$ represents the supernodes of n bioassays and $E_c = \{(v_{ci}, v_{cj})\ 1 \leq i, j \leq n\}$ represents data and biological dependency between all pairs of bioassays ba_i and ba_j. A supernode v_{ci} comprises a directed sequencing graph $G_s = (V_s, E_s)$, where $V_s = \{v_{s1}, v_{s2}, ..., v_{sm}\}$ represents m bioassay operations and $E_s = \{(v_{sl}, v_{sk})\ 1 \leq l, k \leq m\}$ represents dependencies between all pairs of operations l and k that belong to the bioassay v_{ci}.
(2) The digital microfluidic library, which describes the types and locations of the on-chip modules.
(3) The resource preferences of every bioassay ba_i, which describes the initial resource requirement. The values of the bioassay's completion time CT_i^j and the degradation level DG_i^j as a function of the granted resources gr_j are also specified.
(4) The resource-allocation constraints (Section 2.4).
Output: Allocation of chip modules to the bioassays such that the constraints on resource-allocation are satisfied.

Resource-Allocation Sequence and Constraints

We have developed a shared-resource allocation scheme based on a timewheel that is controlled by the *coordinator*, as shown in Figure 2.10. The sequence of actions is indicated by the numbers 1–6. Whenever the coordinator receives a command from the firmware layer about the decision for a certain pathway, it firsts stores the command in a global queue until all preceding requests are fulfilled. When the bioassay's command is ready to be processed (essential resources are available), the coordinator forwards it to the *resource allocator* agent. The resource allocator, in turn, checks the preferences of the bioassay and the scheme-specific constraints on resource allocation. Then, the resource allocation is determined and transferred to the *actor* which, in turn, invokes online synthesis for this particular bioassay.

When the restricted resource-sharing scheme is adopted, the resource allocator must ensure that no non-essential shared resource is allocated to the requesting bioassay; *i.e.*, $sr_i = gr_j$. For instance, a bioassay for the preparation of master mixes cannot get access to the heaters, which are not essential for its execution, but it can get access to additional optical detectors to shorten time-to-completion. Therefore, the worst-case computational complexity for allocating resources to a bioassay ba_i is $O(|sr_i|)$.

Conversely, the resource allocator in unrestricted resource sharing enables the option of using non-essential shared resources. In this case, these resources will be used for sample processing. For instance, an mRNA extraction bioassay can get access to a heater to perform sample processing in order to shorten its completion time. As a result, the worst-case computational complexity for allocating resources to a bioassay ba_i is $O(|R_{sh}|)$.

Finally, degradation-aware resource-allocation method initially allocates shared resources to requesting bioassays without restrictions. It also keeps track of the actual degradation levels at all shared sources resulting from the synthesized bioassays. This is achieved via direct communication with the synthesis tool. As a result, when the reliable operation time for the electrodes at a certain shared resource is exceeded, the resource allocator imposes restrictions

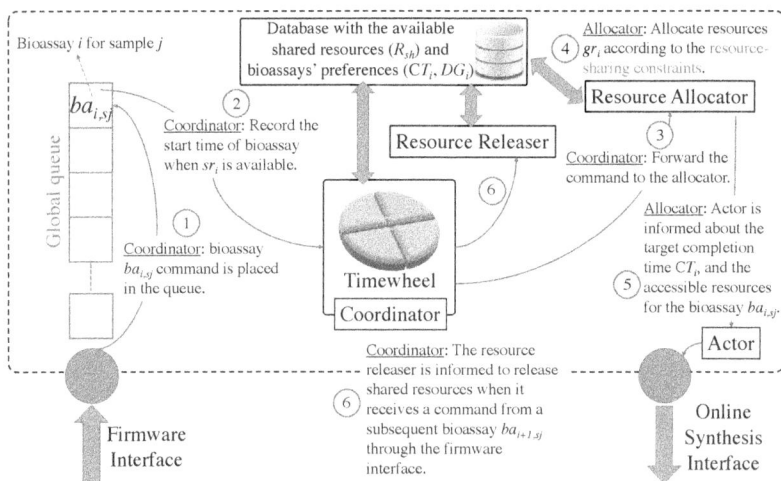

FIGURE 2.10

The components of the shared-resource coordination and allocation.

on accessing this resource; *i.e.*, it switches to restricted resource sharing for this particular resource. Since electrode degradation is considered in resource allocation, resources are sorted according to their degradation level whenever a bioassay ba_i requires resource allocation. Hence, the worst-case computational complexity of this scheme is $O(|R_{sh}|^2)$.

While our methodology is the first to consider dynamic resource allocation in DMFBs, the literature of real-time embedded systems is rich with advanced and cost-effective middleware designs that provide dynamic adaptation [25, 8, 216]. Objectives such as fault tolerance, reduction of power consumption, and reliability of safety-critical applications have been the focus of these designs. Similarly, we foresee many research opportunities for providing optimized dynamic-adaptation techniques in DMFBs.

2.5 Firmware for Quantitative Analysis

The firmware is an interactive software layer that needs to be incorporated in the architecture of cyber-physical DMFBs. This layer is constructed based on a set of C++ libraries that can collect data from on-chip sensors via a signal tracer library, use the collected data to perform real-time data analysis, and feed the upper layer with appropriate decisions. Previous studies overlooked the incorporation of such a layer since the demonstrated designs of cyber-physical DMFBs solely focused on error recovery. The feedback system therefore uses simple signal conditioning and checking schemes to provide decisions. However, advancing cyber-physical DMFBs towards automated quantitative analysis requires taking into consideration the challenges imposed by the distributed sensors. In this section, we focus on the design of the firmware component required for evaluating protocol efficiency and the level of gene expression.

Firmware Components

Figure 2.11 shows an example of a block diagram for the firmware layer that is used for qPCR-based quantitative gene-expression analysis. This layer consists of two libraries: (1) Tracing and queuing (TQ) library; (2) Analysis and decision-making (AD) library. Since multiple sensors are used in the quantification, the TQ library is responsible for interacting with the hardware. Using the sensor-sample map that is generated by the resource allocator, this library can trace the real-time readout of each sensor at every amplification cycle. The synchronization between the library and the underlying sensors is achieved via a designated protocol stack that is much simpler than its counterpart in wireless sensor networks. The circuitry for each sensor must transmit the readout, packetized along with the sensor identification, to the firmware.

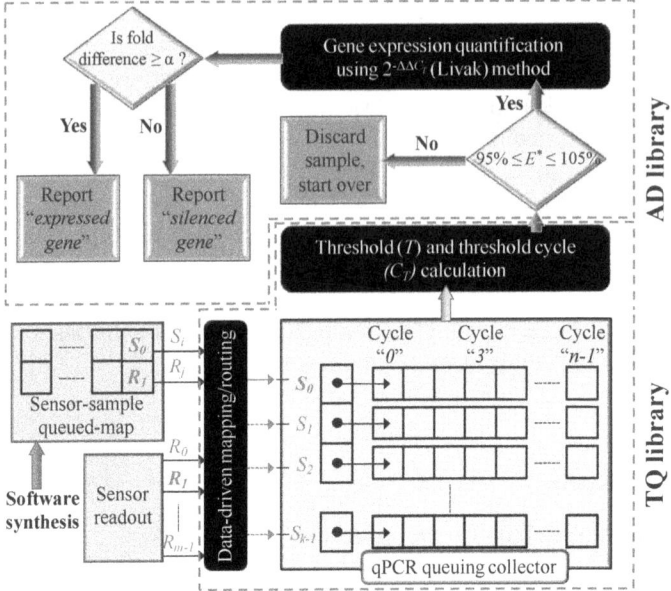

FIGURE 2.11
The structure of the firmware layer for qPCR-based gene-expression analysis.

Next, the firmware matches the received identification information with the sensor-sample map to recognize the detected sample. The received signal is then logged into the library's queuing system at the appropriate location.

When the amplification data is collected, there is a need to use this data to provide useful information. Therefore, the TQ library generates the standard curve for the assessment of the amplification quality. Next, TQ library invokes the AD library to trigger the start of analysis and decision-making. The targets of this library is to execute gene-expression quantification methods. As shown in Figure 2.11, the captured threshold cycles of the samples, communicated from the TQ library, are used to draw the standard curve and determine the efficiency factor E^* (formally defined in Section 2.1). Optimized methods of standard-curve analysis include fit-point method and derivative method [161]. Based on the resulting value of the efficiency factor, the firmware will command the upper layer (that forms the feedback system) to transition the platform into a new start or to report the final result. If the first path is chosen, the command will trigger online synthesis at the upper layer. On the other hand, if the latter path is chosen, the AD library invokes the Livak procedure [153] to quantify the expression of the gene of interest based on the communicated threshold cycles.

Firmware Operation with Amplification Variations

The firmware is a key layer, not only because of its role in collecting sensor data and conveying decisions, but it is also responsible for ensuring that the reported data is meaningful and precise to a large extent. Impreciseness in reporting the level of gene expression might arise due to the following reasons: (1) indeterministic component interactions are ubiquitous in biological processes; as a result, DNA amplification response for the same sample may witness a limited random deviation over several trials [237]; (2) similar to the IC design process, the fabrication process of a DMFB and the integrated devices (e.g., heaters and detectors) are subject to process variations and faults [250]. Unreliable design of the firmware will undoubtedly propagate these inefficiencies to the user side through imprecise gene-expression results.

An effective solution to this problem is based on a probabilistic approach. The protocol experiment is iterated several times, and the results with the highest probability of being precise are reported. A simple probabilistic online approach is to use the mean of the collected data to directly report the results [187]. However, a more accurate scheme will consider the fact that, like several other biological processes [90], the trend of deviation in DNA amplification follows a certain distribution function. With several offline trials, the parameters of such a function are identified. Next, the outcomes of an online experiment (e.g., the level of gene expression) can be probabilistically determined. The firmware design incorporates a tool, called RAmp[1], that carries out such optimization and reports the results based on a probabilistic distribution. Figure 2.12 illustrates the working principle of RAmp assuming a normal distribution over C_T. The acceptable range of $C_{T,on}$[2] can be calculated given a pre-determined threshold on the cumulative distribution function $F(C_T)$. For example, the acceptable range can be associated with the following condition: $F(C_{T,2}) - F(C_{T,1}) = 0.5$. Note that the parameters of the constructed model can be fixed along the spectrum of C_T; this approach is referred to as Static-Variation Model ($\mathcal{SM}$), or they can be curve-fitted to a polynomial function over C_T; we refer to this approach as Dynamic-Variation Model ($\mathcal{DM}$). Additional research and experiments are needed to investigate such probabilistic models through decoupling the effect of indeterminism in biological processes and the impact of process variation in biochip fabrication and operation.

2.6 Simulation Results

We implemented the proposed resource allocation schemes using C++. All evaluations were carried out using a 2.4 GHz Intel Core i5 CPU with 4 GB

---

[1]RAmp stands for $\underline{R}$eliable computation with $\underline{Amp}$lification variations.

[2]We use $C_{T,on}$ to distinguish online C_T values.

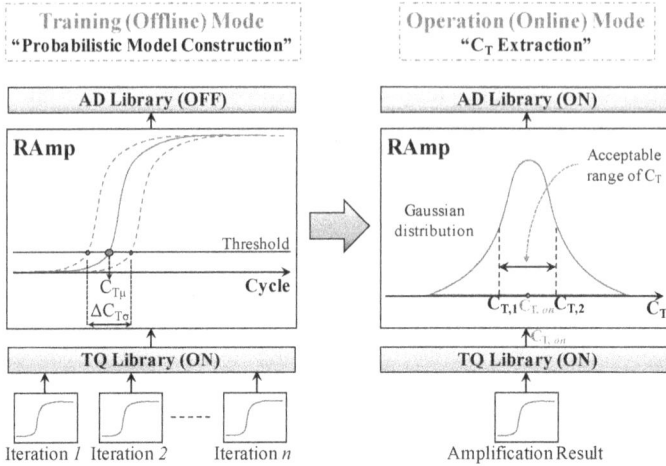

FIGURE 2.12
Using RAmp to: (1) construct a probabilistic model based on amplification variations (offline mode); (2) specify an acceptable range of experimental C_T based on the constructed model (online mode).

RAM. In order to obtain the resource preferences for each bioassay before resource allocation begins, synthesis results were obtained from offline simulation using [86]. The same tool is used for online synthesis when the simulation of real-time quantitative analysis starts. Since this is the first work on optimization for multiple sample pathways, a comparison with prior work is not feasible.

Evaluation of Resource-Allocation Schemes

Two chip arrays have been utilized to simulate the quantitative protocol; these chips are labeled C_{NR} (non-reconfigurable chip) and C_R (reconfigurable chip). In both arrays, the chip has a dedicated port for each liquid and each bioassay also has its own sample processing (SP) region. However, reconfigurable modules, namely heaters (H), magnet (M), CCD camera region (CCD), and optical detectors (OD), are integrated in the shared area of C_R. This area is accessible to droplets from any bioassay via a ring electrode-bus. The chip sizes are 18×18 and 17×17 for C_{NR} and C_R, respectively; these represent the lower bounds on the size required to execute the bioassays. We expect that increasing the array size for C_{NR} and C_R will result in a proportional decrease in completion time. Performance assessment for different chip sizes is left for future work.

We evaluate four resource-allocation schemes: (1) Non-reconfigurable scheme ($\mathcal{NON}$); (2) Restricted resource sharing ($\mathcal{RR}$); (3) Unrestricted resource sharing ($\mathcal{NR}$); (4) Adaptive degradation-aware resource sharing ($\mathcal{DA}$). Using these schemes, we run simulations on the quantitative analysis protocol described in Section 2.2. We consider three samples being concurrently subjected to fluidic operations. The samples are $\mathcal{S}_1$ (GFP gene-targeted sample), $\mathcal{S}_2$ (YFP gene-targeted sample, and $\mathcal{S}_3$ (actin gene-targeted sample). Using the notation in Table 2.2, we consider three different cases in terms of the sample pathways: (1) Short homogeneous pathways (Case I): an optimistic case in which all the three samples follow the same shortest pathway (**CL-mE-mP-MM-SD-TC**); (2) Long homogeneous pathways (Case II): a pessimistic case in which all the three samples follow the same long pathway (**CL-mE-mP-CL-mE-mP-MM-SD-TC**); (3) Heterogeneous pathways (Case III): a realistic case in which the pathways of these samples are different. The considered pathways are as follows: **CL-mE-mP-MM-SD-TC**, **CL-mE-mP-CL-mE-mP-MM-SD-TC**, and **CL-CL-mE-mP-MM-MM-SD-TC**, respectively. Table 2.2 lists the minimum resource requirements for the bioassays.

The metrics of comparison include: (1) The total completion time of the protocol execution (in time steps); (2) the occupancy time (degradation level) for the shared resources (heaters and magnet modules). A time-step refers to the clock period, typically in the range of 0.1 to 1 second [68]. For fair comparison between non-reconfigurable and resource-sharing schemes, the chip array used with the non-reconfigurable scheme (*i.e.*, $\mathcal{C}_{NR}$) is designed such that the resources are sufficient only for a single pathway; therefore, the resources are not replicated with the number of samples. In addition, due to the absence of resource sharing in the non-reconfigurable scheme, the occupancy time is not considered in our analysis. Table 2.3 compares the four schemes based on the chip configuration. As expected, due to the absence of resource sharing in the non-reconfigurable scheme, the chip modules are replicated. For example, three heaters are sufficient for the resource-sharing schemes (in $\mathcal{C}_R$) for thermal manipulation in different bioassays. For $\mathcal{C}_{NR}$, both cell lysis and thermal cycling bioassays require their own heaters, thus increasing chip fabrication cost.

TABLE 2.2

Bioassay notation and resource requirement.

Bioassay	Notation	Minimum resource requirement
Cell Lysis	**CL**	SP, H, and CCD
mRNA Extraction	**mE**	SP and M
mRNA Purification	**mP**	SP and OD
RT Master Mix	**MM**	SP and OD
Serial Dilution	**SD**	SP
Thermal Cycling	**TC**	H

TABLE 2.3

Number of on-chip modules and array electrodes in $\mathcal{C}_{NR}$ and $\mathcal{C}_R$.

Resource	# Modules		# Electrodes	
	$\mathcal{C}_{NR}$	$\mathcal{C}_R$	$\mathcal{C}_{NR}$	$\mathcal{C}_R$
SP	1 (CL), 2 (mE), 2 (mP), 2 (MM), 4 (SD)		176	176
H	2 (CL), 3 (TC)	3	80	48
M	1 (mE)	1	18	18
OD	2 (mP), 2 (MM)	2	36	18
CCD	1 (CL)	1	9	9
Chip size	22	18	18×18	17×17

Based on our simulations, the CPU time, which includes the time for resource allocation and online synthesis, averaged over the three cases (I, II, and III), is less than 1 ms for all allocation schemes. Thus, compared to the protocol completion time (in the order of minutes), the CPU time is negligible.

Case I: Short Homogeneous Pathways

This case arises when all the samples are perfectly grown in a well-controlled medium. In addition, the reagents are contamination-free. We compare the four methods in terms of completion times; see Figure 2.13(a). As expected, non-reconfigurable resource allocation leads to the shortest completion time. Restricted resource sharing, on the other hand, shows the worst completion time. We also note that we can use adaptive, degradation-aware resource allocation ($\mathcal{DA}$) to achieve a short completion time, while the chip resources are not severely degraded (degradation is measured in terms of the occupancy time); see Figure 2.13(b).

Case II: Long Homogeneous Pathways

We next study the case where all the samples are re-suspended for additional bioassays. As shown in Figure 2.13(c) and Figure 2.13(d), the proposed schemes show the same profiles of completion times and degradation levels as in Case I, but with higher values. However, the completion time for $\mathcal{DA}$ is closer to that obtained with restricted resource sharing.

Case III: Heterogeneous Pathways

This is a realistic case that emerges due to the inherent uncertainty about the biological contents of each sample. It introduces the challenge of deciding the allocation of resources among the heterogeneous pathways at runtime. Again, we evaluate the allocation schemes based on the completion time (Figure 2.13(e)) and the degradation level (Figure 2.13(f)).

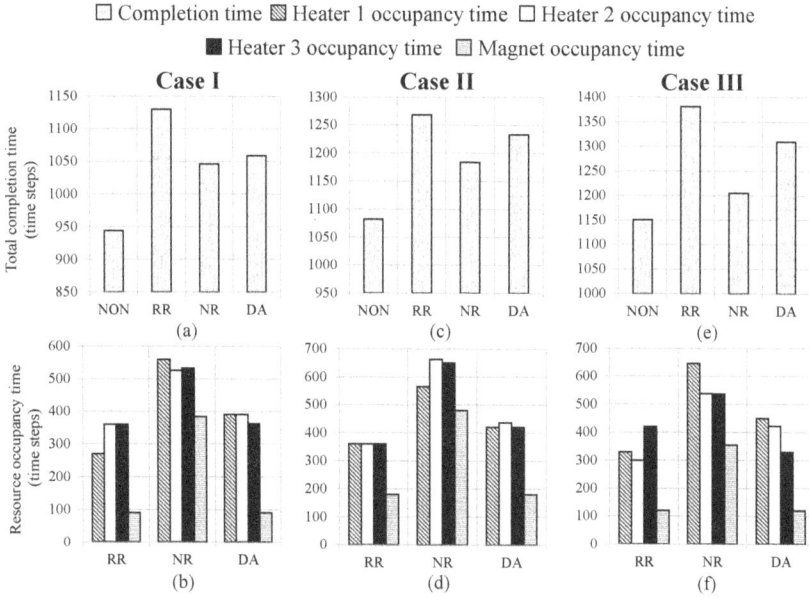

□ Completion time ▨ Heater 1 occupancy time □ Heater 2 occupancy time
■ Heater 3 occupancy time ▨ Magnet occupancy time

FIGURE 2.13
Comparison between the four resource-allocation schemes—Non-reconfigurable ($\mathcal{NON}$), restricted resource sharing ($\mathcal{RR}$), unrestricted resource sharing ($\mathcal{NR}$), and degradation-aware resource sharing ($\mathcal{DA}$)—executing three pathways.

Convergence of Degradation-Aware Resource Allocation

Finally, we study the convergence of the $\mathcal{DA}$ method. We analyze the completion time and the average degradation level of the shared resource-allocation schemes with various lengths of homogeneous pathways, as shown in Figure 2.14. We observe that the completion time for $\mathcal{DA}$ begins to converge to its counterpart in $\mathcal{RR}$ when the length of the pathway is increased. However, due to the restrictions imposed by both $\mathcal{RR}$ and $\mathcal{DA}$ on resource allocation, the degradation levels remain below a certain limit even when we increase the pathway length substantially, as shown in Figure 2.14(b). In real-life scenarios, chip users tend to make optimizations in order to make sure that the protocol is finished as early as possible to avoid droplet evaporation [123]. These scenarios make the $\mathcal{DA}$ method especially attractive in practice.

Variants of Probabilistic Models in RAmp

We demonstrate two variants for probabilistic-model construction and the use of the firmware: (1) Static-Variation Model ($\mathcal{SM}$); (2) Dynamic-Variation

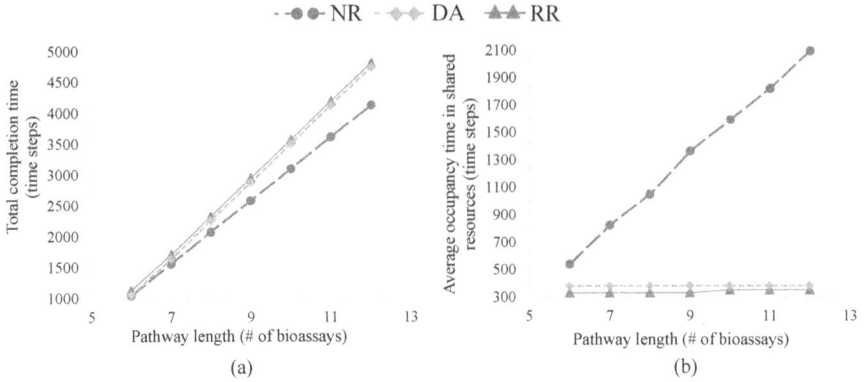

FIGURE 2.14
Performance of the shared resource-allocation schemes with various lengths of homogeneous pathways in terms of: (a) total completion time, (b) average occupancy time.

Model ($\mathcal{DM}$). The following details describe the methodology for training and operation for RAmp:

Input data set (for training): We used the C_T values of β-actin that were extracted from the benchtop experiment (Figure 2.3) as the input data set. Additional values were randomly generated, but clustered to be near the measured data, and included to the set of extracted points in order to enlarge the training set. Therefore, the amplification of each sample droplet is characterized by a "cluster" of possible readings.

Probabilistic models: The input data set for each cluster was fit to a local normal function, as shown in Figure 2.15(a). Next, the standard deviation of the constructed local functions were averaged over C_T spectrum to create a global normal distribution function that is used during operation mode ($\mathcal{SM}$ approach); see Figure 2.15(b). In other words, this global function ($\sigma = 0.7139$) is always used to represent the acceptable range for any input amplification result $C_{T,on}$. On the other hand, to generate $\mathcal{DM}$, the obtained values of standard deviation were curve-fitted furthermore to a function (in our case, it is a first-order polynomial function), as shown in Figure 2.15(c). Based on the training, the standard deviation of the global distribution function will change as $C_{T,on}$ varies, depending on the deviation trend of the original data clusters. Figure 2.15(d) compares the operation outcomes of $\mathcal{SM}$ and $\mathcal{DM}$ for two amplification results ($C_{T,on} = 17$ and 30). At $C_{T,on} = 17$, there is a high chance that this result matches the original input data at $C_T = 17.29$. Accordingly, using $\mathcal{DM}$ leads to a steepened global curve, reducing the standard deviation at this point ($\sigma = 0.6211$). However, at $C_{T,on} = 30$, the $\mathcal{DM}$ flattens the curve ($\sigma = 0.9199$) in an attempt to include the original input data points at $C_T = 27.84$ and 31.56 in the reported probability distribu-

------- Local distributions (clusters) over extracted C_T values ----- Fixed global distribution function for $C_{T,on}$ in $\mathcal{SM}$

------- Linear regression result for $\mathcal{DM}$ ——— Variable global distribution function for $C_{T,on}$ in $\mathcal{DM}$

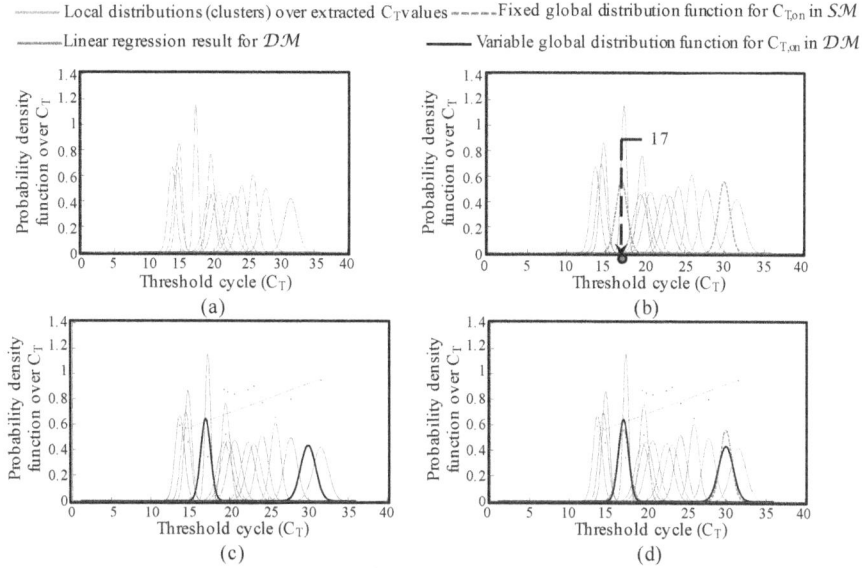

FIGURE 2.15
Illustration of training and operation methodologies in RAmp.

tion. Note that the above points represent corner cases, typical cases such as the range $C_{T,on} : 20 \rightarrow 25$ will exhibit the same behavior for both $\mathcal{SM}$ and $\mathcal{DM}$. Additional statistical research is needed to improve the characteristics of the curve-fitted function. There is a need to reflect the disturbances in qPCR parameters and kinetics of the reaction on this function via stochastic modeling [217].

2.7 Chapter Summary

We have introduced an automated design method for a cyber-physical DMFB that performs quantitative gene-expression analysis. This design is based on a spatial-reconfiguration technique that incorporates resource-sharing specifications into the software synthesis flow. A firmware has been developed to collect data from sensors during runtime, perform analysis, and make decisions about the multiple pathways for samples. We have also presented an adaptive scheme for shared-resource allocation that efficiently utilizes on-chip modules. This method manages resource access between the bioassays. The proposed scheme has been evaluated based on the completion time and the electrode degradation level for realistic multi-sample test cases. We have reported results on the performance of the allocation scheme for various pathway lengths.

3

Synthesis of Protocols with Temporal Constraints: Epigenetic Analysis

In Chapter 2, a synthesis methodology was introduced to support non-trivial biology-on-a-chip applications (e.g., quantitative gene-expression analysis). In this work, a dynamic reconfiguration technique is used to spatially map resource specifications of multiple sample pathways to the biochip devices. This technique is embedded in an adaptive framework that facilitates real-time resource sharing among bioassays and reduces protocol completion time without sacrificing the chip's lifetime.

However, a drawback of the previous method is that it makes spatial reconfiguration decisions at the bioassay level (i.e., "locally"), and it does not capture interactions between multiple sample pathways at the protocol level. In other words, resource sharing among different sample pathways is achieved at a coarse-grained level, thus biochip devices are not efficiently exploited, particularly when further complicated protocols such as gene-regulation analysis (referred to here as epigenetics) are considered. Furthermore, this work does not consider the upper-bound temporal constraints imposed by the application domain; these constraints may arise due to physical phenomena such as droplet evaporation and deadlines imposed by the target chemistry, e.g., degradation of samples and reagents. It was found that the droplets that are subjected to overly long-lasting reactions (especially in a medium of air) are susceptible to evaporation, which impacts the efficiency of enzymatic reactions and alters protocol outcomes [123]. In previous methods, the reaction time is specified using a microfluidic library, which considers the overly pessimistic worst-case timing of bioassay reactions. Moreover, these methods overlook droplet evaporation.

In this chapter, we overcome the above drawbacks by presenting a system design and a design-automation method that carries out task assignment and scheduling for cyber-physical DMFBs for quantitative analysis, e.g., the study of alterations in gene expression or cellular phenotype of epigenetics [112, 107]. The proposed method scales efficiently to multiple independent biological samples and supports on-the-fly adaptation under temporal and spatial constraints. The main contributions of this chapter are as follows:

- We first present the outcomes of our benchtop experiment to motivate the study of epigenetic regulation in fission yeast. Motivated by the experimental results and the $\mathcal{C}^5$ architecture (Section 1.4), we introduce a layered system design for DMFBs. The experimental outcomes are also used to guide the synthesis of the underlying protocol.

- We map the synthesis of epigenetics to real-time multiprocessor scheduling and formulate it as an Integer Programming problem. Since the scheduling is NP-hard, we develop a heuristic for dynamic fluidic task scheduling to respond to protocol-flow decisions. The proposed algorithm provides resource sharing and handles droplet evaporation.

- To promote component-based design in DMFBs [197], the interaction between the proposed algorithm and other system components (actuation and firmware) is demonstrated using an embedded micro-controller board.

The remainder of the chapter is organized as follows. An introduction to epigenetic regulation and the layered system design are presented in Section 3.1. In Section 3.2, we introduce the system model and associated constraints. Next, an algorithm for task scheduling is presented in Section 3.3. Finally, results of our experimental evaluation are presented in Section 3.4 and conclusions are drawn in Section 3.5.

3.1   Miniaturization of Epigenetic-Regulation Analysis

One of the important uses of gene-expression analysis described in Chapter 2 (flowcharts corresponding to the benchtop protocol and the miniaturized implementation are shown in Figure 2.4(a) and Figure 2.4(b), respectively) is in epigenetics, which identifies changes in the regulation of gene expression that are not dependent on gene sequence. Often, these changes occur in response to the way the gene is packaged into chromatin in the nucleus. For example, a gene can be unfolded ("expressed"), be completely condensed ("silenced"), or be somewhere in between. Each distinct state is characterized by chromatin modifications that affect gene behavior [36]. An improved understanding of the *in vivo* cellular and molecular pathways that govern epigenetic changes is needed to define how this process alters gene function and contributes to human disease [279].

Based on our benchtop study of gene-expression analysis (Section 2.1), we assessed the relevance of chromatin structure on regulation of the gene function. A second benchtop experiment was carried out to image yeast chromatin samples under a transmission-electron microscope (TEM) [36]. Figure 3.1 relates the outcome of the experiment to gene-regulation behavior based on chromatin structure. With multiple samples and with several causative factors affecting chromatin behavior, implementing epigenetic-regulation analysis using a benchtop setting is tedious and error-prone; thus this preliminary

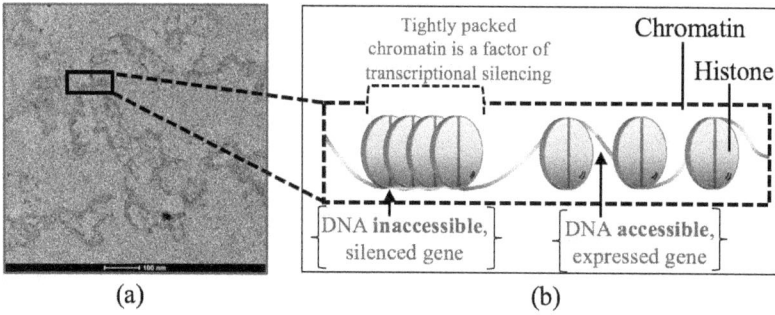

(a)   (b)

FIGURE 3.1
(a) TEM image of chromatin; (b) correlation with chromatic control of epigenetic gene regulation.

benchtop study motivates the need to miniaturize epigenetic-regulation analysis. The benchtop study also provides important guidance on the design of the miniaturized protocol for a DMFB. Figure 3.2(a) depicts a flowchart of the benchtop protocol. The protocol consists of two stages:

1. Up-stream stage in which the transcriptional profile (gene expression) of a GFP reporter gene is investigated. Control samples (GFP not under epigenetic/drug control) and experimental strains (under epigenetic control) were analyzed by qPCR. The goal was to explore how chromatin-folding alterations influence gene expression.

2. Down-stream stage in which novel modifiers of epigenetic gene regulation are identified. Samples are mutagenized, for example by ultraviolet radiation, and cells whose transcriptional activity has been enhanced or suppressed are analyzed further. Following quantitative gene expression analysis, the causative mutation can be identified by whole genome sequencing [16]. The role of the genes in epigenetic processes can be verified by additional studies, including TEM analysis, ChIP, and other assays.

Similar to Chapter 2, reliable concurrent manipulation of independent samples requires the incorporation of sample-dependent decision-making into the protocol. We have developed a miniaturized protocol for epigenetic-regulation analysis; see Figure 3.2(b). Using DMFBs, both up-stream and down-stream stages are carried out using the protocol for gene-expression analysis, followed by DNA pyrosequencing [31], to identify the sequences of the generated mutations. To provide a successful transformation of the complex epigenetic protocol into a biochip setting, a layered structure of a digital-microfluidic system is deployed.

Figure 3.3 illustrates the components required for on-chip implementation of the protocol, and the interactions among them. The control software consists of a system model, embedded in a firmware layer, and a real-time resource

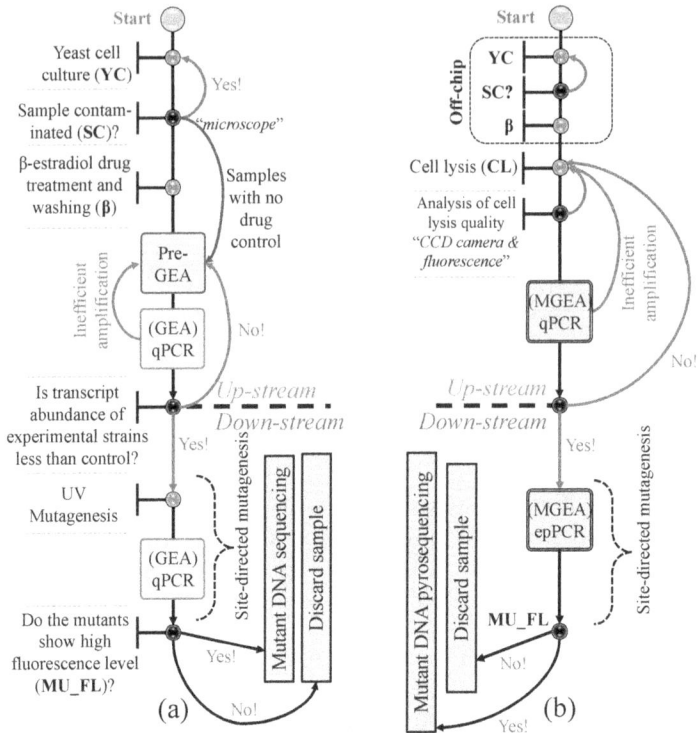

FIGURE 3.2
Protocol for epigenetic gene-regulation analysis using (a) benchtop setup; (b) DMFBs.

assignment and scheduling layer. The protocol is initially synthesized to process the fluids across independent sample pathways. At the end of a bioassay, a sample is subjected to a detection operation that triggers the start of a decision-making process. The firmware receives the sensor readout, analyzes the data, and produces a decision to the scheduler. Real-time synthesis is then employed to provide the needed actuation sequences to adapt the system to the new situation.

In principle, the real-time synthesis is sufficient to capture the dynamics of fluid-handling operations within multiple sample pathways. However, an integrated system for epigenetics must take into account the impact of other active components: electrode actuation and firmware. In other words, the system scheduler is responsible for the timely coordination between the periodic loading of actuation sequences (CM_1), firmware computation (CM_2), and synthesis execution (CM_3).

Therefore, we represent our digital microfluidic (DMF) platform as a hierarchy of components (Figure 3.4); CM_1 triggers a set of *periodic* tasks to stimulate the CPU to transfer actuation sequences from the controller memory to the biochip control pins at the start of every actuation period. The tasks triggered by CM_2 are invoked *sporadically* at every decision-point within the protocol. Note that the durations of tasks generated by CM_1 and CM_2 are fixed and can be easily determined using offline simulation.

Our focus in this chapter is on the control-software design and optimization, specifically, system modeling and real-time scheduling. A fully integrated hardware/software demo involving a fabricated biochip is a part of ongoing work, and beyond the scope of this work. In Section 3.4, we present a preliminary demo using a micro-controller and simulation data. In the following two sections, we present details about the system model and real-time scheduling.

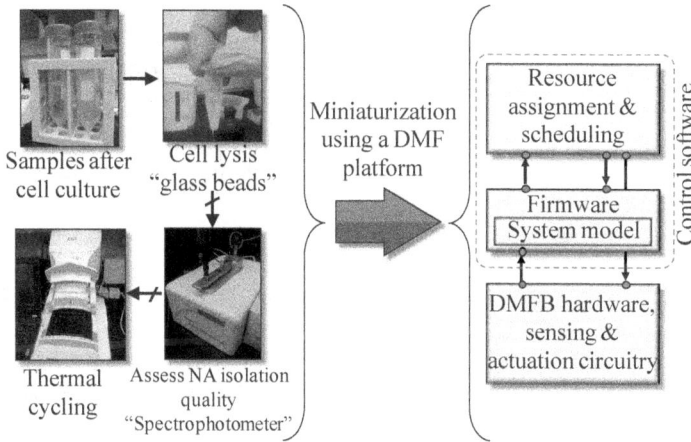

FIGURE 3.3
Proposed layered structure of a DMF platform to miniaturize a quantitative protocol. In this chapter, we design the system model and the real-time scheduler.

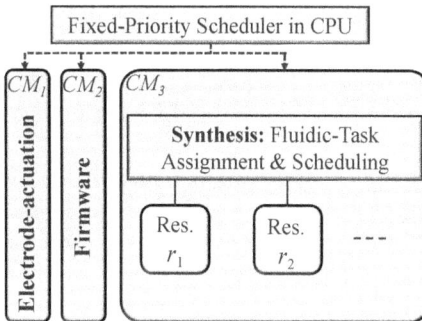

FIGURE 3.4
Hierarchical scheduling of DMF system components.

3.2   System Model

We model the digital microfluidic (DMF) system for gene-regulation analysis in terms of real-time computing systems.

Biochip Resources

The DMF platform includes three categories of resources: (1) *physical, non-reconfigurable* resources ($\mathcal{PN}$) such as I/O ports, (2) *physical, reconfigurable* resources ($\mathcal{PR}$) such as heaters, detectors, and regions to manipulate magnetic beads, and (3) *virtual, reconfigurable* resources ($\mathcal{VR}$) such as mixers. The set of chip resources $\mathcal{R}$ is defined as $\mathcal{R} = \mathcal{PN} \cup \mathcal{PR} \cup \mathcal{VR}$. Unlike $\mathcal{VR}$, the resources in $\mathcal{PR}$ and $\mathcal{PN}$ are spatially fixed, but a resource in $\mathcal{PR}$ can be reconfigured to leverage the electrodes located within its region for sample processing in addition to its original function. For example, a magnet resource can be used either for magnetic-bead snapping or for sample processing, but not both at the same time.

Consequently, a biochip resource $r_i \in \mathcal{R}$ is characterized by $r_i = (rt_i, rx_i, ry_i)$ where rt_i is the resource type, rx_i and ry_i are the x and y coordinates of the resource interface, respectively. The resource interface is represented by an electrode that connects the resource to the global, unidirectional routing bus. Figure 3.5 shows an example of DMF resources. A dedicated dispensing reservoir is used to store a replenishment solution to counter droplet evaporation [123]. The proposed chip layout can implement the epigenetic regulation protocol, and it facilitates the real-time coordination of multiple droplets along the global routing bus.

Fluidic Operations in Multiple Pathways

In prior work, bioassay operations and the interdependencies among them have been modeled as a directed sequencing graph $G_s = (V_s, E_s)$ [245]. A node $v_{si} \in V_s$ signifies an operation and an edge $e_{si} = (v_{s1}, v_{s2}) \in E_s$ represents precedence relation between operations v_{s1} and v_{s2}, respectively. This model is sufficient to handle synthesis for a single sample pathway, but inadequate for multiple sample pathways.

When multiple sample pathways are involved, modeling the synthesis problem based on real-time system theory enables us to assess system performance and robustness under various conditions and constraints, e.g., through compositionality and schedulability analyses [197]. In analogy with real-time multiprocessor scheduling, we introduce the following key terms to model the synthesis problem:

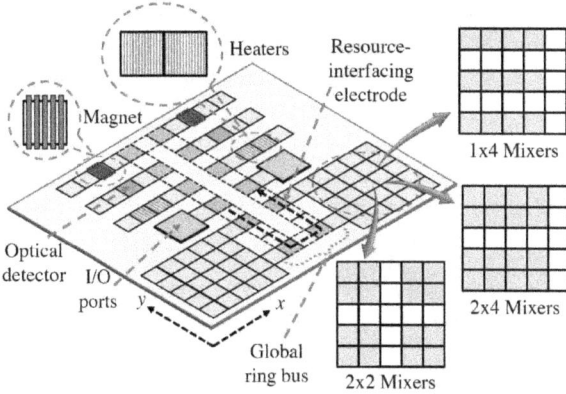

FIGURE 3.5
Resources of a DMF system used for implementing the epigenetic regulation protocol.

- *Fluidic-task set (𝒯):* Similar to computational tasks in computer systems, a fluidic task $\tau_i \in \mathcal{T}$ represents a bioassay in a target protocol. A fluidic subtask τ_i^j represents a fluid-handling operation within a bioassay τ_i.

- *Fluidic-processor set (ℛ):* A biochip resource is also referred to as a fluidic processor. Hence, a DMFB is comprised of a set of heterogeneous processors corresponding to the chip resources $\mathcal{R}$.

In real-time systems, precedence-related subtasks can be assigned and scheduled on dedicated processors, where inter-processor communication induces delays in the release of subsequent subtasks [20]. Similarly, we build on the directed-acyclic graph (DAG) model [32] to capture the characteristics of a quantitative-analysis protocol. This model can then be used to compute task-assignment and scheduling solutions in our DMF system. The details of our task model are described below.

- The protocol consists of a set of bioassays $\mathcal{B} = \{ba_1, ba_2, ..., ba_n\}$. Each bioassay $ba_i \in \mathcal{B}$ is a fluidic task τ_i that is characterized as a 3-tuple (tl_i, G_{si}, td_i), where tl_i is the task release time, G_{si} is a DAG, and td_i is a positive integer representing the relative deadline of the task. DAG G_{si}, whose number of nodes will be denoted by m_i, is specified as $G_{si} = (V_{si}, E_{si}, PC_i, LC_i)$, where V_{si} is a set of vertices representing subtasks $\{\tau_i^1, \tau_i^2, ..., \tau_i^{m_i}\}$, $E_{si} \in [0, 1]^{m_i \times m_i}$ is an adjacency matrix that models the directed-edge set of G_{si}, and $PC_i \in [0, 1]^{m_i \times m_i}$ is a matrix derived from E_{si} that represents precedence relationships between the vertices V_{si}; formally, $PC_i(k, l) = 1$ if and only if subtask τ_i^k has to be completed before subtask τ_i^l starts. Finally, $LC_i \in \mathbb{N}^{m_i \times m_i \times |\mathcal{R}| \times |\mathcal{R}|}$ represents the *lower-bound* routing-cost matrix between the vertices V_{si}, considering all resource combinations used for subtask executions. Specifically, if resources used to execute τ_i^j and τ_i^k are determined to be r_l and r_a, respectively, $LC_i(j, k, l, a) = Z$ indicates that the lower-bound value for the routing distance from the interfacing electrode of r_l to that of r_a is equal to Z. Note that $LC_i(j, k, l, a)$ is a function of coordinates (rx_l, ry_l) and (rx_a, ry_a), and it is calculated based on the unidirectional ring bus shown in Figure 3.5. Furthermore, note that $LC_i(j, k, l, a)$ does not have to be equal to $LC_i(j, k, a, l)$.

- A fluidic subtask $\tau_i^j \in \{\tau_i^1, \tau_i^2, ..., \tau_i^{m_i}\}$ is characterized by $\tau_i^j = (O_i^j, \alpha_i^j, st_i^j, ft_i^j)$, where: **(i)** O_i^j is a vector used to specify the operation times needed by the chip resources to execute subtask τ_i^j;[1] **(ii)** $\alpha_i^j \in \{0, 1\}^{|\mathcal{R}|}$ is a vector that specifies the subtask assignment to biochip resources $\mathcal{R}$ (i.e., $\alpha_i^j(l) = 1$ if subtask τ_i^j is allocated to the resource r_l); **(iii)** st_i^j represents the start time of the subtask; **(iv)** ft_i^j represents the end time of the subtask.

Thus, to satisfy the requirements of the bioassay ba_i, the following constraints must be satisfied:

$$\forall i, j, k : \tau_i^j, \tau_i^k \in \{\tau_i^1, ..., \tau_i^{m_i}\}; l, a : r_l, r_a \in \mathcal{R},$$

(C1) $\sum_l \alpha_i^j(l) = 1$ {allocation};

(C2) $ft_i^j \geq st_i^j + O_i^j(l) \cdot \alpha_i^j(l)$ {task duration};

(C3) $st_i^j - ft_i^k \geq LC_i(k, j, l, a) - K(2 - \alpha_i^k(l) - \alpha_i^j(a)) - K(1 - PC_i(k, j))$ {task precedence and inter-processor communication}, where the constant K is a large positive constant that linearizes the AND Boolean terms in the task-precedence satisfaction problem, described in Equation (3.1).

$$\left(st_i^j - ft_i^k \geq LC_i(k, j, l, a)\right) \text{ if } \left(\alpha_i^j(a)\right) \wedge \left(\alpha_i^k(l)\right) \wedge \left(PC_i(k, j)\right) \qquad (3.1)$$

- In order to counter droplet evaporation, replenishment steps are incorporated before the start of every new bioassay [123]. Note that each bioassay ba_i (composed of a set of subtasks τ_i^j) is characterized by a deadline, td_i, on its completion time after which a sample must be run through a just-in-time replenishment process. Note that fulfilling a bioassay deadline is a soft real-time requirement. In addition, the need to fulfill all protocol deadlines imposes temporal constraints on our design, since resorting to extra replenishment steps during protocol execution significantly impacts completion time. In other words, if a fluidic task violates its deadline (creating non-zero tardiness), extra replenishment steps are applied to the associated sample pathway to counter droplet evaporation, leading to an increase in completion time.

---

[1]$O_i^j(l) = \infty$ indicates that subtask τ_i^j cannot execute on r_l.

To address this problem, each subtask τ_i^j is also characterized by a Boolean variable $\Omega_i^j \in \{0, 1\}$, where $\Omega_i^j = 1$ indicates that subtask τ_i^j is planned to start execution after the deadline td_i. In addition, ts_i represents the start time of a bioassay ba_i and Q models the number of time steps[2] needed to complete sample replenishment. The following additional constraints must be satisfied:

$$\forall i, j, k : \tau_i^j, \tau_i^k \in \{\tau_i^1, ..., \tau_i^{m_i}\}; l, a : r_l, r_a \in \mathcal{R},$$

(C4) $st_i^j \geq ts_i$ {bioassay start time};

The value of the variable Ω_i^j for every subtask τ_i^j can be specified using the difference between the subtask start time st_i^j and the absolute deadline $(ts_i + td_i)$ of the bioassay, as follows:

(C5) $U \cdot \Omega_i^j \geq st_i^j - (ts_i + td_i);$ $\Omega_i^j \geq 0;$ $\Omega_i^j \leq 1$ {tasks beyond deadline}, where U is a large positive constant that can be used to upper-bound the allowable range of tardiness;

(C6) $ft_i^j \leq T_{PF}$ {protocol finish time};

In addition, the inequality in **(C3)** is modified to incorporate replenishment as follows. Note that just-in-time replenishment is applied before subtask τ_i^j only when $\Omega_i^j = 1$ and $\Omega_i^k = 0$, such that $E_{si}(k, j) = 1$. In other words, the bioassay deadline td_i is violated during or immediately after the execution of τ_i^k.

(C3′) $st_i^j - ft_i^k \geq LC_i(k, j, l, a) - K(2 - \alpha_i^k(l) - \alpha_i^j(a)) - K(1 - PC_i(k, j)) + Q \cdot \xi_i^{(k,j)};$ where $\xi_i^{(k,j)}$ is a Boolean variable specified through the expression $\left(\xi_i^{(k,j)} = \neg \Omega_i^k \wedge \Omega_i^j \wedge E_{si}(k, j)\right)$, and it can be formulated as in **(C7)**;

(C7) $\xi_i^{(k,j)} \geq \Omega_i^j - \Omega_i^k + E_{si}(k, j) - 1; \xi_i^{(k,j)} \leq \Omega_i^j; \xi_i^{(k,j)} \leq 1 - \Omega_i^k; \xi_i^{(k,j)} \leq E_{si}(k, j); \xi_i^{(k,j)} \geq 0$ {replenishment requirement}.

- Finally, to achieve mutual exclusion in system resources, each resource $r_l \in \mathcal{R}$ is also characterized by a Boolean variable $\omega_{(j,i)}^l(t)$, where $\omega_{(j,i)}^l(t) = 1$ indicates that r_l is being utilized by subtask τ_i^j at time t. Therefore, the following constraint must be satisfied:

(C8) $\sum_{(j,i)} \omega_{(j,i)}^l(t) = 1$ {mutual exclusion};

The value of $\omega_{(j,i)}^l(t)$ can be specified as follows:

(C9) $\omega_{(j,i)}^l(t) \geq 1 - K(3 - \alpha_i^j(l) - \kappa_i^j(t) - \eta_i^j(t));$ where $\kappa_i^j(t)$ and $\eta_i^j(t)$ are Boolean variables that are specified through the following formulation:

(C10) $U \cdot \kappa_i^j(t) \geq t - st_i^j;$ $\kappa_i^j(t) \geq 0;$ $\kappa_i^j(t) \leq 1;$

(C11) $U \cdot \eta_i^j(t) \geq ft_i^j - t;$ $\eta_i^j(t) \geq 0;$ $\eta_i^j(t) \leq 1.$

---

[2]A time-step refers to the clock period, typically in the range of 0.1 to 1 second [194].

3.3   Task Assignment and Scheduling

In this section, we present our algorithm for fluidic task assignment and scheduling. Table 3.1 summarizes the notation used in this chapter.

Problem Formulation

Inputs: (i) A set of bioassays $\mathcal{B}$, where each bioassay $ba_i \in \mathcal{B}$ is characterized by a task $\tau_i \in \mathcal{T} = \{\tau_1, \tau_2, ..., \tau_n\}$, bioassay deadlines $TD = \{td_1, td_2, ..., td_n\}$, and adjacency matrices $\{E_{s1}, E_{s2}, ..., E_{sn}\}$; (ii) precedence-relationship matrices $\{PC_1, PC_2, ..., PC_n\}$; (iii) a set of subtasks $\{\tau_i^1, \tau_i^2, ..., \tau_i^{m_i}\}$ for each task $\tau_i \in \mathcal{T}$; (iv) the routing-cost matrix LC_i for each task τ_i; (v) the processing-time vector O_i^j for each subtask τ_i^j; (vi) biochip resources $\mathcal{R}$.

Output: (i) Assignment of fluidic subtasks to resources $\alpha_i^j(l)$; (ii) start time st_i^j and finish time ft_i^j for each subtask τ_i^j.

TABLE 3.1
Notation used in Chapter 3.

$\mathcal{R}$	The complete set of the biochip resources
r_i	A biochip resource; $r_i \in \mathcal{R}$
rt_i	The type of resource r_i
rx_i	The x coordinate of r_i interface
ry_i	The y coordinate of r_i interface
$\mathcal{B}$	The complete set of bioassays
ba_i	A bioassay; $ba_i \in \mathcal{B}$
$\mathcal{T}$	The set of fluidic task that represents the bioassays $\mathcal{B}$
τ_i	A fluidic task that represents ba_i; $\tau_i \in \mathcal{T}$
tl_i	The release time of τ_i
ts_i	The actual start time of τ_i
G_{si}	The sequencing graph of τ_i
td_i	The relative deadline of τ_i
PC_{si}	The precedence matrix of τ_i
LC_{si}	The lower-bound routing-cost matrix of τ_i
τ_i^j	A fluidic subtask representing a fluid-handling operation performed as part of τ_i
O_i^j	A vector of operation times for τ_i^j
α_i^j	A binary vector specifying resource assignment of τ_i^j
st_i^j	The start time of τ_i^j
ft_i^j	The finish time of τ_i^j
Q	Number of time steps needed for replenishment

Objective: Reduce the number of tardy tasks to avoid repetition of droplet-replenish-ment procedures.

Heuristic Algorithm

The scheduling problem addressed here is mapped to the problem of scheduling DAG tasks on heterogeneous multiprocessors. Recently, schedulability of DAG tasks in uniform multiprocessors has been investigated and it has been shown that this problem is NP-hard [32, 220]. Not much is known about schedulability of DAG tasks in heterogeneous multiprocessors, but the problem is likely to be computationally intractable [20]. We handle this problem as described below.

The pseudo-code for the algorithm is described in Algorithm 3.1. The algorithm expects a user-specified input that characterizes the upper-bound (Υ) on the number of iterations (Line 4) to terminate the optimization process.

Algorithm 3.1 Task Assignment and Scheduling

1: **procedure** MAIN( )
2:     $Sol \leftarrow \emptyset$; $util_{min} \leftarrow \infty$;
3:     $iter \leftarrow 0$; $\alpha_{prev} \leftarrow \emptyset$;
4:     **while** $iter \leq \Upsilon$ **do**
5:         $iter \leftarrow 0$;
6:         $\{\alpha_i^j\} \leftarrow$ PROCESSORASSIGN();
7:         $T_{PF}, st_i^j, ft_i^j \leftarrow$ COORDINATE();
8:         $nv \leftarrow$ CALCULATEVIOLATIONS();
9:         $util \leftarrow \beta_1 \cdot T_{PF} + \beta_2 \cdot nv$;
10:         **if** $util < util_{min}$ **then**
11:             $util_{min} \leftarrow util$;
12:             $Sol \leftarrow$ GETRESULTS();
13:         $\alpha_{prev} \leftarrow \alpha_{prev} \cup \{\alpha_i^j\}$; $iter \leftarrow iter + 1$;
14:     **return** Sol;
15: **procedure** COORDINATE( )
16:     $t \leftarrow 0$;
17:     **while** $true$ **do**
18:         $RQ \leftarrow$ PUSHREADYTASKSORT(τ_i, PC_i, LC_i);
19:         **if** $(t == td_i)$ AND TASKUNFINISHED(τ_i, E_{si}) **then**
20:             REPLENISH(τ_i, Q);
21:         $\{st_i^j, ft_i^j\} \leftarrow$ SCHEDULEIFPOSSIBLE(RQ);
22:         $t \leftarrow t + 1$;
23:         **if** all $\mathcal{T}$ Scheduled **then** break;
24:     **return** $\{T_{PF}, st_i^j, ft_i^j\}$;

In addition, β_1 and β_2 are integers that are used as weighting factors for the utility function (Line 9), which is used to determine the goodness of a solution.

At every iteration, the algorithm performs task assignment and scheduling (Lines 6-7); subtasks are randomly assigned to the fluidic processors according to the required resources (Line 6), whereas a scheduling algorithm is used for real-time scheduling (Lines 18-30). The algorithm evaluates the utility of the produced solution based on the number of bioassay-deadline violations (nv) and the total completion time (T_{PF}) for the protocol (Lines 8-9). The solution with the best utility (Lines 10-13) is finally selected. We introduce two policies to guide the behavior of the scheduler (Line 21) below for updating the ready subtask queue RQ. Note that a scheduling policy must preserve precedence relationships among subtasks. In addition, a scheduling decision taken by a policy must take into consideration the droplet-routing cost (inter-processor communication cost).

1. *First-Come-First-Served (FCFS):* A static policy in which the ready subtasks are prioritized based on their resource-request time. Note that the resource-access time for a subtask τ_i^j depends on the finish time of the preceding subtasks. This static approach is oblivious to bioassay deadlines.

2. *Least-Progression-First (LPF):* A dynamic policy in which the task that belongs to the least-progressing bioassay is selected first. The *least-progressing bioassay* is a bioassay that is most likely to miss its deadline and its tasks urgently need to be advanced. This policy is similar to the Least-Laxity-First policy [177]. Quantifying the progression of bioassay execution is performed through a utility function $f_u(td_i, t, ns_{comp}, ns_{tot}) = (td_i - t)(1 - \frac{ns_{comp}}{ns_{tot}})$, where td_i is the bioassay deadline, t is the elapsed time since the bioassay has started, ns_{comp} is the number of time steps completed by this bioassay, and ns_{tot} is the summation of the number of time steps for all the bioassay operations. Note that the least-progressing bioassay has the lowest utility value.

We are given a set RQ of ready subtasks. We determine the processing time, the start time, and the finish time of every subtask in RQ and ensure that a subtask can get hold of the pre-assigned resource at time t (Line 25). Note that a task τ_i that is not completed by the deadline is suspended until sample replenishment is carried out (Lines 22-24). Based on the scheduling choices, the synthesis algorithm (using the one-pass algorithm in [280]) is invoked to generate the actuation sequences (Line 12).

The scheduling scheme developed in this work is based on a timewheel that is controlled by an entity known as the coordinator and a priority queue, that is used to enforce the policy. The sequence of actions involved in our scheduling scheme is illustrated in Figure 3.6.

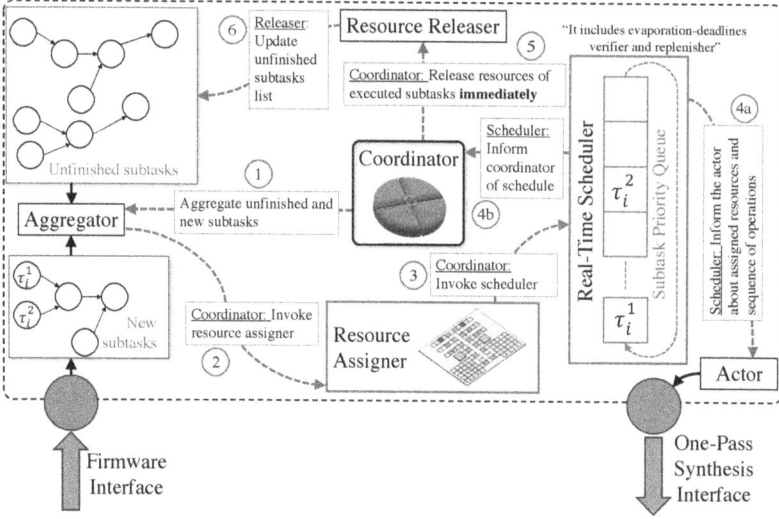

FIGURE 3.6
The sequence of actions in the proposed real-time scheduling scheme.

Scheduling-Policy Analysis

We analyze the scheduling policies explained above using the example in Figure 3.7. We consider two cases: (i) Case A: a case where a limited-resource chip is given; (ii) Case B: a case where an unlimited-resource chip is given. In Case A, we consider a biochip with a single mixer, a single optical detector, and a single waste reservoir that is used to discard droplets. In both Case A and Case B, we consider a dedicated reservoir for each dispensing operation

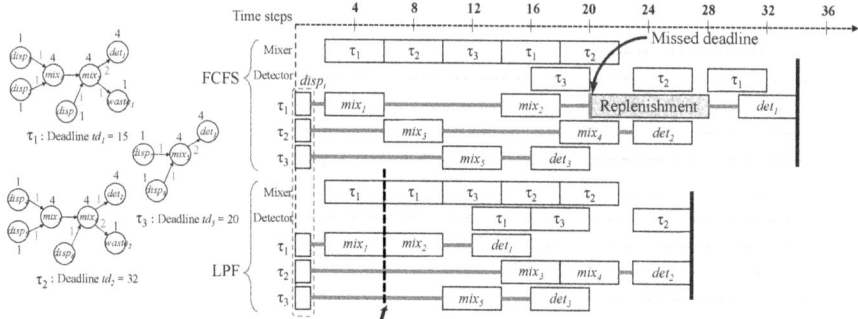

FIGURE 3.7
Illustration of scheduling fluidic tasks using FCFS and LPF polices with a limited-biochip setting. Horizontal thick lines indicate inter-processor communication costs (in time steps).

$disp_i$. The task set consists of three tasks τ_1, τ_2, and τ_3, which have release times $tl_1 = tl_2 = tl_3 = 0$. The DAG representation for these tasks is shown in Figure 3.7. The numbers above the nodes (shown in black) represent the worst-case processing time resulting from task assignment, and the numbers shown on the arrows represent the droplet-routing cost. We consider that the replenishment steps are carried out using dedicated resources in both Case A and Case B, and the time needed to complete droplet replenishment is 8 time steps.

In Figure 3.7, we demonstrate the timeline of the scheduling output for FCFS and LPF when Case A is considered. The results show that LPF outperforms FCFS for a limited-biochip setting. The reason is that LPF dynamically adapts the priorities of tasks based on their progression. For example, at time t (shown in Figure 3.7), all tasks are competing for the mixer. The utility value f_u of τ_1, τ_2, and τ_3 at this time are 5.625, 22.75, and 11.2, respectively. As a result, τ_1, with the least utility, is selected to process upon the mixer starting at t. The completion times for FCFS and LPF are 34 and 27, respectively.

Nevertheless, we expect that both approaches converge to an equivalent lower-bound completion time when we increase the number of on-chip resources. Given the same set of tasks and considering Case B, the number of biochip resources is $\sum_{i=1}^{n} m_i = 7 + 7 + 4 = 18$; i.e., there is a dedicated resource for each subtask τ_i^j. In this case, it is easy to demonstrate that task scheduling depends only on the timing characteristics of a given task set, and that the computed completion time represents the lower-bound, based on a specific task assignment. We introduce the following definition that aids in our explanation [32]:

In Case B, the completion times obtained by FCFS and LPF are equal and they can be defined using the following equation: $T_{PF}^* = \max\limits_{\forall i: G_{si}} len(G_{si})$. According the task set given[3] in Figure 3.7, $T_{PF}^* = max(17, 17, 12) = 17$, which is the lower-bound completion time for FCFS and LPF when the number of resources are increased; this finding is corroborated using simulations in Section 3.4.

The worst-case time complexity of the above algorithm is $O(\Upsilon \cdot ns_{tot} \cdot |\mathcal{R}|)$ when FCFS is used and $O(\Upsilon \cdot ns_{tot}^2 \cdot |\mathcal{R}|)$ when LPF is used. The algorithm is invoked whenever there is a decision that has been taken, necessitating a change in the task assignments and schedules for the sample pathways. Note that that the timing overhead of this algorithm is based on the value of the optimization parameter Υ, the number of on-chip resources, and the total number of protocol time steps. Since both the number of resources $|\mathcal{R}|$ and time steps ns_{tot} are specified in advance for a given platform, a biochip user can modify the response characteristics of a microfluidic system only by changing the value of Υ. In our micro-controller demonstration, described in Section 3.4, we set Υ to 1 to ensure faster responses to protocol decisions.

---

[3]Consider a unit routing cost between the two mix_i operations in τ_1 and τ_2.

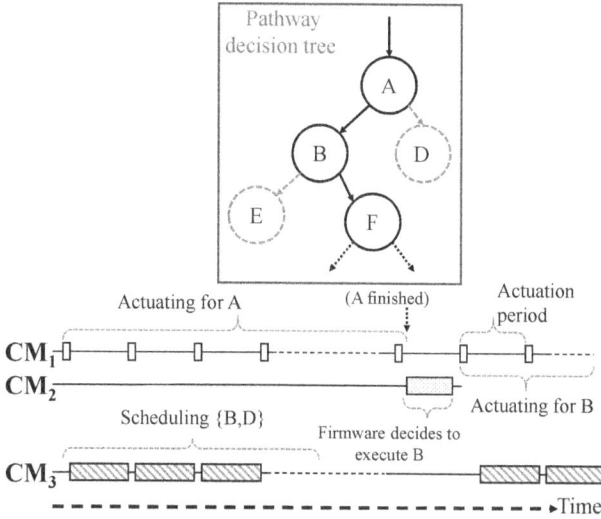

FIGURE 3.8
Lookahead-based execution for a sample pathway. The progression of the sample is modeled as a decision tree.

To make use of the hierarchical system structure in Figure 3.4, the execution of the algorithm needs to be interleaved with electrode actuation and firmware computation. An approach for employing this scheme is to run the task-assignment and scheduling algorithm in a *lookahead* manner. The progression of a sample pathway through a sequence of bioassays, based on the flow decisions, is modeled as a decision tree. While CM_1 executes a bioassay at the qth level, CM_3 concurrently carries out task assignment and scheduling considering all choices at the $(q+1)^{th}$ level. Subsequently, the appropriate assignments and schedules are selected based on the detection results obtained from CM_2. Note that CM_3 must complete computation of the $(q+1)^{th}$ level before CM_1 finishes the execution at the qth level. Figure 3.8 depicts the timeline of a lookahead-based execution for a pathway. This execution mechanism can significantly reduce the impact of the timing overhead; the algorithm is executed concurrently with biochip actuation, which is usually configured to function at a rate of 1 to 100 Hz.

3.4 Simulation Results and Experimental Demonstration

We implemented the proposed heuristic algorithm using C++. The set of bioassays of the quantitative gene-regulation protocol (described in

Section 3.1) were used as a benchmark. A comprehensive analysis of the algorithm is performed based on three groups of evaluations: (1) evaluation of scheduling policies using simulation-based analysis; (2) comparison with previous resource-allocation techniques; (3) real-time experimental demonstration using an embedded micro-controller. Simulation-based assessment was carried out using an Intel Core i7, 3 GHz CPU with 16 GB RAM.

Evaluation of Scheduling Policies

Since most of the previous design-automation methods have not considered multiple sample pathways, we derive a baseline in which the protocol is executed for each sample separately. Therefore, we evaluate the performance for three task-scheduling schemes: (1) the baseline; (2) FCFS; (3) LPF. The metrics of comparison include: (1) the total completion time for the protocol (including replenishment time); (2) the time overhead incurred by replenishment procedures due to deadline violation; all measured in time steps. The results were obtained for various chip sizes, which are represented in terms of the number of biochip resources (e.g., heaters, mixers, and detectors); see Figure 3.5. The deadlines for bioassays were obtained using offline simulation—each bioassay was simulated for various chip sizes and then the longest completion time was considered as a deadline.

We consider three samples being concurrently subjected to fluidic operations. The samples are S_1 (GFP gene-targeted sample), S_2 (YFP gene-targeted sample, and S_3 (actin gene-targeted sample). We consider three different cases in terms of the sample pathways: (1) Short homogeneous pathways (Case I): a case where all three samples follow the same shortest pathway (12 bioassays in each pathway); (2) Long homogeneous pathways (Case II): a case where all three samples follow the same long pathway (16 bioassays in each pathway); (3) Heterogeneous pathways (Case III): a case where these samples are different (the three pathways are comprised of 12, 14, and 16 bioassays, respectively).

Figure 3.9 compares the three scheduling schemes in terms of completion times for the three cases. Although the baseline scheme does not provide resource sharing (hence no deadline violation occurs), it leads to the highest completion times for all chip sizes. We also observe that LPF provides lower completion time compared to FCFS for tight resource constraints. This result corroborates our analysis in Section 3.3, in which we demonstrated that LPF is a deadline-driven approach and it achieves less tardiness, compared to FCFS.

We also estimate the overhead incurred due to replenishment when FCFS and LPF are used. Without loss of generality, we assume that a replenishment procedure takes 10 time steps in the worst case. It is obvious that LPF incurs less replenishment overhead for all chip sizes, as shown in Figure 3.9. As a result, compared to FCFS, the priority scheme of the LPF scheduler is more effective in countering droplet evaporation while the protocol completion time is not significantly increased. Also, it is noteworthy to mention that

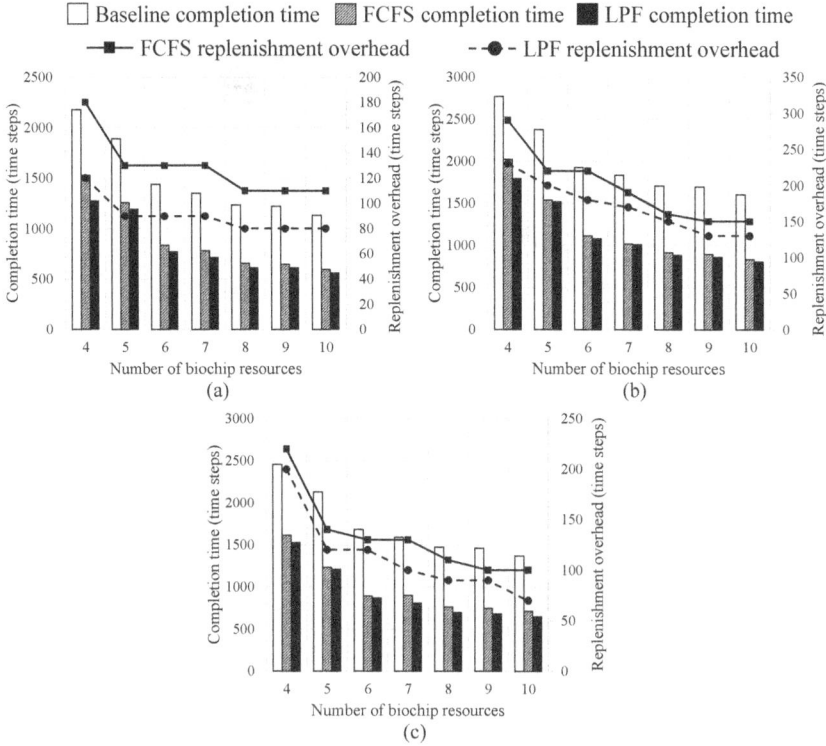

FIGURE 3.9

Completion times and replenishment overhead for three task-scheduling schemes—baseline, FCFS, and LPF—running (a) short homogeneous pathways, (b) long homogeneous pathways, (c) heterogeneous pathways.

the difference in protocol completion time between LPF and FCFS gradually decreases when the number of biochip resources is increased; see Figure 3.9. This finding also adheres to our analytical study performed in Section 3.3.

Next, we explore scalability of the three scheduling schemes. We analyze the completion time as the number of homogeneous pathways is varied; see Figure 3.10. We observe that, as expected, the baseline scheme does not scale with the number of samples, since its completion time increases considerably when the number of samples is increased.

Comparison with Previous Results

We next compare the completion time and resource utilization of our real-time scheduling approach (using LPF) with resource-allocation techniques proposed in Chapter 2. This work introduced two schemes to derive upper and lower bounds on protocol completion time considering resource sharing.

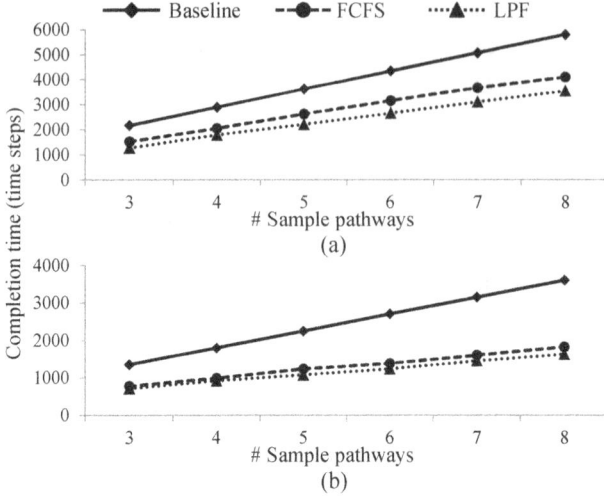

FIGURE 3.10
Scalability of the scheduling schemes with a varying number of homogeneous pathways; (a) using a biochip with four resources, (b) using a biochip with seven resources.

These schemes are: (i) restricted resource sharing ($\mathcal{RR}$): a scheme that fully restricts the reconfigurability of shared resources among bioassays; (ii) unrestricted resource sharing ($\mathcal{NR}$): a scheme that does not consider any restrictions on resource sharing.

For comparison, we use a simulation environment similar to that in Chapter 2, in which a gene-expression analysis protocol is used. The minimum resource requirements as well as fluid-handling operations are re-used from Chapter 2. We utilize three short homogeneous sample pathways, but the conclusions of this simulation holds also for other cases. The biochip architecture consists of seven resources: two mixers, two optical detectors, one magnet, one heater, and one CCD camera. The utilization profile of a resource r_i over the course of the protocol execution is denoted by $ur_i(t)$, where $ur_i(t) = 1$ indicates that resource r_i is used for execution during the time step $[t, t+1]$, whereas $ur_i(t) = 0$ indicates that the resource is idle during the same time step. The summation of the utilization profiles of all chip resources is referred to as *cumulative system utilization* $U_{cm}(t)$; thus $U_{cm}(t) = \sum_{i=1}^{7} ur_i(t)$. This function gives an indication of how efficiently chip resources are used by a resource-allocation scheme over time. We use this function to derive the mean overall system utilization $\overline{U}_{cm}$ to compare system utilization of resource-allocation schemes using numerical values. The paremeter $\overline{U}_{cm}$ is calculated as follows:

$$\overline{U}_{cm} = \frac{\int_0^{T_{PF}} U_{cm}(t) \cdot dt}{T_{PF}} \qquad (3.2)$$

where T_{PF} is the protocol completion time. An upper-bound for $\overline{U}_{cm}$ is equal to the number of chip resources; i.e., $\overline{U}_{cm} \in [0, 7]$. Note that higher $\overline{U}_{cm}$ signifies better resource utilization. Although our real-time scheduler takes into consideration droplet-routing overhead, we ignore this cost in our comparison.

Figure 3.11(a) compares the three resource-management schemes based on the completion time for gene-expression analysis. It is obvious that the proposed real-time scheduler, i.e., the LPF scheduler, achieves the shortest completion time, compared to the resource-allocation schemes in Chapter 2. As expected, $\mathcal{RR}$ sharing leads to the worst-case completion time due to the restrictions imposed on resource sharing.

Next, we analyze the resource utilization that results from the resource-management schemes, based on two metrics: (1) percentage of resource time used effectively by protocol execution; (2) $\overline{U}_{cm}$. The results for the former metric is illustrated in Figure 3.11(b). Using the LPF scheduler, 32.13% of the overall resource time is effectively used to execute protocol fluid-handling operations, whereas only 25.53% and 28.34% of the overall resource time are effectively used by $\mathcal{RR}$ sharing and $\mathcal{NR}$ sharing, respectively. Note that resource allocation in Chapter 2 is carried out at the bioassay level; hence a set of resources might be reserved to execute a bioassay, but not used until the associated fluid-handling operations are invoked. As a result, both $\mathcal{RR}$ sharing and $\mathcal{NR}$ sharing incur a time cost due to reserving, but not using, chip resources; see Figure 3.11(b).

Finally, results involving the parameter $\overline{U}_{cm}$ for resource utilization are shown in Figure 3.11(c). The proposed LPF scheduler achieves the best cumulative system utilization ($\overline{U}_{cm} = 2.5$), compared to $\mathcal{RR}$ sharing ($\overline{U}_{cm} = 1.786$) and $\mathcal{NR}$ sharing ($\overline{U}_{cm} = 1.98$), respectively. These results indicate that the proposed real-time scheduler is more cost-effective and it is applicable for tighter resource-budget cases.

Experimental Demonstration

We next demonstrate the application of hierarchical scheduling using a commercial off-the-shelf micro-controller (TI 16-bit MSP430 100-pin target board; see Figure 3.12(b)). An oscilloscope is used to probe signals generated from the components CM_1, CM_2, and CM_3—the experimental setup is shown in Figure 3.12(a). The generated waveform is shown in Figure 3.12(c), where the signal waveform at the bottom indicates the synthesized time steps based on the task scheduler (CM_3), the signal waveform at the middle represents the actuation pulses (CM_1), and the signal pulse at the top reflects firmware computation (CM_2) based on a detection operation.

Our experimental target was to execute gene-expression analysis using two sample pathways. For testing and verification purposes, we generate a

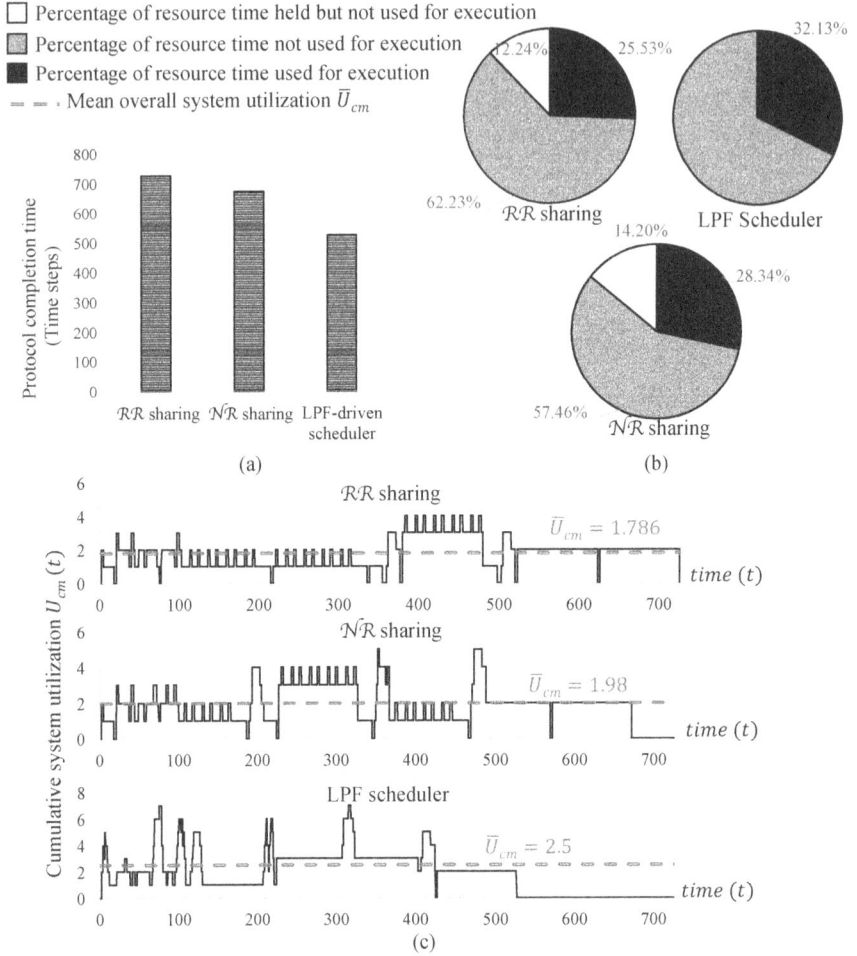

FIGURE 3.11

Comparison between three resource-management schemes: (i) restricted resource sharing ($\mathcal{RR}$); (ii) unrestricted resource sharing ($\mathcal{NR}$); (iii) LPF-based real-time scheduler (proposed). Comparison is based on: (a) protocol completion time, (b) percentage of time effectively used for execution, (c) $U_{cm}(t)$ and $\overline{U}_{cm}$.

hypothetical, but feasible, stream of data to mimic the data transfer between the hardware and the control software. This data was obtained by simulating the gene-expression analysis protocol with two sample pathways.

A timer-interrupt module was utilized to trigger periodic actuation tasks while the task scheduler was running. This mechanism provides flexible tuning for the actuation clock depending on the developed application [194]. The

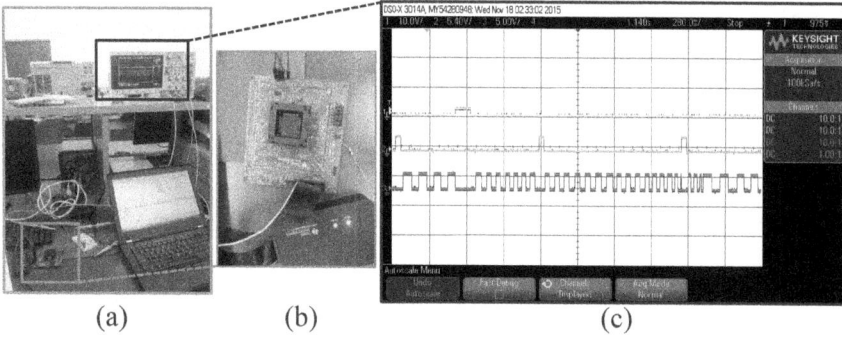

(a)   (b)   (c)

FIGURE 3.12
Experimental demonstration of hierarchical scheduling. (a) Experimental setup. (b) The 16-bit MSP430 100-pin target board. (c) Probed signals from CM_1 (middle signal waveform), CM_2 (top signal waveform), and CM_3 (bottom signal waveform).

firmware tasks were also realized through a set of interrupt service routines (ISRs) and invoked after each (simulated) detection operation. The selection of an ISR depends on the type of detection method assumed, e.g., CCD-camera monitoring of cell culture or fluorescence detection of amplified nucleic acid.

The outcome of the experiment matches the lookahead timing model described in Section 3.3. This experiment lays the foundations for developing and testing the key hardware/software co-design components for real-time DMFBs that can be used in realistic microbiology protocols. Hence, the work proposed in this chapter adds another key component, which is the real-time component, to the synthesis infrastructure established in Chapter 2. As a result, similar to multi-core system-on-chips, mapping the resource allocation of multiple samples (Chapter 2) to the real-time context offers modular and channel-based integration between a DMFB and other onboard analysis modules such as DNA sequencers.

3.5   Chapter Summary

We have introduced a design-automation method for a cyber-physical DMFB that performs quantitative epigenetic gene-regulation analysis. The design is based on real-time multiprocessor scheduling; it provides dynamic adaptation to protocol decisions under spatio-temporal constraints. We have also presented a hierarchical structure to illustrate the coordination between the synthesis of multiple pathways, electrode actuation, and firmware computation. A heuristic algorithm has been presented for the NP-hard task-scheduling problem. An experimental demonstration has been provided using a micro-controller board to show the interaction between the scheduling algorithm and other components.

Our ongoing work is focused on the integration of a fabricated biochip with the proposed design methodology. The work presented in this book will be an important component of such an integrated hardware/software demonstration of a quantitative-analysis protocol.

4

A Microfluidics-Driven Cloud Service: Genomic Association Studies

In this chapter, we extend our synthesis toolkit for biomolecular quantitative analysis by introducing a framework that enables adaptive and complex single-cell genomic studies [117]. The proposed framework is suited for a large-scale biochemistry experimental setting that cannot be handled by a single microfluidic platform and requires coordination among multiple biomolecular studies. An important application of this framework is cancer genomics, where researchers study the interplay between single-cell biology and *omics technologies* (i.e., DNA sequences, RNA expression levels, proteomics, and other epigenetic markers) and their impact on disease development and evolution (i.e., association studies) [82]. To realize this framework, we discuss the integration of different technologies—microfluidic biochips, biochemical analysis protocols, cyber-physical adaptation, as well big data and cloud computing—and introduce the fundamentals of a cyber-physical distributed infrastructure for integrative genomic studies based on microfluidics as a service (MaaS). The proposed framework presents the first discussion on networked and collaborative lab-on-chip environments and it provides the following benefits:

- It will coordinate the operation of a large number of microfluidic devices (referred to as nodes) to dynamically process iterations of single-cell analysis with high-throughput sequencing control. This approach will pave the way for Internet-of-Things (IoT)-enabled real-time collaborative experiments, whereby large number of labs and researchers will be able to coordinate experiments, guide each other, and immediately make decisions on follow-up biochemical protocols or procedures.

- It will leverage the capability of a big-data infrastructure to cumulatively build and enhance the accuracy of biomarker-influence and lineage networks besides cell-type clustering.

- By coupling cyber-physical integration and big-data infrastructure, it will introduce a physical-aware (self-adaptive) microfluidic system, which can reconfigure its nodes (i.e., refocus its analysis scope) based on the dynamic restructuring of computational models—such reconfiguration can be performed either automatically or using human-in-the-loop. This design facilitates sharing genomes and related omics data among researchers, and enables the coordination of thousands of nodes.

The rest of the chapter is organized as follows. Section 4.1 presents necessary background that motivates this work. An overview of biological systems and omic data is introduced in Section 4.2. A case study from cancer research is presented in Section 4.3 to demonstrate the significance of the proposed framework, BioCyBig. Next, Section 4.4 depicts the overall architecture of BioCyBig, whereas Sections 4.5, 4.6, and 4.7 discuss the anticipated design challenges and research opportunities at the application level, the microfluidics level, and the middleware level, respectively. Finally, conclusions are drawn in Section 4.8.

4.1   Background

Microfluidics is a key technology that enables advances in personalized medicine. Breakthroughs in microfluidics and genome technologies can significantly advance personalized cancer treatment and transform clinical diagnostics from the bench to the bedside. Ultimately, with portable microfluidic devices, patients with breast, lung, and colorectal cancers, for instance, will be able to routinely undergo point-of-care molecular testing as part of patient care, enabling physicians to precisely select treatments that improve the chances of cure. These repeated test results, coupled with timestamps, situational information (time, location, and environment of such tests), and personal information (age, weight, height, gender, etc.), will form the data fabric that cannot only highlight the medical condition of an individual but also collectively inform the evolution of the state of the population, for instance, early detection of potential outbreak of highly contagious diseases. This microfluidic service has the potential to perform complicated genomic studies through assisting in quantitative analysis (e.g., epigenetics [29]). These studies therefore can accelerate cancer research and provide point-of-care clinical diagnostics.

One of the most important application areas in cancer-genomics research is the interplay between single-cell biology and omics technologies. Single-cell biology utilizes several microfluidic techniques, some of which are described in Chapters 5 and 6, and computational tools with an ultimate goal of constructing a genome-wide catalog of genetic, epigenetic, and transcriptomic elements [34, 40]. As shown in Figure 4.1, single-cell microfluidic techniques are used to generate omic data from cancer cells; this data is managed and analyzed by computational methods to identify clusters, lineages, and networks, which in turn generate new biological hypotheses.

Hence, the contributions to genomic data analysis include two aspects: (1) improved, large-scale machine learning techniques through big genomic data, which will result in more powerful algorithms; (2) human-centric collaborative environment to facilitate communication, collaboration, and synchronization

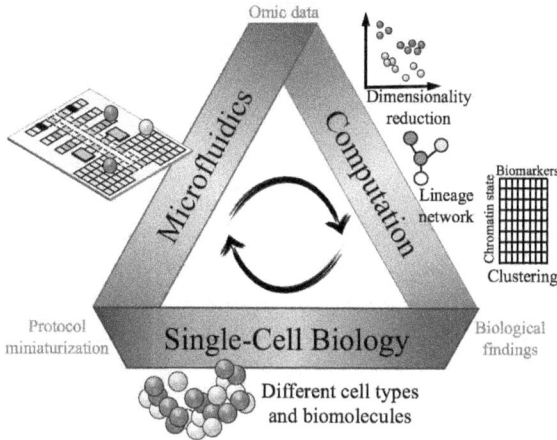

FIGURE 4.1
Iterative genomic-association analysis using microfluidic and computational methods.

of diverse, microfluidics-based research facilities. Next, biological findings, in turn, guide the development of new microfluidic experiments and computational studies [218]. Specifically, the circular process shown in Figure 4.1 needs to be selectively iterated hundreds to thousands of times using cells from a variety of cell populations and tissues. Such an iterative approach enables us to draw precise conclusions about the types and states of these cells, the effective biomarkers that influence genomic or transcriptomic behavior at different loci, and to finally re-focus the scope of subsequent iterations of single-cell analysis. Nevertheless, the application of genome-wide association studies at the single-cell level is extremely complicated and it is hindered by several technical limitations, one of which is the need for a reliable, self-adaptive, and high-throughput control scheme that readily coordinates the experiments of single-cell analysis at a large scale.

Current research methods for single-cell genomic-association studies belong to one of the following categories: (1) extrapolated single-cell methods, which rely on the *in-vivo* findings of a certain level of biological systems or a single omic data type, such as DNA sequences or RNA expression levels, to extrapolate the results or conclusions of cell subpopulations; this approach (similar to epigenetics analysis in Chapters 2 and 3) is pursued by experimentalists using high-throughput microfluidics on real biological systems; however, the analysis that assesses the variation of only a single omic data type can miss complex models that require variation across multiple levels of biological regulation, (2) static data integration techniques, which are applied to large omics data generated from multi-source experimental setting [82]. The development of such analytical methods aims to harness the utility of these comprehensive high-throughput data to elucidate important biomakers. Nevertheless, decoupling such analytical methods from experimental biology expertise does not lead to efficient search techniques for patterns over very large collections of omic data in very high dimension [244]; i.e., raw massive genomic data is not efficiently exploited.

It is evident that the gap between the two approaches represent a technological barrier for researchers against developing an integrative, highly adaptive analysis framework that can relate changes in molecular measurements to disease development, behavior, and evolution. We describe our vision to close this gap by investigating an interactive CPS for a cloud-based microfluidic service in the Internet of Things (IoT) framework, referred to as BioCyBig. This CPS framework will introduce MaaS for genomic association studies. We investigate several design challenges, at different architectural layers, and potential solutions that can make BioCyBig a reality.

4.2   Biological Pathway of Gene Expression and Omic Data

Genomic association mechanisms and the associated omic data are linked to different stages of the gene-expression pathway. In the gene-expression process (Figure 4.2), a particular segment of DNA is enzymatically processed, or transcribed, into an RNA molecule. Then, a specific product of RNA, namely messenger RNA (mRNA), can be expressed and contribute to the translation process. This step leads to proteins that form the functions of our life. Notably, the DNA sequences involved in the establishment of proteins are said to be "expressed genes." On the other hand, the DNA sequences that do not elucidate a high level of expression (i.e., sequences that are not regularly transcribed or are under the influence of "epigenetic transcriptional control") are said to be "silenced/suppressed genes."

Heterogeneous omic data exist within and between stages of gene-expression pathway [211]. For example, analysis of single-nucleotide polymorphism (SNP) and copy-number variation (CNV) can be applied at the genome level. Next, DNA methylation, histone modification, and chromatin accessibility are investigated at the epigenome level. Note that genes with similar DNA sequence (genome level) may not necessarily show the same expression behavior due to variation in chromatin accessibility (epigenome level); see Figure 4.2. In fact, silencing of a certain gene through tight chromatin packaging is enforced by a protein expressed from another gene located upstream or downstream from the suppressed gene. The search for such complicated interaction among thousands of genes is what motivates the need for BioCyBig.

At the transcriptome level, gene-expression analysis can be carried out, whereas analysis of protein expression or post-translational modification can be conducted at the proteome level. Finally, at the far end of the biological pathway, metabolite profiling in serum, plasma, etc., can be employed. To extract a certain type of omic data, dedicated bioassay protocols have been implemented using microfluidics in an isolated environment—Table 4.1 lists examples of such miniaturized protocols. However, only an integrative, multi-omics study of biological systems from genome, epigenome, transcriptome, proteome, and metabolome can lead to the identification of phenome; that is key to recognize serious diagnostic conditions, such as cancer or metabolic syndrome [199]. Our cyber-physical framework facilitates the coordination and management of thousands of bioassay protocols running interactively to generate usable multi-omic data in an efficient manner. The details of the

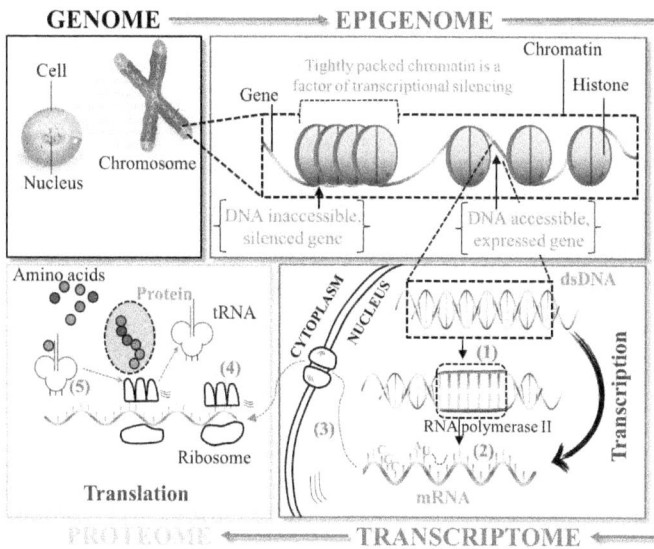

FIGURE 4.2
The biological pathway inside a cell, and the phases of omic-data generation: genome, epigenome, transcriptome, and proteome.

TABLE 4.1
Miniaturized protocols for genomic-association studies.

Pathway Omics	Biological Property	Microfluidic Protocol
Genome	SNP genotyping	Mass spectrometry [299]
Epigenome	DNA CpG methylation	Bisulfite sequencing [128]
Transcriptome	Gene expression	qPCR [212]
Proteome	Protein-DNA interactions	ChIP-seq [40]
Metabolome	Lactate release	Fluorescence-based [178]

computational techniques associated with each omic data type and the integrative approaches for multi-omics are beyond the scope of this chapter.

4.3   Case Study: Integrative Multi-Omic Investigation of Breast Cancer

Breast cancer, like all cancer diseases, is triggered through abnormal changes in a combination of heterogeneous, yet inter-related, biological processes, including gene mutations, DNA methylation, and modifications in gene regulation and metabolism. Changes in each mechanism arise due to the activity of specific genes, which need to be identified. Combining this data leads to a genomic network that explains the multivariate association model.

Our approach not only enables progressive disease models with higher resolution over time, but it also improves our understanding of the adaptive evolutionary changes of cancer diseases [84], specially when geographically scattered microfluidic devices are involved. A special case of our progressive methodology has been proposed for studying metabolic models in biological systems [73].

Herein, we present a simplified case study that elucidates the need for an integrative multi-omic analysis for investigating breast cancer [211]. We also explain BioCyBig's role in exploring complex models of such a disease.

Multi-Omics of Breast Cancer

The goal of the study is to pinpoint the root causes of breast cancer (i.e., biomarkers). As a first step, thousands of cancerous cells must be extracted from fresh tumor tissue. Next, these cells need to be run while observing different aspects of the biological pathway (e.g., targeting genomic, epigenomic, proteomic, or metabolic associations) to construct a precise disease model using the generated multi-omic data. Note that it is difficult to obtain a large number of samples from a fresh tissue at the same site; such a limitation represents one of the bottlenecks for today's analysis techniques. Obviously, an IoT-enabled, microfluidics-driven service facilitates data integration and coordination among multiple sites.

To identify the breast-cancer model, we need to measure and integrate four types of omic data: common genetic variants (genome level), DNA methylation (epigenome level), gene expression (transcriptome level), and protein expression (proteome level). The problem objective is to construct a representative breast-cancer model based on these omics data, where gene expression is co-regulated by both DNA methylation and genetic variants. This model can be used as a disease signature to identify patients with similar tumor characteristics via clustering techniques. Thus, the model description is given below:

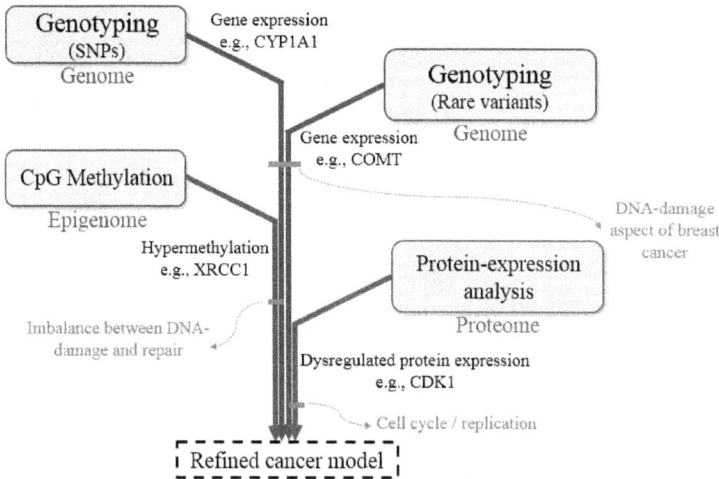

FIGURE 4.3
Flow of integrative genomic-association analysis for breast cancer.

- [**Constraints**] Number of cancerous samples extracted from fresh tumor tissue per site.

- [**Variables**] gx : Selected gene probes; gy : SNPs around each gene probe per window size[1] (genomic data); gz : CpGs around each gene probe per window size (epigenomic data); gw : protein expression (proteomic data).

- [**Output**] gf : Gene expression (transcriptomic data).

- [**Integrative Analysis**] Multi-staged, concatenation-based regression techniques[2].

Figure 4.3 shows the multi-omic analysis flow for breast cancer investigation [211]—we apply this flow to both cancer and normal cells for comparison. First, genotyping of tumor samples is performed to select gene probes and to determine the associated SNPs per each gene probe within a pre-specified window size (e.g., 1 MB window). Next, regression techniques are applied to assess the association between each expression probe and the SNPs in single and multivariate models (e.g., SNP-CpG [199]). The SNPs of probes with increasing expression activity, such as CYP1B1 gene, *may* result in high risks of carcinogenic instances. Likewise, genetic rare variants (or SNPs) in COMT gene can reduce the metabolism of carcinogenic product, resulting in a higher level of DNA damage. Even so, these variations may not increase the risk of

---

[1]A window encompassing the gene of interest is measured in terms of megabases (MBs).
[2]A valid assumption here is that the relationship between genotype and phenotype can be modeled in a linear manner, as is the case for SNPs associated with metabolites [211].

cancer if the DNA-damage repair can adequately absorb carcinogenic metabolites. In other words, using variations in genetic and transcriptomic association solely as a signature for breast cancer could be misleading.

To refine the model, epigenomic and proteomic data must be integrated in the analysis flow; thus, CpG methylation data is generated and associated with gene expression. Accordingly, higher levels of methylation at XRCC1 gene and variation in the gene expression of XRCC3 result in reduced transcription levels, and the repair mechanism may no longer be able to adequately keep DNA repair at necessary levels. Even though inadequate rate of DNA-damage repair likely indicates a carcinogenic tissue, dysregulated protein expression of genes in the cell cycle pathway (e.g., CDK1) may result in a rate of cell replication that is higher than average and therefore reduces the impact of damaged cells. Hence, protein-expression analysis is equally important.

While it is evident that a study of all of the variation mentioned above is required to assess cancer development, constructing such a model requires significant quantities of samples, major effort in experimental work and interactive research, and sophisticated computation utility. These requirements can be realized using our framework.

The Use of BioCyBig

The adoption of BioCyBig as a solution for breast-cancer analysis brings the following advantages.

- BioCyBig provides unification of research goals, which enables efficient exploitation of multi-site benchtop resources (e.g., tissue samples, reagents, and workers). Such a coordination allows precise modeling of cancer, for example, through directing a research site to focus their study on specific genome loci; enabling them to increase the number of gene probes per locus and thereby the system precision. In analogy with electronic systems, this is similar to increasing the number of representation bits of an analog signal during analog-to-digital conversion.

- The big-data infrastructure can be seamlessly exploited for high-dimensional machine learning and data mining, giving a significant advantage for cancer researchers. For example, sophisticated Bayesian inference can be employed for assessing cancer risk or for predicting patient survivorship.

- Deploying BioCyBig as an open framework and reporting on constructive progress of multi-omic disease models will encourage researchers to contribute under the umbrella of BioCyBig.

Figure 4.4 shows the timeline of a typical scenario for the interactions between BioCyBig and breast-cancer researchers, following the logical sequence in Figure 4.3. Note that a microfluidics-based facility can communicate with BioCyBig, via a handshaking mechanism, to run a bioassay protocol and augment the genomic model of a disease (i.e., "write" mode) based on a "call" from BioCyBig. Alternatively, a researcher can inquire about the current status of the model (i.e., "read" mode) for diagnosis purposes.

Working Example: CanLib

To facilitate the understanding of the BioCyBig architecture and explain the system components (Sections 4.5-4.7), we consider an example derived from the above case study of breast-cancer research. In this example, referred to as CanLib, BioCyBig is utilized to investigate the association between three types of omic data: (1) common genetic variants (via SNPs); (2) DNA methylation (via CpGs); (3) gene-expression level. The measurement of gene-expression level is performed by microfluidics-based fluorescence detection (Chapter 2). BioCyBig collects expression data, integratively builds an association model for cancer (referred to as SNP+CpG) through a penalized regression method, assesses the significance of the model and provides cyber-physical adaptation with the aid of visual analytics. The objective is to study the expression of 21,000 gene probes, and the initial SNPs/CpGs window size for each probe is 1 Megabase. The results of the first iteration are collected from 3000 sam-

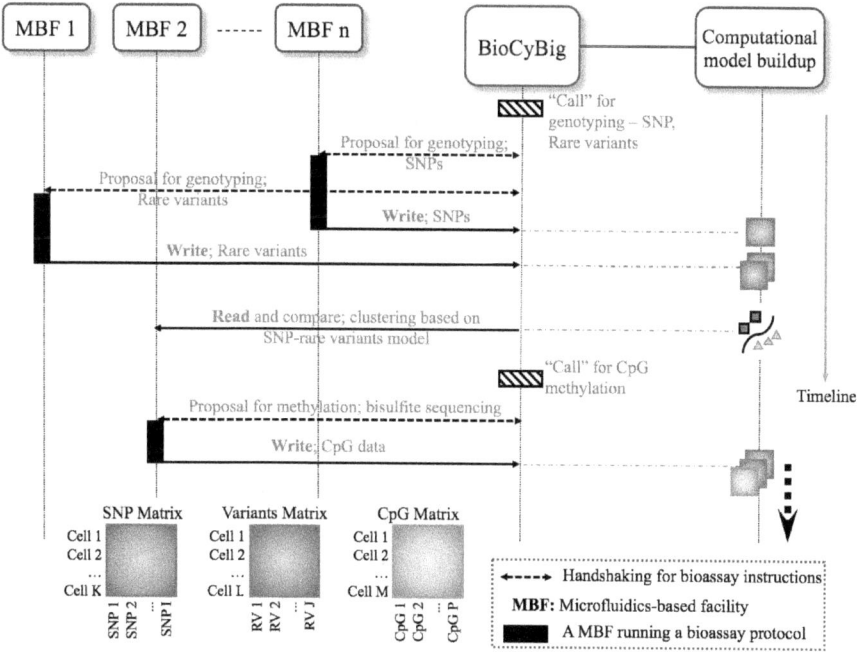

FIGURE 4.4
Flow of integrative genomic-association analysis for breast cancer.

ples located at multiple sites. According to [199], it is expected that nearly 1 million SNPs and 400,000 CpGs might contribute data due to this window size. The contributing SNPs and CpGs are also referred to as transcription factors. Based on the above setting, we estimate that the size of the raw data generated in the first round to be approximately $21,000 \times 3000 \times 1,400,000 \times 4$ Bytes per entry ≈ 353 Terabytes. This value only considers static data, i.e., it does not take into account the dynamic data generated at runtime due to statistical analysis. BioCyBig is clearly motivated by the complexity of the lab procedures for this example, which considers the large number of model parameters but is nevertheless limited to only three types of omics. In practice, the association between more than three types of omic data will need to be studied.

4.4   The Proposed Framework: BioCyBig

In this section, we discuss the system architecture of the proposed framework.

Overall System Architecture Design

Motivated by the $\mathcal{C}^5$ architecture in Section 1.4, significant rethinking in system design is needed to leverage big-data infrastructure, CPS adaptation, and human-system interaction for distributed bioassay protocols. Figure 4.5 shows a high-level view of the system components. The realization of a cloud-based microfluidic service requires the development and integration of four main components: [C1] cloud software infrastructure; [C2] distributed-system component; [C3] microfluidic biochip (node) component; [C4] human-interface component. The component [C1] hosts all the adaptive data-mining and machine-leaning models in the cloud. It also distributes the computational effort for constructing the multi-omics models using a cloud framework such as Apache Spark [76]. The coordination among the components [C1], [C3] and [C4] are carried out at [C2]. The implementation of the actual microbiology protocols, either based on a feedback from [C1] or a human operator acting through [C4], is performed through [C3]. Finally, the component [C4] incorporates human interaction, which enables a human operator to visualize the obtained results at [C1], launch quick analytics procedures at runtime, and instantaneously control the distribution of biochemical assay (tasks) among [C3] nodes, based on technical or budgetary constraints. It is necessary to integrate these components to enable the seamless on-chip execution of complicated biochemical protocols across multiple devices, using the power of big-data analytics.

Since this is a big-data solution and intended for wide community use, the entire stack needs to be built with open-source big-data software including scalable machine learning environment such as Apache Spark, scalable, highly-available, fault-tolerant data store such as Apache Cassandra [77], and visual analytics toolings such as Standford Seaborn [277]. Figure 4.6 outlines our vision for the software stack matched with the system-component schematic in Figure 4.5. The four boxes at the top of the software stack represent the components on the cloud side, whereas the two boxes at the bottom indicate the client components. Apache Cassandra is chosen as a distributed-data store that can directly interface with the sea of client microfluidic nodes. LoC-to-cloud interface utilities (adapters) are required to enable a microfluidic node to directly write data into Cassandra (Omic-Logging Utility). As shown in Figure 4.6, Apache Spark cluster overlays on top of Cassandra to allow Spark to efficiently process complex, real-time streaming data stored in Cassandra. Apache Spark serves as our general-purpose distributed compute workhorse where many diverse multi-omics computational applications (e.g., graph data, machine learning, etc.) can be executed. To serve the visual analytics vertical application for human-system interaction, Spark will synthesize the diverse, dynamic data collected by Cassandra and write the results into Seaborn guided by a set of Python APIs that will describe visual analytics objectives. Seaborn dashboards will interactively display relevant biological and operational reports to an end user, and assist in making decisions.

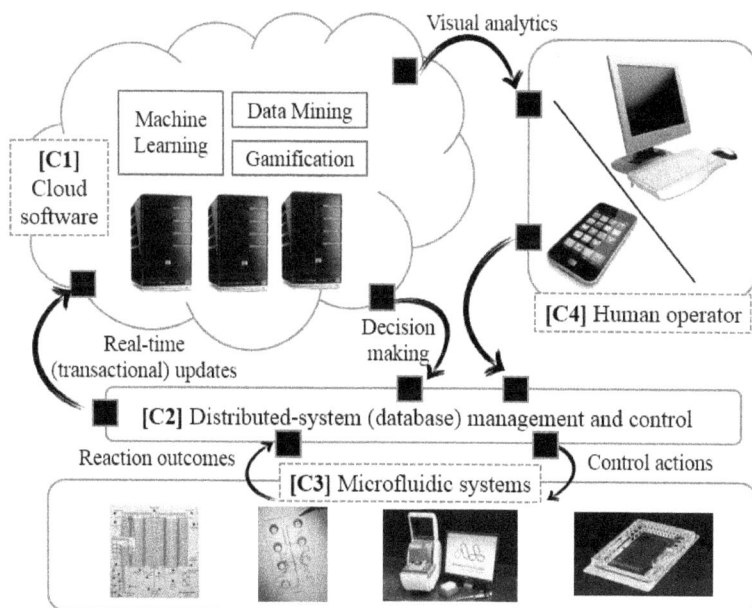

FIGURE 4.5
Illustration of components and data flow in BioCyBig.

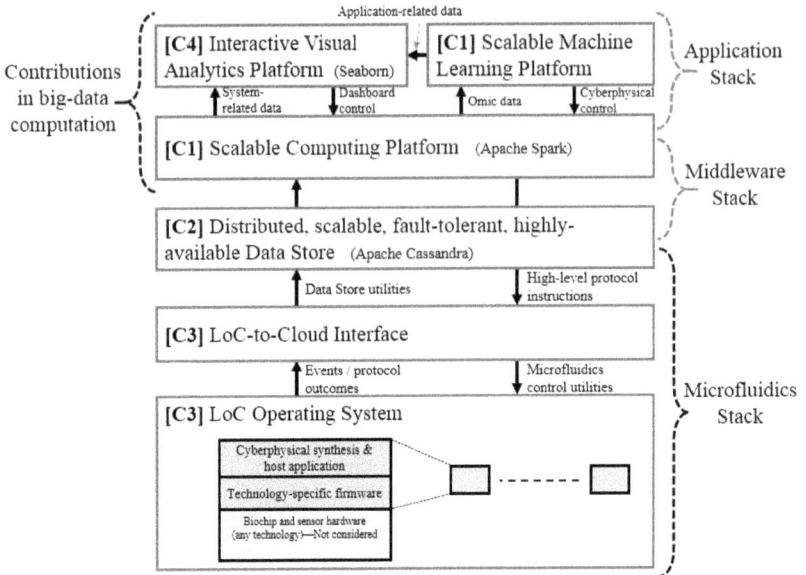

FIGURE 4.6
Software stack of BioCyBig.

Cloud Software and Human-Interface System

To improve our understanding of genomic association mechanisms, the development of novel computational tools has become an integral part of large-scale data analysis; such tools aim at converting raw data signals obtained from experimental setting to quantitative biological information. Significant effort has been devoted to addressing several challenges arising with omic data analysis [218]. For instance, due to the high dimensionality of single-cell data, enabling data visualization requires the application of specific dimensionality-reduction approaches that map the data points into a lower-dimensional space while maintaining the single-cell resolution [162]. Unsupervised clustering is widely used to group samples with similar genomic properties, which can be used to identify previously unknown subpopulations from multi-omic data. Specifying the set of genes that discriminate these subpopulations has also been studied using relevant computational tools [132]. Network modeling can provide mechanistic insights into lineage relationships and the coordination of gene activities to help in understanding the overall dynamics of the biological system. Taken together, applications of multi-omic data analysis greatly enhance the power of systematic characterization of cancer heterogeneity.

Distributed-System Architecture

An efficient coordination among multiple (heterogeneous) microfluidic devices and the cloud triggers the need for a new sensor-actor coordination model, which takes into account the specific properties of integrative multi-omics analysis. Related concepts of coordination and communication have been previously studied in systems powered with wireless sensor networks [210]. Similarly, a distributed-system design can be leveraged to collect omic data out of the microfluidic devices (sensing action). In addition, the control decisions communicated from the cloud software or the human interface are managed at this level (actuation action). At the level of large-scale genomic association analysis, the process of management and communication of control decisions is critical and it needs to be carefully studied. Several management criteria can be employed for allocating and prioritizing biochemical analysis at different microfluidic nodes. For example, a situational criterion can be used to focus biochemical analysis at certain locations in order to depict the evolution state of subpopulations, leading to an early detection of potential outbreak of highly contagious disease. Another example is the level of expertise, in which elaborate microfluidic-based biochemical analysis of liver and breast cancerous cells are performed separately at the associated research centers. Gathering information from different specialized institutions/centers will support researchers with a big picture of cancer and other diseases, and ultimately lead to a better understanding of the mechanisms behind drug resistance.

Plug-and-Play Control of Microfluidic Devices

The key technology behind the proposed cloud-based analysis system is microfluidics, which offers significant advantages for performing high-throughput screens and sensitive assays. Various microfluidic technologies (e.g., flow-based, droplet-based, and digital microfluidics) have been presented in the literature for genomic association analysis, targeting epigenome [40], transcriptome, proteome [281], etc. To seamlessly coordinate microfluidic nodes, it is necessary to develop a control methodology that is acquainted with various microfluidic technologies. A universal (canonical) control interface can be designed and customizations to specific technologies can be realized through "adapters". The design of the control interface and the adapters will include software synthesis to complement the available hardware. This standardized design of such a control interface will facilitate the plug-and-play addition of microfluidic nodes.

Logging utility tools (e.g., Omic-Logging Utility) can enable both the Lagrangian traces that record the complete fulfillment flow (e.g., which microfluidic devices worked on this biochemical experiment, when and with what outcome) and the Eulerian traces that record all state changes of a device (e.g., when this component is up, down, faulty, performing which type of operations on which experiment, and associated resource usages). This utility tool can be invoked by any biochip firmware to enable full tracing coverage in the host application.

Relationship to "Big Data" Community Goals

BioCyBig is aligned with the "Big Data" community's key goal of fostering smart and connected communities and utilizing IoT to benefit society. Our framework also offers fundamental advances into medical research by leveraging machine learning, cloud computing, and recent advances in lab-on-chip as IoT devices. It explores a completely new opportunity to rethink the principles and methods of systems engineering that are built on the foundations of real-time control, data analytics, and cyber-physical microfluidic biochips. The proposed solution investigates new engineering principles that are needed to advance personalized medicine and patient care for diseases such as cancer. It provides the means for controlling/coordinating distributed biochemistry experiments. Our choice of using open-source tools in the design is to facilitate easy adoption and technology transition. Unlike previous cyber-physical designs of microfluidic biochips, the asssessment of BioCyBig requires a combination of microfluidics-related benchmarks (e.g., technology parameters and protocols) and functional genomic data that are publicly available [27].

BioCyBig relies on three inter-related layers; namely application, middleware, and microfluidic layers. In the following three sections, we discuss design aspects and solutions of each layer. The realization of CanLib components is used as a motivating example in these sections.

4.5 BioCyBig Application Stack

We anticipate that multi-omic data will be collected from potentially thousands of microfluidic devices, which run biochemical experiments asynchronously. Therefore, due to the emerging complexity of the collected data, not only because of the data size but also the asynchronicity of the communicated signals, genomic-association applications (e.g., lineage inference, gene-regulatory networks, and unsupervised cell clustering) need to be populated on the cloud and accessed/delivered in the form of Software-as-a-Service (SaaS) model. To improve system productivity and engage more participants, techniques such as gamification can be used. A breakdown of the design aspects involved in this stack is given in this section.

Development of Scalable, Integrative Multi-Omic Applications on a Cloud Service

The development of novel computational tools is an integral part of genomic-association analysis; such tools nowadays come in one of two forms: (1) methods for studying the correlation between a single omic data type and gene expression; e.g., transcriptomic clustering techniques [243]; (2) integrative methods to combine different types of omics methodologies into a unified toolbox; e.g., adapted regression methods [199]. Even though such computational tools have emerged recently as an extension to single-cell benchtop work, the underlying machine learning models (highlighted below) are static and these methods rely on offline learning mechanisms. Thus there are concerns about their scalability for genome-wide association analysis and their applicability for cyber-physical adaptation. As a first step, it is necessary to analyze scalability of these techniques when multi-omic data streams (from different sources) are communicated. The evaluation metrics are computation time, computation accuracy, and system response time. Second, it is required to develop a methodology to port these offline tools into online frameworks, which interact with data from multiple sources in near-real time. Online machine-learning tools became prominent with the introduction of big data, and their role needs to be explored for genomic-association applications. Third, there is a need to design an automated delivery system that feeds data streams, received from the middleware, into the online learning models.

Figure 4.7 shows a typical flowchart for single-cell genomic-association applications that need to be imported to the cloud using Apache Spark. Preprocessing and quantification are the first steps in any large-scale data analysis. The purpose of these steps is to convert raw data to quantitative biological information. In addition, significant effort is paid to the estimation and removal of systematic biases due to technical variability. A major issue in genomic analysis is that technical variation is always confounded with biological variation. Methods that aim at building error models [132] to account for biological biases, or constructing normalization techniques to correct the biases at an early stage [154], have been presented for the specific loci of a specific omic instance and they have been developed to run offline. Our solution aims at looking into algorithmic techniques to scale these methods and to incorporate cyber-physical, online adaptation feature into the application.

The high dimensionality of omic data provides a challenge for visual analytics. Several dimensionality-reduction approaches are available to map data points into a lower-dimensional space while maintaining single-cell resolution. Methods such as principal component analysis (PCA) and t-distributed stochastic neighbor embedding (t-SNE) [162] can be used to visualize omic data in different contexts. However, with technological breakthroughs in integrative, multi-omic association analysis, new distributed high-dimensional techniques are needed to keep pace with the unforeseen scaling of the cell-subpopulation features. Our proposed big-data infrastructure represents a

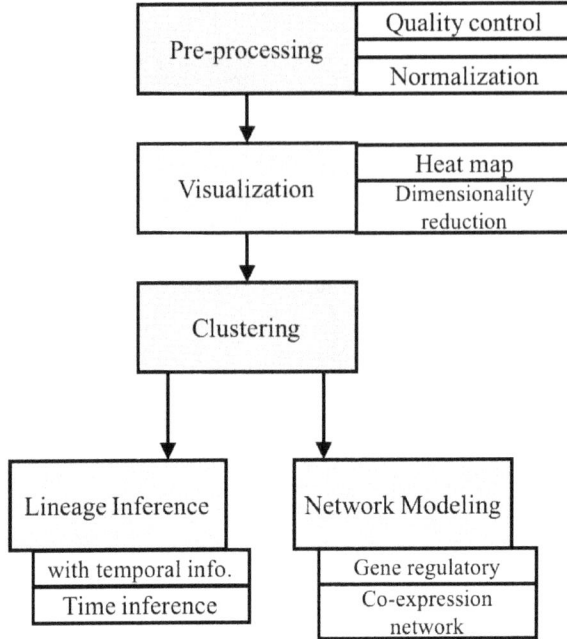

FIGURE 4.7
Flowchart of computational single-cell analysis methods.

timely advance and it provides an ideal opportunity to develop such analysis methods with unprecedented resolution. BioCyBig aims to couple algorithmic innovations with responsive big-data ecosystem leading to self-contained, dynamic computational models—this approach tends to resolve one of the biggest challenges in genomic-association studies [218].

CanLib: In addition to addressing the high dimensionality problem, CanLib must also utilize scalable machine-learning techniques that take into consideration the very small number of samples compared to the number of parameters. The most common solution to this problem is to select a subset of important explanatory variables, where the subset selection is the key of such a problem. According to [199], penalized regression allows us to accomplish this goal in a stable and computationally efficient fashion. The following multivariate model (SNP+CpG) is used: *Gene expression*$_i = c_1 \cdot SNP_1 + c_2 \cdot SNP_2 + ... + c_1 \cdot CpG_1 + c_2 \cdot CpG_2 + ...; i = 1...gm.$ The symbol *gm* represents the number of gene probes.

To apply integrative analysis, the following steps are performed recursively: (1) microfluidic nodes start to perform gene-expression analysis on specified probes where a 1 MB window of SNPs and CpGs are selected from each probe; (2) the above model is modified accordingly, and penalized regression is applied to each probe and model, thereby providing the deviance per model[3]; (3) the significance of gene probes is tested and used to guide the next iteration of probe selection. In the subsequent iterations, the window size of SNPs and CpGs is increased and only the significant probes are re-examined to update the model. This approach expands the model horizontally.

The above discussion demonstrates that cyber-physical integration and the associated big-data infrastructure offer powerful means to study the diversity and evolution of single cancer cells, which can ultimately be applied to the clinic from an early detection stage to identifying therapeutic strategies for cancer patients.

Gamification for Improving System Productivity

Improving system productivity and promoting client participation are key requirements that can play a critical role in enhancing the precision of model learning and ultimately cancer diagnosis. Therefore, inspired by Games in Health [182], the incorporation of game-design elements and strategies into the cloud side will lead to the establishment of models that will act as a key player in decision making. The cloud software needs to include the game components that will motivate the researchers/clients to gather more omic data and daily make reasonable decisions influencing positively the scope of analysis and cancer diagnosis. Gamification creates the atmosphere where more microfluidic experts become eager to participate with their experimental outcomes. As a result, a gamification software architecture (Figure 4.8) must be considered as a major building block in BioCyBig.

There are several challenges that need to be addressed as part of this design aspect, as listed below:

Reward scheme: It is required to design a game where the participating users (e.g., microbiologists) are awarded based on their experimental effort, as a form of incentives. The reward scheme may be intellectual-based such as recognition in terms of academic publications and awards, or monetary-based in terms of commercialization. Automated social/academic network involvement can be used to share participants' achievements.

Mechanism design: It is needed to design the rules of the game—that can for example be taken from the principles of microeconomics to achieve fairness and efficiency among participants [206]. Features such as Pareto efficiency, envy-freeness, and sharing incentives will be potential characteristics in this game. For example, the problem of incentivizing effort and rewarding in online systems based on user contributions has evolved from the economics literature on "tournaments" [64]. Tournaments, as a broad class of game-theoretic mechanisms, are used for diverse purposes such as choosing winners in sporting events, procuring innovations, or rewarding workers. Our innovation lies

---

[3]Deviance is often used in conjunction with regression to quantify the quality of fit for a model.

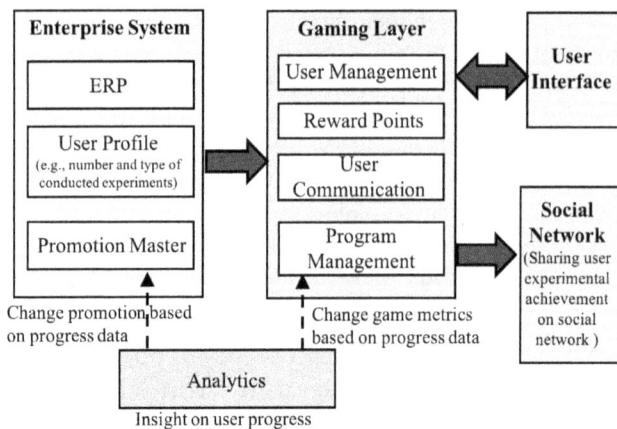

FIGURE 4.8
Proposed gamification architecture to encourage user progression and improve service productivity.

in the application of these tournaments in the cyber-physical microfluidics domain, where the players are biochemists/microbiologists utilizing microfluidics.

Game coordination and analytics: A secure and trusted software party/agent needs to be developed to coordinate/manage the game and control participants' promotion. It is necessary to investigate a game-theoretic approach for communication security (or cyber-security) to guarantee system integrity and authenticity. As a first step, potential cyber-attacks need to be explored and the actions of attackers and defenders are studied. Second, behavioral game theory can be utilized to investigate the role of certain actions taken by both parties in a set of simulated scenarios [39]. Third, reinforcement learning is used to represent a simulated attacker and a defender in cyber-security game. Such a methodology will provide us with insights about efficient techniques for ensuring cyber-security. Finally, as a part of game coordination, gamification analytics needs to be applied to our genomic analysis system.

Development of Visual Analytics and Decision Making

These tools have two main functionalities: (1) applying analytics methods to the obtained models in order to initiate automated decision making; (2) enabling users/ participants to visualize and extract knowledge from the models to aid in decision making. Seaborn visualization dashboards will be adapted to be highly programmable so that they can fit different use cases. These dynamic dashboards will allow users to monitor the progression of experimental work and make protocol decisions based on their own perspective. A key

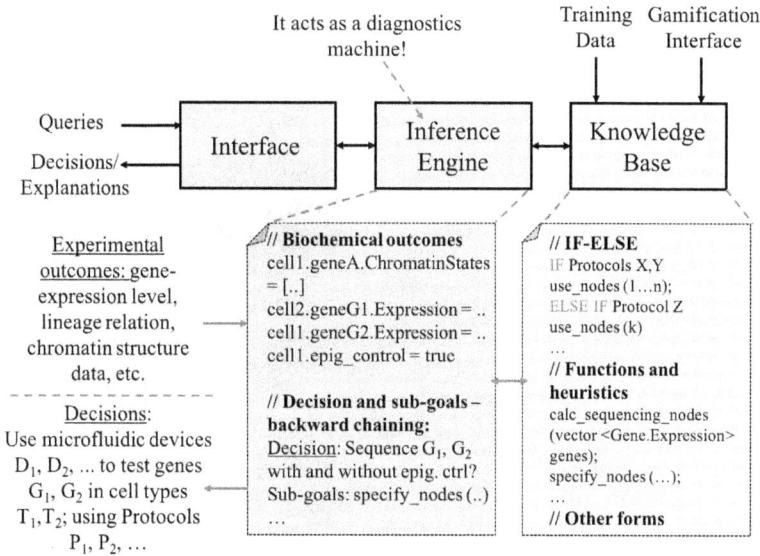

It acts as a diagnostics machine!

Training Data

Gamification Interface

Queries

Decisions/ Explanations

Interface

Inference Engine

Knowledge Base

Experimental outcomes: gene-expression level, lineage relation, chromatin structure data, etc.

Decisions:
Use microfluidic devices D_1, D_2, ... to test genes G_1, G_2 in cell types T_1, T_2; using Protocols P_1, P_2, ...

// Biochemical outcomes
cell1.geneA.ChromatinStates = [..]
cell2.geneG1.Expression = ..
cell1.geneG2.Expression = ..
cell1.epig_control = true

// Decision and sub-goals – backward chaining:
Decision: Sequence G_1, G_2 with and without epig. ctrl?
Sub-goals: specify_nodes (..)
...

// IF-ELSE
IF Protocols X,Y
use_nodes (1...n);
ELSE IF Protocol Z
use_nodes (k)

// Functions and heuristics
calc_sequencing_nodes
(vector <Gene.Expression>
genes);
specify_nodes (...);
...
// Other forms

FIGURE 4.9
Components of a decision-support system.

challenge that needs to be addressed in this context is as follows:

Decision-making models: Decision-support systems (DSS) are used in many applications [205], and they rely on knowledge-based reasoning systems (KBS); thus KBS is adopted in BioCyBig, as shown in Figure 4.9. The knowledge base will be constructed and iteratively trained (with online machine learning methods) to support decisions about protocol implementation and selection of microfluidic platforms.

The knowledge base can be encoded as IF-THEN, for example based on specific microfluidic technology, or it may incorporate heuristics or probabilities, for example when algorithmic partitioning of deep sequencing protocols among multiple platforms is considered. The knowledge base will be mainly stored and handled via Spark. Note that these models will run in conjunction with the designed game described above; such a pairing needs to be investigated. In other words, we are interested in investigating the influence of a decision-support mechanism on a tournament. The inference engine (or reasoning mechanism), in turn, can be developed using known concepts from artificial intelligence. It will be designed such that it receives description/analysis findings from the middleware and it may request additional information from the system user if needed. The engine will interpret the knowledge base, draw conclusions, and ultimately make decisions about protocols. In order to apply the inference engine in our setting, a backward-chaining inference methodology needs to be developed [181]—this methodology is applicable for diagnostic problems since it is a goal-directed inference, i.e., inferences about protocols or microfluidic facilities are not carried out until the system is able to reach a particular goal (e.g., execute certain protocols on specific types of cells). Figure 4.9 shows an example of a cyber-physical, computer-aided decision-making scenario.

CanLib Visual Analytics: Recall that cyber-physical adaptation expands the CanLib model horizontally by re-examining the influence of transcription factors on gene probes. However, to reach a meaningful conclusion, the previous gene-expression study must be performed on different cell types, wherein cells vary based on the activity of the transcription factors. Therefore, considering multiple cell types in the study expands the CanLib model vertically. With this expansion, a user can view clustering of cells and transcriptional factors based on the level of gene expression [176]. As shown in Figure 4.10(a), gene-expression analysis, obtained from multiple cell types over a specific window of transcription factors, suggests that there are two major classes of cells: C_{e1} and C_{e2}. Based on Figure 4.10(a), a system user/administrator can infer the following: (i) analysis over a subset of C_{e1} cells is sufficient to predict the overall behavior of C_{e1} cells; (ii) unlike C_{e2} cells, C_{e1} cells hypothetically exhibit gene-expression activity at the north extension of transcription factors. As a result, a decision can be made to extend the association analysis for a subset of C_{e1} by expanding the window towards the north direction.

Similarly, the synthesis of a regulatory network of transcription factors for CanLib allows researchers to infer the influence chain of each factor; see Figure 4.10(b). Graph-theoretical algorithms such as finding cycles of nodes can also be used to narrow down the analysis scope.

4.6   Design of Microfluidics for Genomic Association Studies

Support for diverse microfluidic technologies is an important design characteristic that must be considered. Therefore, a universal control interface connected with the distributed-system software (middleware) is needed. Such an interface will enable tracing the outcomes of individual biochemical pathways and also the delivery of synthesis specifications in a technology-independent manner. Customization to specific technology, if needed, will be carried out using an "adapter" system. Figure 4.11 depicts the main components of the microfluidic control and sensing interfaces.

Design of Microfluidic Control and Sensing Interface

A well-defined interface will be constructed to enable tracing and logging of experimental data. Control plans will also be conveyed through a

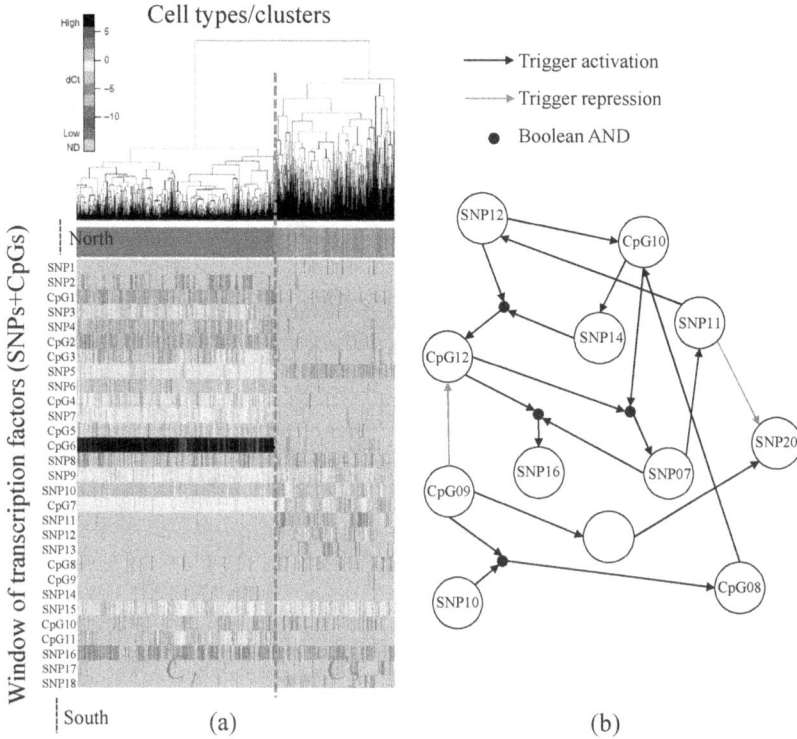

FIGURE 4.10
Visual analytics for CanLib. (a) Hierarchical clustering of cells showing the correlation between gene-expression analysis and transcription factors. (b) A synthesized Boolean network model of transcription factors illustrating the interaction between these factors (i.e., activation and repression).

technology-independent host controller. This level of abstraction will achieve horizontal scalability, since it can be employed at any microfluidic system. The following are some challenges that we need to consider:

Microfluidic System Model Composition: From the control perspective, our solution follows a model-verification approach at every microfluidic node to verify whether the required control decisions can be fulfilled given the advertised node model, which embeds information related to available reagents, fault-tolerance threshold, timing specifications, and others (see Figure 4.11). Every participant (or researcher) will declare a unified model for the microfluidic systems they possess. Such a model will incorporate, for example, the types of tissues they work on, microfluidic technologies, and corresponding reagents, and many other features. Thus, a tool is built to turn the expert specifications into a file format that is compliant with a model-checker tool. Note that at this level, the details of the biochemical procedures are not

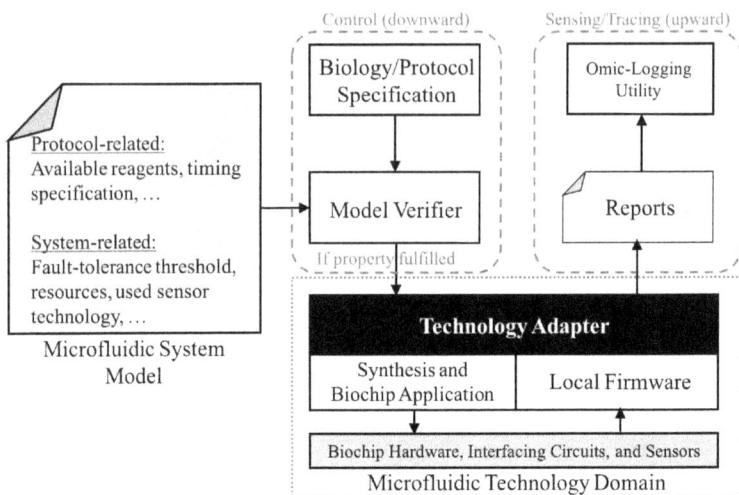

FIGURE 4.11
Adaptation of microfluidic platforms (clients) for BioCyBig-compliance.

known yet. Therefore, model verification is considered as a first step in the communication protocol between the cloud infrastructure and the underlying microfluidic systems.

Reporting: From the sensing/tracing perspective, the solution uses a reporting engine to aggregate all the details of the experimental outcomes. This data is then used by logging tools.

Omics-Driven Biochip Synthesis and Firmware

Breakthroughs in microfluidics technologies have allowed the realization of high-throughput single-cell genomic studies. The massive amount of data generated by these platforms allows mapping of complex heterogeneous tissues and most likely uncovers previously unrecognized cell types and states. The communication established between thousands of microfluidic biochips and the cloud infrastructure will interactively make use of this data. However, for automated, high-throughput, cost-effective execution of single-cell tests (e.g., deep DNA sequencing) using microfluidic biochips, there are several design and algorithmic challenges that need to be tackled first:

Scalable Protocol-DNA Co-Modeling: In Chapter 2, we presented a graph-theoretic modeling approach for quantitative-analysis protocols. While this modeling approach is capable of capturing the characteristics of a wide class of protocols (i.e., support for multiple sample pathways and sample-dependent decision making), the introduction of single-cell analysis creates new challenges (opportunities) for protocol modeling. More specifically, since a sample droplet encapsulates an individual cell, there is need to rethink protocol modeling to combine the representation of fluid-handling operations (e.g., mixing and heating) with modeling information about a particular DNA (or protein) and its associated omic data. Chapter 5 presents a co-modeling and synthesis solution for single-cell analysis using a hybrid microfluidic platform. This model fits the requirement of many single-cell genomic-association protocols such as chromatin immunoprecipitation (ChIP) that aims to investigate DNA-protein interactions.

Sequencing Depth vs. Number of Droplets/Cells [Staged or Sequencing-Driven Synthesis]: Given the above protocol model as input, we propose to develop a technology-specific synthesis framework, where the progression of fluid-handling operations within the sequencing network is not known *a priori*. One of the challenging questions in next-generation sequencing associated with single-cell analysis is whether it is possible to predict the amount of sequencing that is required, both to answer a biological question and, at the same time, to prevent excessive sequencing. Figure 4.12 illustrates the tradeoff between the number of cells (or replicates) required and sufficient (saturated) sequencing. Current approaches are offline; i.e., they rely on statistical analysis at the design stage. Therefore, the first problem to be handled with our framework is how CPS can enable biochip-level decision making to iteratively determine the extent of sequencing. In this case, biochip detection is applied at every step in the sequencing network. Since multiple droplets (or pathways) are involved in the protocol, the synthesis framework must consider

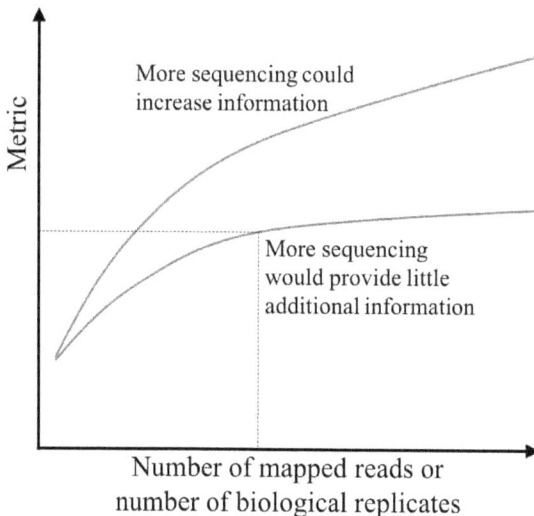

FIGURE 4.12
Motivation for staged sequencing to predict sequencing requirements.

multi-sample decision making, as presented in Chapters 2 and 3. In conclusion, our synthesis framework will combine cyber-physical control of reaction (sequencing) termination with multi-sample decision making.

The second problem arises when cell clustering and lineage network applications at the cloud layer come into play. Based on online models, a cloud-level decision making will be communicated to the biochip; the objective is to carry out deep sequencing up to a pre-determined limit. The design will aim to traverse the minimum set of fluid-handling operations to achieve the specified goal in a statistical fashion. As a first step, optimization techniques based on network-flow algorithms can be investigated to solve this problem.

Barcoding-Aware Synthesis and Firmware Design: Since thousands of cells can be involved in one experimental run in genomic association analysis, researchers have recently developed a high-throughput droplet-microfluidic approach for barcoding the RNA from thousands of individual cells for subsequent analysis by next-generation sequencing [136]. With such data, we can track heterogeneous cell sub-populations, and infer regulatory relationships between genes and pathways.

While this method provides an innovative solution to the challenging problem of cell heterogeneity and its dynamics during early differentiation, random barcoding does not allow individual cell identities (marked by gene-expression, lineage, or location) to be associated with a given barcode. In this case, sequencing efforts (to identify de novo barcoded cells) are either completely ad hoc or exhaustive, leading to extremely high completion time. Therefore, it is necessary to implement a barcoding strategy that facilitates sequencing and cell analysis at a later stage; such strategy must be incorporated in the synthesis engine and it can dynamically change the ordering of dispensed droplets accordingly. This requires appropriate reservoir/port management facility associated with microfluidic biochips. A firmware layer is also required to collect and analyze sequencing information, and to guide the next steps in the protocol.

The input to barcoding-aware synthesis will be a protocol model and a barcoding library that represents the barcoded hydrogels. The library may contain hundreds of hydrogel barcodes, pipetted in separate reservoirs. This implies the following constraints: (1) A single port may be time-multiplexed among multiple reservoirs; (2) A multi-reservoir pressure modulator is used to control pressures at multiple reservoirs. It is obvious that such an architectural configuration may lead to significant time overhead despite being scalable and cost-effective. Our goal is to leverage this architecture, but develop a new synthesis framework (we call it barcoding-aware synthesis) that takes into account the following characteristics:

- Barcoding specifications must be fulfilled. For example, if cell differentiation is based on the gene-expression level for a gene (high, medium, and low), then we will assign a unique barcode for the cells in each category. With fine-grained gene-expression discretization and with the need to

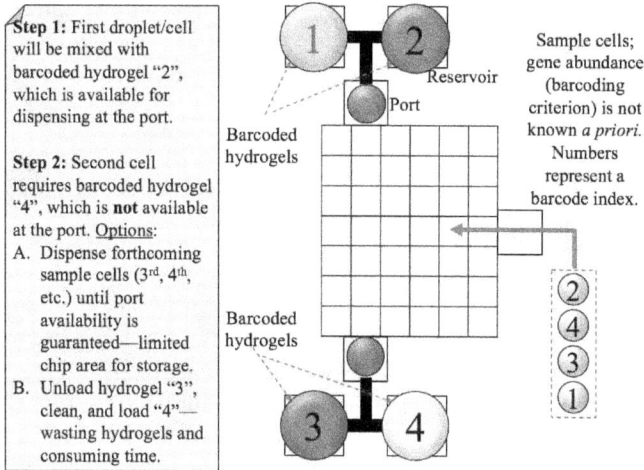

Step 1: First droplet/cell will be mixed with barcoded hydrogel "2", which is available for dispensing at the port.

Step 2: Second cell requires barcoded hydrogel "4", which is **not** available at the port. Options:
A. Dispense forthcoming sample cells (3rd, 4th, etc.) until port availability is guaranteed—limited chip area for storage.
B. Unload hydrogel "3", clean, and load "4"— wasting hydrogels and consuming time.

Barcoded hydrogels

Barcoded hydrogels

Reservoir

Port

Sample cells; gene abundance (barcoding criterion) is not known *a priori*. Numbers represent a barcode index.

FIGURE 4.13
Motivation for barcoding-aware synthesis.

consider hundreds of genes simultaneously, barcoding becomes extremely complicated.

- Completion time must be minimized. This requires a synthesis solution for fluid handling, where multi-reservoir pressure modulators are efficiently utilized. It is impractical to design a pressure modulator for each reservoir; the biochip may contain hundreds of reservoirs to cover a wide spectrum of barcodes. Figure 4.13 illustrates the motivation for barcoding-aware synthesis.

In Chapter 5, we describe an efficient framework for single-cell barcoding and analysis using an integrated microfluidic platform and an associated synthesis solution.

4.7 Distributed-System Interfacing and Integration

The distributed-system (middleware) software will handle incoming and outgoing data streams between the microfluidic devices and the cloud, ideally in a real-time fashion. Developing the middleware stack is a big-data design problem for which similar solutions have been engineered in the past [188]. Nevertheless, this stack is important for our system and must be adopted to integrate, test, and evaluate our MaaS solution. Open-source tools, specifically Apache Spark and Cassandra, can be used for database management and analytics realization. The objective is to investigate an efficient method for data structuring with Cassandra.

Deploying Spark on Cassandra

Apache Cassandra is a masterless, NoSQL online database architecture with no single point of failure (i.e., fault tolerant). Apache Spark is a centralized scheme that designed to handle a large amount of data by simultaneously processing it at scale. For example, to develop scalable regression models for CanLib, Spark Machine Learning Library (MLlib) can be deployed and used. In BioCyBig, we tightly integrate Spark and Cassandra, which gives us the capability to use Spark to analyze the data stored at Cassandra; this data is generated online by the individual microfluidic platforms that may be geographically distributed. This integration provides horizontal scaling, fault tolerance, operational-level reporting, and analytics-friendly environment, all in one package. To achieve this integration, it is imperative to study how to extract biochemical outcomes from Cassandra and incrementally move the updates to Spark in a real-time fashion.

Data Structuring and Storage in Cassandra for Real-Time Transactions

The Cassandra data model is schema-optional and column-oriented. This means that, unlike for a relational database, we do not need to model all of the columns required by the application upfront, as each row is not required to have the same set of columns. The primary language for communicating with the Cassandra database is the Cassandra Query Language (CQL). As a first step, we study how the Cassandra data model can be leveraged to store experimental reports driven by microfluidic nodes. Second, we use Cassandra APIs to write, read, and tune data replication (for consistency), via CQL. Node and cluster configurations are also an important part of our design. Our main challenge lies in the fact that biochemical reports must be structured and prepared such that real-time data writing and reading across Cassandra (via CQL) can be performed.

Integration and Evaluation: The above components need to be integrated to evaluate BioCyBig. For evaluation, publicly available information about microfluidic protocols and single-cell data can be used. Examples include a rich set of high-throughput functional genomic data a from the Bioconductor project [27]. Framework scalability can be assessed by tuning the number of microfluidic devices, labs, and users. It is required to systematically consider a range of simultaneous users and devices (e.g., in the range of tens to thousands). The responsiveness of BioCyBig can be evaluated against varying frequencies of adaptation requests. These evaluations are considered for future work.

4.8 Chapter Summary

We presented our vision for a microfluidic-driven framework, referred to as BioCyBig, that enables the interpretation of genomic sequences and how DNA mutations, expression changes, or other molecular measurements relate to disease, development, behavior, or evolution. We illustrated the system components and their functionalities, and we explained, through a case study, how the integration of biological domain expertise, large-scale computational techniques, and a computing infrastructure can support flexible and dynamic queries and system adaption to search for patterns of genomic association over large collections of omic data.

The knowledge gained from applying molecular biology protocols to software-controlled biochips in large-scale and distributed experiments will be a big step forward towards personalized medicine. Such a cloud-infrastructure based on CPS and MaaS will advance our understanding of a variety of diseases, including cancer. The advances proposed in this book will be applicable to a range of quantitative analysis protocols, such as gene-expression and immunological analysis.

Although this framework was originally envisioned and designed for scalable genomic and cancer research, the layered design methodology of our framework can be leveraged for other CPS areas; especially in smart city domains. For example, coupling big data analytics with cyberphysical adaptation enables management of sustainable mobility and traffic control in a smart city. In this setting, the application layer can be used for traffic data fusion, adaptive traffic-light control, and coordination of driverless transportation buses. The middleware layer, in turn, collects sensor data for traffic and weather conditions. This application and several others can be seamlessly deployed using our framework.

In the following chapters, we solve synthesis problems related to microfluidics-based single-cell analysis; that is, enabling the microfluidic layer for BioCyBig.

Part II

High-Throughput
Single-Cell Analysis

5

Synthesis of Protocols with Indexed Samples: Single-Cell Analysis

Single-cell analysis using affordable microfluidic technologies has now become a reality [95, 272]. Thousands of heterogeneous cells can be explored in a high-throughput manner to investigate the link between gene expression and cell types, thereby providing insights into diseases such as cancer [218]. Microfluidic techniques have recently been developed to conduct each step of the following single-cell experimental flow.

1. **Cell Encapsulation and Differentiation:** Heterogeneous cells are isolated, encapsulated inside droplets, and differentiated according to their identity (type); e.g., their shape, size, cell-cycle stage, or lineage [44].

2. **Droplet Indexing (Barcoding):** Each droplet is manipulated through a sequence of biochemical procedures such as cell lysis and mRNA analysis. At the end of these steps, the *in-situ* type of the encapsulated cell may no longer be available for down-stream analysis [227]. Therefore, indexing of droplets using barcodes is needed to keep track of their identity.

3. **Type-Driven Cell Analysis:** Single-cell bioassays such as chromatin immunoprecipitation (ChIP) are carried out using microfluidics, where the selection of a bioassay relies on the cell type that is identified in Step 1 [95]. To draw meaningful conclusions, the experimental outcomes are associated with droplet barcodes injected in Step 2 [136].

Although the synthesis methods in Chapters 2 and 3 can handle multiple independent sample pathways, i.e., they support concurrent manipulation of cells, there are two barriers that need to be addressed in order to adopt these methods for practical single-cell studies:

- **Integration of Heterogeneous Single-Cell Methods:** Not all the above biochemical steps can be efficiently miniaturized using a single microfluidics technology. Valve-based techniques are used to rapidly separate and isolate biomolecules with high resolution, making them suitable for cell encapsulation (Step 1) [95]. On the other hand, digital-microfluidic biochips (DMFBs) enable real-time decision making for sample processing and genomic-analysis protocols, such as quantitative polymerase chain reaction (qPCR) [212, 190] (Step 3). However, DMFBs are not as effective for interfacing to the external world [124]. Hence, there is a need for a hybrid microfluidic system that combines the advantages of the two domains, and a synthesis method that controls single-cell experiments in a dual-domain microfluidic setting.

- **Scalable Droplet Indexing:** A single-cell analysis flow may involve hundreds of cell types [28], each of which requires a distinct barcode for downstream analysis using digital microfluidics. Therefore, droplet indexing on a DMFB requires either the use of pre-stored droplets that host individual barcoding hydrogels [136] (or dyes)[1]—not feasible when a large number of cells are being investigated—or a specific input reservoir for each cell type. The latter solution increases the fabrication cost dramatically. Furthermore, since reservoir control is not readily automated[229], it is unrealistic to assume that each dispensed droplet contains only one barcoding particle. Therefore, there is a pressing need for a low-cost mechanism for droplet barcoding.

- **Dynamic synthesis:** Due to the inherent uncertainty about cell types, cyber-physical integration can play a key role in streamlining microfluidic cell-type identification and single-cell analysis. However, employing cyber-physical integration for processing every cell requires a dynamic synthesis capability, which can effectively explore resource space and also provide a prompt solution. As a result, the need for such a capability introduces a tradeoff between synthesis performance, e.g., protocol completion time, and system responsiveness—this tradeoff has yet to be investigated.

- **Stochasticity of Protocol Conditions:** Despite the efforts made towards standardization of single-cell protocols, many protocol steps, such as the duration of sonication in ChIP, are deemed to be cell-type-specific. For example, CD4+ whiteblood cells require significantly longer sonication time compared to other red blood cells [17]. Failing to characterize the variation in sonication time and other parameters among cell types may lead to degradation in down-stream immunoprecipitation and thus the overall ChIP performance [78]. Hence, it is important to take into consideration such stochasticity when defining the protocol guidelines.

In Chapters 2 and 3, we introduced a synthesis framework for DMFBs to support multiple sample pathways, but we considered that sample differentiation and indexing (Stage #2 in Figure 1.14) have been carried out in advance before protocol execution. Moreover, stochasticity of protocol conditions was not considered. Therefore, the earlier frameworks overlooked the above challenges.

In this chapter, we address the above challenges by introducing a hybrid microfluidic platform for integrated single-cell analysis [113, 114]. We present

---

[1] Henceforth, we refer to dyes as barcoding droplets.

a synthesis method, referred to as Co-Synthesis (CoSyn), to control the dual-domain platform. We also present a probabilistic model that employs a discrete-time Markov chain (DTMC) to capture protocol settings where experimental parameters vary among cell types. The main contributions of this chapter are as follows:

- We present an architecture of a hybrid microfluidic platform that integrates digital-microfluidic and flow-based domains (using valves) for large-scale single-cell analysis.

- We describe CoSyn, which enables coordinated control of the microfluidic components, and allows dual-domain synthesis for concurrent sample pathways.

- We propose two schemes for valve-based routing (graph-theoretic and incremental methods), which enable dynamic routing of concurrent samples within a reconfigurable valve-based system.

- We construct a DTMC model, which utilizes probabilistic information related to protocol steps and experiment budget to investigate the efficiency of probabilistic protocol decisions.

- We evaluate system performance and reconfigurability while exploring various configurations of the valve-based system.

The rest of the chapter is organized as follows. An overview of the hybrid microfluidic platform and its use for single-cell analysis are presented in Section 5.1. Next, we formalize the single-cell analysis flow in Section 5.2 and describe the proposed synthesis framework (CoSyn) in Section 5.3. Subsequently, details of the valve-based synthesizer are introduced in Section 5.4, and probabilistic protocol modeling is presented in Section 5.5. Our experimental evaluation is presented in Section 5.6 and conclusions are drawn in Section 5.7.

5.1 Hybrid Platform and Single-Cell Analysis

Similar to the cyber-physical microfluidic implementation of gene-expression analysis (Chapter 2), single-cell analysis relies on the concurrent manipulation of sample droplets, where each sample cell is run through the protocol flow discussed earlier. An efficient on-chip implementation of the single-cell analysis protocol is accomplished using a hybrid platform. Figure 5.1 shows the platform components matched with different protocol stages. The two domains are connected through a capillary interface; this technique has been successfully adopted in practice [229].

Cell Encapsulation and Flow Control

As shown in Figure 5.1, on-chip operation starts with the encapsulation of single cells in droplets, which is efficiently accomplished using flow-based microfluidics [229].

A flow-based system can be configured to function as a droplet-in-channel device, allowing a two-phase flow to be generated. More specifically, encapsulation of individual cells is easily accomplished by considering three intersecting flows; an aqueous flow (containing cells) and two oil flows. These flows are pressure-driven by syringe pumps and therefore they can be carefully balanced, allowing aqueous droplets (containing single cells) to be automatically formed with a surrounding oil phase. Next, the resulting two-phase flow is transported to the digital microfluidic device through a capillary interface. Figure 5.2 shows cell encapsulation and droplet generation using a two-phase flow.

On the other hand, the digital microfluidic device consists of two parallel plates. The gap that separates the two plates is flooded with oil, which acts as a filler medium. According to [229], the rate of oil injection between the two plates can be controlled using a feedback system in order to prevent the evaporation of droplets that are collected from the flow-based side. The oil medium also facilitates the injection of the two-phase flow into the digital side through the capillary interface.

To efficiently integrate both sides, the droplet generator uses the syringe pumps such that the flow rate of pressure-driven droplets can be automati-

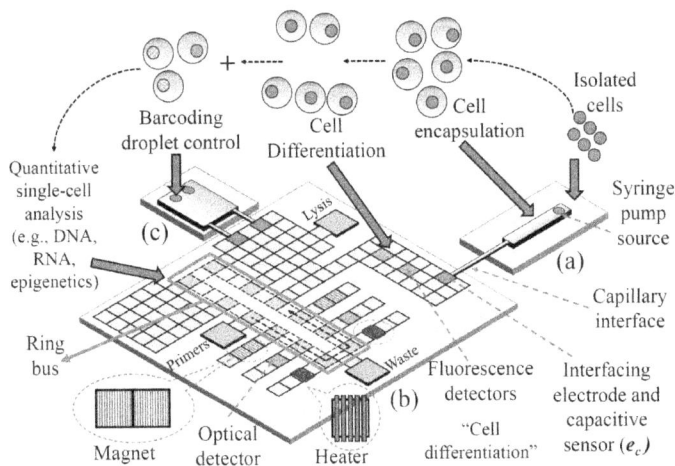

FIGURE 5.1
Schematic of the hybrid platform for single-cell analysis: (a) a flow-based biochip used for cell encapsulation and droplet generation; (b) a DMFB used for quantitative analysis; (c) a reconfigurable valve-based fabric used for barcoding.

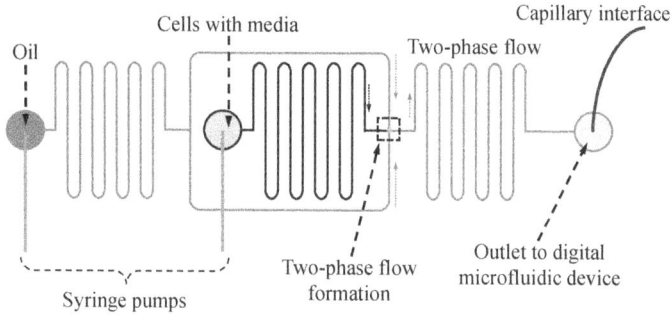

FIGURE 5.2

A flow-based device used to generate aqueous droplets containing single cells as part of a two-phase flow.

cally controlled via feedback. A capacitive sensor is placed at the interfacing electrode (Figure 5.1) on the digital side to sense a droplet [180]. When the digital array is unable to accommodate additional droplets, it stops the flow by switching off the pump.

Note that an actuator is used in the flow-based component, whereas a sensor is placed on the digital side. To synchronize the two domains, the flow-control procedure (capacitive sensing and pump control) is invoked at the same frequency as droplet actuation in the digital domain (1 Hz to 10 Hz).

Cell Differentiation

Automated cell-type identification can be achieved by analyzing signaling events in single cells *in situ*. Similar to the miniaturization of gene-expression analysis, a green fluorescent protein (GFP) reporter is used for cell differentiation. In each cell, the fluorescence intensity from the GFP (detected in real-time using an on-chip fluorescence detector or imaging apparatus) is used to account for differences in expression level among cells; this is equivalent to classifying cells into functional clusters that represent cell types.

Although a valve-based biochip can also be used for cell differentiation, we consider a DMFB for this purpose in CoSyn due to its demonstrated ability to carry out high-throughput fluorescence detection and distinguish between hundreds of cell types [183]; this feature is not supported by valve-based mechanisms.

Droplet Barcoding

Droplet barcoding is essential to maintain cell identity during down-stream analysis. Since thousands of cells (and hundreds of cell types) can be involved in an analysis protocol, a barcoding droplet must be dispensed *on demand*,

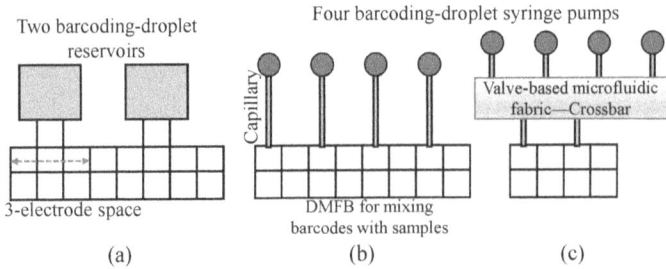

FIGURE 5.3
Droplet barcoding using: (a) a DMFB; (b) a valve-based biochip with one-to-one mapping between syringe pumps and DMFB ports; (c) a valve-based biochip with many-to-many mapping.

and mixed with a sample droplet and other reagents according to the cell type [136]. If we consider a population with n_c cell types, droplet barcoding on a digital-microfluidic array requires n_c reservoirs, each of which typically covers a 3-electrode space. An additional electrode is needed for separation; see Figure 5.3(a). In this case, to accommodate n_c reservoirs, a lower bound on the array perimeter is $4n_c + K$ electrodes, where K is a constant that represents the number of electrodes covered by other reservoirs. This approach is therefore impractical because of the significant increase in chip size with the number of target cell types. It also requires the dispensing of a single hydrogel particle per droplet; this feature has not been implemented yet using reservoir control.

To overcome the above limitations, a valve-based biochip is connected to the DMFB to exploit its pressure-driven ports that have smaller footprints than reservoirs; see Figure 5.3(b). This biochip is used to generate barcoding droplets via a syringe pump; the droplets are routed to appropriate locations on the digital-microfluidic array through a capillary interface [136]. With this hybrid configuration, the lower bound on the digital-array perimeter is reduced to $2n_c + K$ electrodes for n_c cell types. A one-electrode gap is needed to prevent accidental mixing of droplets. This hybrid configuration, however, overprovisions the number of concurrently utilized ports; it is unlikely that all cell types will simultaneously request barcoding.

As a cost-effective solution, we utilize a reconfigurable valve-based fabric [232]; see Figure 5.3(c). This routing fabric acts as a crossbar since it allows routing of barcoding droplets from any of the n_b input ports to any of the m_b output ports, where $n_b >> m_b$. The m_b-output valve-based fabric is then stitched to the DMFB; hence the lower bound on the perimeter is decreased to $2m_b + K$. However, in this situation, p_b valves ($p_b > 0$) are necessary for routing barcoding droplets across the fabric. By unlocking this capability of valve-based crossbars, we shift the scaling complexity from the digital domain

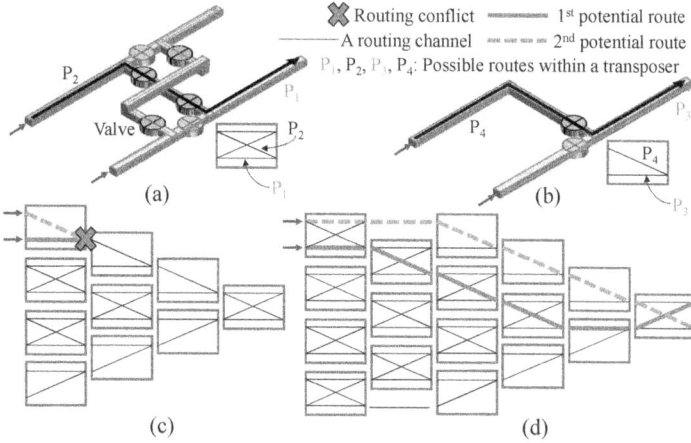

FIGURE 5.4
A valve-based routing fabric for droplet barcoding: (a) a 2-input full transposer; (b) a 2-input half transposer; (c) a 4-level, 8-to-2 routing fabric; (d) a 6-level, 8-to-2 fabric.

to the flow-based domain, which is known to have a cost-effective fabrication process and efficient peripheral components.

In addition to the above advantages, the use of a reconfigurable valve-based fabric is important since it allows us to employ a design methodology that investigates the tradeoff between biochip cost and single-cell throughput and therefore provides more choices for biochip designers. For example, a biochip designer may prefer to utilize a cost-effective design, in which fewer electrodes are used at the digital side. To achieve this goal, i.e., to optimize for cost, sharing of flow channels among several barcoding inputs gains significant importance in order to minimize the number of peripherals to the DMFB. In other words, the role of the reconfigurable valve-based fabric becomes more significant. On the other hand, if a designer opts for a high-throughput platform, in which a large DMFB size is utilized, then sharing of flow channels among barcoding particles can be reduced, and therefore the role of the reconfigurable valve-based system becomes less significant. This observation has been studied and the results are reported in Section 5.6.

We utilize the "transposer" primitive introduced in [232]. As shown in Figure 5.4(a)-(b), a valve-based transposer appears in two forms: (1) a two-input, two-output transposer, which is comprised of six valves, controlled via two pneumatic inputs (full transposer); (2) a two-input, one-output transposer, which consists of two valves controlled via two pneumatic inputs (half transposer). Note that only a full transposer allows simultaneous dispensing of two barcoding droplets, wherein the droplets can be driven "straight" or "crossed."

TABLE 5.1
Design problems for assembling and integrating a n_b-to-m_b valve-based crossbar.

Problem	Objective
Architecture	Analyze the tradeoff between routing flexibility and operation timing and cost to obtain the best transposer configuration.
Modeling	Map the architecture into algorithmic semantics that support real-time routing.
Synthesis	Solve the valve-based routing problem considering real-time reconfiguration and in coordination with resource allocation in DMFBs.

The use of transposers to construct an n_b-to-m_b valve-based crossbar leads to various design problems that must be tackled; see Table 5.1. An architectural design challenge arises because various configurations of transposers can be exploited to achieve the required number of input and output ports. For example, an 8-to-2 crossbar can be constructed using four "vertical" levels, as shown in Figure 5.4(c), or using six levels, as shown in Figure 5.4(d). A six-level crossbar, while incurring higher cost, provides a higher degree of reconfigurability and flexibility in routing. We present our solution for the modeling, synthesis, and architectural design problems in Sections 5.2, 5.3, and 5.4, respectively.

Type-Driven Single-Cell Protocol

After a droplet is barcoded, a single-cell analysis protocol is applied to the constituent cell in the DMFB, where protocol specifications are determined based on the cell type. For example, an investigator might be interested in identifying the gene expression of a specific genomic loci for a certain cell type A. Another cell type B might show unexpected heterochromatic state at a certain loci, and the investigator might be interested in identifying the protein interactions (i.e., causative proteins) or chromatin modifications causing this behavior. For type A, it is sufficient to perform gene-expression analysis using qPCR [212, 115], whereas ChIP protocol followed by qPCR must be used for type B to reveal the DNA strains contributing in the activity of the causative proteins [282].

Ultimately, by supporting type-driven analysis, biologists can launch multiple single-cell applications, aiming for drawing a holistic picture of the interactions between different biomolecules (e.g., DNA, mRNA, protein, etc.). Note that our proposed framework can support the execution of multiple applications, a single application for all cell types, or even multiple applications for a stream of cells that are of the same type (cell barcoding can be used to distinguish these cells at the end of down-stream analysis).

Platform Throughput and Scalability

By combining the two microfluidic formats (flow-based and digital microfluidics), we can maintain scalable single-cell analysis. As described earlier, flow-based microfluidics is efficient in generating thousands of droplets that encapsulate individual cells using a two-phase format. However, on the negative side, flow based microfluidics relies on an etched micro-structure that, despite its capability in droplet preparation, fails to support reconfigurable analysis at the down-stream part. To overcome this drawback, digital microfluidics comes into play—digital microfluidics is well-suited for adding reagents in parallel, and reagents can be mixed on demand without the requirement of optimal and precise flow rates.

A DMFB is scalable in terms of handling multiple cells concurrently. In [212], a 230-electrode chip was used to process 20 cells concurrently. By using our design-automation technique, the number of cells can be drastically increased since we allow resource sharing among pathways. However, note that a DMFB still provides a lower throughput compared to the flow-based devices; this observation does not mean that digital microfluidics cannot process droplets quickly, but it means that a DMFB is burdened with the largest portion of work for single-cell analysis. Further increase in the chip size can also increase the number of cells manipulated concurrently, thus leading to a higher throughput. Nevertheless, it needs to be clear that our objective is not to achieve the optimal throughput; our objective is to provide a cost-effective design methodology that employs the hybrid platform to process a continuous stream of cells with a reasonable throughput.

Based on the above discussion, flow-based microfluidics offers temporal scalability (high-throughput), whereas digital microfluidics offers reconfigurability and spatial scalability (concurrent single-cell analysis). Combining both technologies with an adequate feedback system (to synchronize throughput rates at both domains) offers spatio-temporal scalability for single-cell analysis.

Adaptation of the $\mathcal{C}^5$ Architecture to the Hybrid System

Cyber-physical integration plays a key role in enabling type-driven single-cell analysis. Clearly, an efficient cyber-physical system design bridges the gap between the microfluidic components and the software specifications of the single-cell analysis protocol. Hence, an efficient solution to bridge this gap can be developed based on the the the $\mathcal{C}^5$ architecture described in Section 1.4. The adaptation of this architecture to the single-cell protocol lays the ground for a seamless interaction between a sequence of heterogeneous cells and the

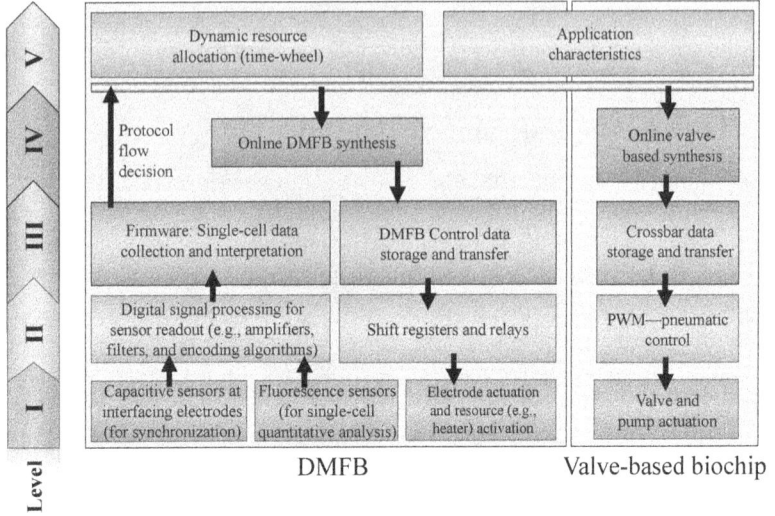

FIGURE 5.5
$\mathcal{C}^5$-compliant architecture design of cyber-physical hybrid microfluidic platform for single-cell analysis.

hybrid microfluidic system. Figure 5.5 shows the $\mathcal{C}^5$-compliant design of cyber-physical hybrid microfluidic system for single-cell analysis.

In this chapter, we focus only on the top two levels, i.e., Level IV and Level V. We specifically study the interaction between the microfluidic components and single-cell analysis (Level V) and dynamic synthesis algorithms for both microfluidic domains (Level IV).

5.2   Mapping to Algorithmic Models

This section provides a mapping of the hybrid platform and the single-cell analysis protocol to algorithmic models.

Modeling of a Valve-Based Crossbar

Recall that an n_b-to-m_b valve-based crossbar can be constructed using different combinations of transposers. We represent the set of transposers and their interconnections as a directed acyclic graph (DAG) $G_b = (V_b, E_b)$, where a vertex $v_{bi} \in V_b$ is a transposer node, and an edge $e_{bi} \in E_b$ represents a connection between two transposers. Within a transposer, the point at which a droplet can be routed either straight or crossed is defined as a decision point.

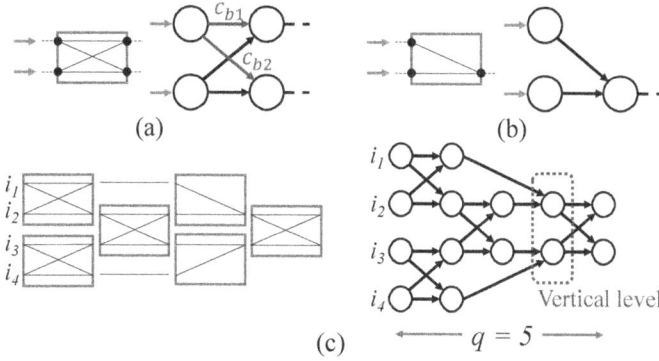

(a)   (b)

(c) $q = 5$

FIGURE 5.6
Mapping a valve-based crossbar to a graph model: (a) a full transposer; (b) a half transposer; (c) a 4-to-2 crossbar.

We map an n_b-to-m_b valve-based crossbar (with a transposer network G_b) into a DAG $\mathcal{F}_{n_b \times m_b} = (\mathcal{D}_{n_b \times m_b}, \mathcal{Y}_{n_b \times m_b})$, where a vertex $d_{bi} \in \mathcal{D}_{n_b \times m_b}$ is a flow-decision node, and an edge $y_{bi} \in \mathcal{Y}_{n_b \times m_b}$ represents a channel that connects two decision nodes. To simplify the discussion, we do not include G_b in the notation for the crossbar DAG. We can view a full (half) transposer as a 2-to-2 (2-to-1) valve-based crossbar; thus, we represent fluid-flow control in a full (half) transposer as a DAG $\mathcal{F}_{2 \times 2}$ ($\mathcal{F}_{2 \times 1}$); see Figure 5.6(a)-(b). The cost c_{bi} of y_{bi} represents the time needed to transport fluid between the two connected nodes, measured in flow time steps (T_b). We assume that the routing time of a droplet on a straight channel between two decision nodes is a unit of T_b. For example, as shown in Figure 5.6(a), c_{b1} is equal to T_b, whereas c_{b2} is equal to $2 T_b$, since even though a diagonal is shown in Figure 5.4 as a fluidic path, routing of such paths in a transposer is implemented only along the x- and y-directions and the distances along these dimensions are equal [232]. Figure 5.6(c) depicts the graph $\mathcal{F}_{4 \times 2}$ for a 4-to-2 crossbar with 4 levels of transposers and 5 levels of nodes (denoted henceforth by q; $q = 5$ in this case).

Modeling of a Digital-Microfluidic Biochip

While DMFBs are highly reconfigurable and can support a diverse set of transport paths, we reduce the burden of managing droplet transport in real-time by considering a unidirectional ring-based architecture, as shown in Figure 5.1. Connected to this ring are on-chip resources. Since there is always a route between any pair of on-chip resources, a ring-based DMFB is modeled as a strongly connected DAG $G_f = (V_f, E_f)$, where a vertex $v_{fi} \in V_f$ represents the fluid-handling operation offered by an on-chip resource, and a directed edge $e_{fi} \in E_f$ represents a path (over the ring) that connects two resources.

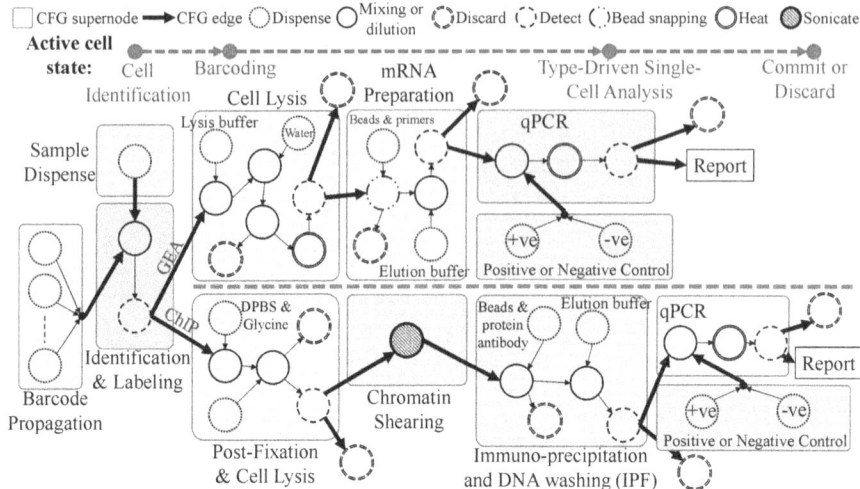

FIGURE 5.7
CFG of type-driven analysis for a single cell pathway.

The cost c_{fi} of e_{fi} indicates the number of digital time steps (T_f) needed to transport a droplet. We assume that the durations corresponding to T_b and T_f are equal, which can be achieved in practice by tuning the actuation frequency (Hz) and the flow rate (mL/min).

Protocol Model and Cell State Machine

To solve the synthesis problem for single-cell analysis, we take into account the complexity imposed by the barcoding mechanism. Similar to the design methodology in Chapter 2, we represent the protocol as a control flow graph (CFG) $G_c = (V_c, E_c)$, in which every node $v_{ci} \in V_c$ (referred to as a *supernode*) models a bioassay such as qPCR; see Figure 5.7. A directed edge $e_{ci} \in E_c$ linking two supernodes $\{v_{cj}, v_{ck}\}$ indicates that a potential decision can be made at runtime to direct the protocol flow to execute the bioassay v_{ck} after v_{cj}. A supernode v_{ci}, in turn, encapsulates the sequencing graph that describes the fluid-handling operations of a bioassay and the interdependencies among them. Since there is inherent uncertainty about the type of barcoding droplets for a sample cell at design time, we extend the basic CFG model by incorporating an internal supernode (barcode propagation) that describes all possible dispensing options of barcoding droplets. Note that this model is agnostic about the type of the microfluidic technology used for implementing the protocol. Yet, the synthesis of each supernode is accomplished in a technology-aware manner using CoSyn.

The model shown in Figure 5.7 represents a protocol where a dispensed cell can be processed using one of two biochemical procedures, namely gene-expression analysis (GEA) and ChIP procedures. As shown in the CFG model, the execution of the protocol starts with dispensing an aqueous sample droplet (Sample Dispense supernode). The type of the sample is then identified and mixed with an associated barcoding droplet that is routed through the reconfigurable valve-based fabric (Identification and Labeling supernode). Next, according to the cell type, either GEA or ChIP procedures will be executed. If GEA is selected, then fluid-handling operations of cell-lysis, mRNA preparation, control preparation, and qPCR will be carried out. At the end of each bioassay, a detection operation is performed to ensure that the efficiency of the resulting solution is above a certain threshold, as described in Section 2.5. Similarly, if ChIP is selected, then fluid-handling operations of post-fixation, cell-lysis, chromatin shearing, immuno-precipitation, DNA washing, control preparation, and qPCR will be performed.

In addition to the CFG model, a state machine is utilized to model the progression of each cell along the single-cell pipeline. Typically, the hybrid platform can iteratively process thousands of cells; such cells might be scattered across the platform domains at any given point in time. Therefore, this state machine (Figure 5.7) is necessary to keep track of the cells that are being processed simultaneously.

5.3 Co-Synthesis Methodology

In this section, we describe the synthesis problem and present an overview of the proposed co-synthesis methodology. Table 5.2 summarizes the notation used in this chapter.

Problem Formulation

Our optimization problem is as follows:

Inputs: (i) The protocol CFG G_c. (ii) A matrix C; each vector $C_i \in C$ corresponds to a cell, and consists of integers that encode cell state machine, cell type, and the assigned bioassays in G_c. (iii) The configuration of the valve-based system; this information includes the graph $\mathcal{F}_{n_b \times m_b}$, the number of inputs n_b, and the number of outputs m_b. (iv) The types of resources corresponding to the DMFB, their operation time, and the routing distance between each pair of resources.

Output: Allocation of chip modules by the individual cells, protocol completion time T_{comp}.

Objective: Minimize T_{comp} to provide high throughput and minimize T_{resp} to improve system responsiveness.

TABLE 5.2
Notation used in Chapter 5.

G_b	The graph model of an n_b-to-m_b crossbar
$\mathcal{F}_{n_b \times m_b}$	The flow-decision model of an n_b-to-m_b crossbar
$\mathcal{D}_{n_b \times m_b}$	The set of vertices in $\mathcal{F}_{n_b \times m_b}$
$\overline{\mathcal{D}}_{n_b \times m_b}$	The set of unoccupied vertices in $\mathcal{F}_{n_b \times m_b}$
P_b	A set of vertices in $\mathcal{F}_{n_b \times m_b}$ representing a routing path
θ_{P_b}	A Boolean variable: True if a path P_b is complete
G_f	The graph model of a DMFB
$\mathcal{R}$	The set of DMFB resources
$\overline{\mathcal{R}}$	The set of unoccupied DMFB resources ($\overline{\mathcal{R}} \subset \mathcal{R}$)
$\hat{s}$	Minimum value of the cost function associated with $\overline{\mathcal{R}}$
C_i	Cell metadata
$\mathcal{A}$	The set of fluidic-operation types
G_c	The control-flow graph of a protocol
$\mathcal{M}$	The DTMC model of a protocol
$\mathcal{W}$	The set of states of $\mathcal{M}$
$\mathcal{U}$	The transition-probability matrix of $\mathcal{M}$

Proposed Solution

The proposed co-synthesis scheme, depicted in Figure 5.8, consists of four components: (1) valve-based synthesizer, which is used to route barcoding droplets through the valve-based crossbar; (2) DMFB synthesizer, which is utilized for allocating DMFB resources, e.g., mixers and heaters, to sample pathways; (3) biology-sample model, which records the progress of a sample (cell) within the protocol CFG and also provides updated resource preferences; (4) time-wheel engine, which seamlessly coordinates real-time interactions between the individual sample models and the synthesizers of the hybrid system. Note that the stages of the single-cell pipeline, simulated by this scheme, match the states of the cell state machine (Figure 5.7).

The time-wheel interacts with other components through APIs. For example, whenever the time-wheel locates an available fluorescence detector at the DMFB, a new sample model is initialized (Figure 5.8) and the associated cell is allocated to the detector in order to perform type identification. Next, when the cell type is identified and there are available valve-based routes to route the associated barcoding droplet, the time-wheel triggers the valve-based synthesizer to start the pipelined routing process of the barcoding droplet; valve-based routing is performed through iterations until the droplet reaches the electrode interface at the digital-microfluidic side and is mixed with the cell. We discuss two methods for valve-based routing in Section 5.4.

When a DMFB resource is available to further process the cell, the previously reserved valve-based channels are released by the time-wheel. Hence,

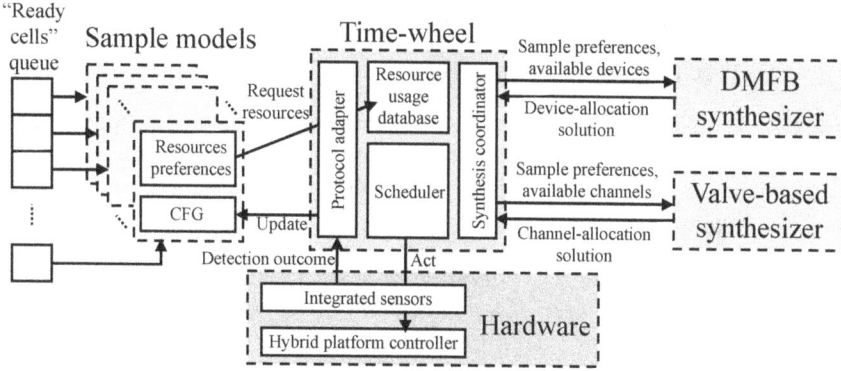

FIGURE 5.8
Overview of the proposed co-synthesis methodology.

real-time resource allocation for the DMFB is also initiated by the time-wheel, which in turn, commits a cell pathway whenever its particular single-cell bioassays have executed. Based on an intermediate decisions whose outcome is communicated to the sample model, the cell might also be discarded during analysis.

DMFB Synthesizer

We use a greedy method to solve the resource-allocation problem in the DMFB; the pseduocode is shown in Algorithm 5.1. We denote a DMFB resource by $r_i \in \mathcal{R}$, where $\mathcal{R}$ encapsulates all DMFB resources. Thus, the cost of allocating resource r_i to execute a fluidic operation of type $a_k \in \mathcal{A}$ (the set $\mathcal{A}$ incorporates all operation types) is $\varpi(r_j, r_i, y_k) = OT(r_i, a_k) + CR(r_j, r_i)$, where $OT(r_i, a_k)$ is the operation time on r_i and $CR(r_j, r_i)$ is the routing distance from r_j (the currently occupied resource) to r_i. The worst-case computational complexity of this algorithm is $O(|V_f|)$.

Algorithm 5.1 DMFB Resource Allocation
***
Input: C_i, G_f, current simulation time "t"
Output: Assigned Resource "r_i"
1: $\overline{\mathcal{R}} \leftarrow$ GetCurrentlyUnoccupiedResources(G_f, t);
2: **if** ($\overline{\mathcal{R}}$ is *empty*) **then return** NULL;
3: $a_j \leftarrow$ GetOperationType(C_i);
4: $\hat{s} \leftarrow$ CalculateMinimumCostAllAvailableResources(a_j, $\overline{\mathcal{R}}$);
5: **if** ($\hat{s} = \infty$) **then return** NULL; ▷ No suitable resource
6: $r_i \leftarrow$ GetSelectedResource($\hat{s}$); **return** r_i;

5.4 Valve-Based Synthesizer

Our goal is to design a fully connected fabric such that a droplet can be forwarded from any of the n_b inputs to any of the m_b output ports. We present a sufficient criterion for achieving a fully connected fabric. The proof can be found in Appendix A.

Theorem 5.1 *An n_b-to-m_b, q-level valve-based crossbar is a fully connected fabric if n_b and m_b are even integers, and $q \geq \frac{m_b+n_b}{2}$.*

Using this theorem, we can automatically generate the graph model $\mathcal{F}_{n_b \times m_b}$, thereby guaranteeing that any barcoding input can reach all m_b outputs. The algorithm is described in Appendix B. Using this model, a systematic methodology for droplet routing can be developed. We introduce the formulation of the routing problem and the solution approach below.

Problem Formulation and Solution Approach

Valve-based routing is described as follows:
Inputs: (i) The fabric model $\mathcal{F}_{n_b \times m_b}$, the number of inputs n_b, and the number of outputs m_b. (ii) The matrix C that encodes the cell state machine; each vector $C_i \in C$ corresponds to a cell (Section 5.3).
Constraints: (i) A droplet must be routed through a path from its specified input port to any output port. (ii) A droplet requires at least one time step to be transported from an intermediate vertex to another directly connected vertex (i.e., no jumps allowed). (iii) At any time t, the routing paths of different droplets cannot overlap.
Output: Allocation of the graph vertices to barcoding droplets at all time steps.

An n_b-to-m_b valve-based crossbar allows only m_b barcoding droplets to be delivered simultaneously to the DMFB. We increase throughput by allowing pipelined routing of droplets. With pipelining, the routing algorithm allows a droplet to be routed even though a complete path to an output is unavailable. In this case, a droplet is immobilized at the furthest intermediate decision point that is not reserved by other droplets (a pipeline stage), then allowed to move forward when a path is freed. Figure 5.9 illustrates pipelined and non-pipelined routing.

In the following subsections, we introduce two schemes for solving the routing problem. The first scheme uses graph search to efficiently route barcoding droplets, whereas the second scheme aims to improve system responsiveness (i.e., reduce computation time) via an incremental routing procedure.

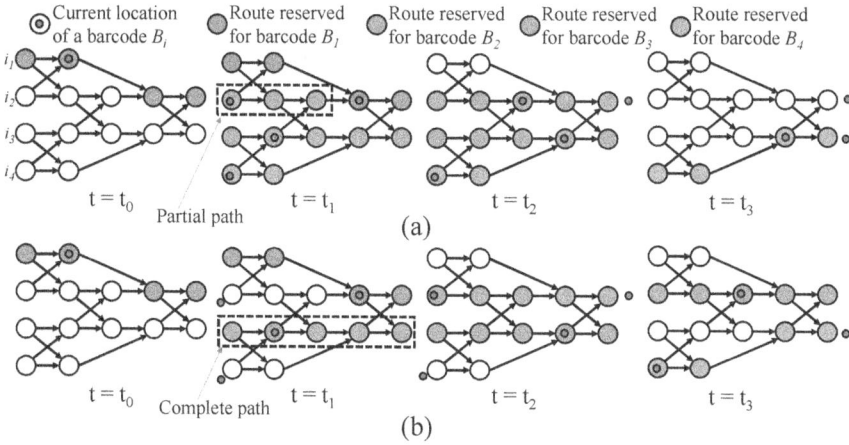

FIGURE 5.9
Valve-based routing of 4 barcoding droplets (4-to-2 biochip): (a) with pipelining; (b) without pipelining.

Method 1: Graph-Theoretic Routing

We utilize a graph-theoretic algorithm to find vertex-disjoint shortest paths [58]; see Algorithm 5.2. By computing disjoint paths, we ensure that different barcoding droplets do not interfere with each other during routing. The routing algorithm is invoked whenever a cell transitions from the identification state to the barcoding state. If all the m_b outputs of the chip are currently reserved, the algorithm generates a partially disjoint shortest-length path from the input source to the furthest node (Figure 5.9). This is equivalent to routing the associated barcoding droplet up to an intermediate point, and holding the droplet until another disjoint path (partial or complete) can be computed to advance the droplet. The channels currently reserved for routing a barcoding droplet cannot be accessed by any other droplet until the droplet being held moves out of the valve-based crossbar.

Algorithm 5.2 Pipelined Valve-Based Routing
___

Input: C_i, $\mathcal{F}_{n_b \times m_b}$, current simulation time "t"
Output: Generated routing path "P_b", is P_b complete "θ_{P_b}"
  1: $\hat{\mathcal{D}}_{n_b \times m_b} \leftarrow$ GetCurrentlyUnoccupiedSubGraph($\mathcal{F}_{n_b \times m_b}$, t);
  2: $vb \leftarrow$ GetVertexCurrentlyHoldingBarcode(C_i, $\mathcal{F}_{n_b \times m_b}$);
  3: $P_b \leftarrow$ GenerateVertexDisjointShortestPath(vb, $\hat{\mathcal{D}}_{n_b \times m_b}$);
  4: **if** (P_b is *empty*) **then return** {NULL, false};
  5: **else if** (P_b is *complete*) **then return** {P_b, true};
  6: **else return** {P_b, false};

Since the vertices of $\mathcal{F}_{n_b \times m_b}$ are generated in topological order, the computation of shortest paths can be simplified; the worst-case complexity of this algorithm is $O(|\mathcal{D}_{n_b \times m_b}| + |\mathcal{Y}_{n_b \times m_b}|)$.

Method 2: Incremental Routing

Method 1 is computationally expensive, despite the use of topological ordering, because it performs exhaustive search to find a vertex-disjoint shortest-length path whenever a barcoding droplet is allowed to move forward. Therefore, to reduce the computation time, we need to limit the size of the search space so that the routing of a droplet can be determined quickly. For this purpose, we replace the graph-theoretic method (Line 3 in Algorithm 5.2) with an incremental routing procedure, which computes the route of a droplet only for the next time step. Hence, by using this approach, the search space for routing a droplet is limited to three choices only: move straight, move crossed, or stay immobilized. The reduction in the search space can potentially reduce the overall computation time, although this procedure will be executed more often compared to the graph-theoretic method.

To facilitate the making of a routing decision, the valve-based synthesizer adopts the following priority scheme: (1) a barcoding droplet must stay immobilized if both the straight and crossed channels are occupied; (2) a droplet must move straight (crossed) if only the straight (crossed) channel is unoccupied; (3) a droplet must move straight if both the straight and crossed channels are unoccupied. Figure 5.10 illustrates incremental routing using a 4-to-2 crossbar. The computational complexity of this procedure is $O(1)$.

Droplet Routing Using Variants of Crossbar Architecture

Our discussion so far has focused on valve-based routing using a fully connected crossbar $\mathcal{F}_{n_b \times m_b}$, which exhibits the highest degree of routing flexibility and fault tolerance. According to Theorem 5.1, this design requires at least $\frac{n_b + m_b}{2}$ vertical levels. A drawback of using this number of vertical levels is that it increases the number of time steps needed to transport a barcoding droplet from an inlet to the DMFB side, which in turn may increase the total completion time. Therefore, we need to investigate the tradeoff between routing flexibility (or fault tolerance) and system performance.

To perform this study, we explore various configurations of an n_b-to-m_b crossbar; these configurations differ in the number of vertical levels and therefore channel connectivity. For example, a 16-to-8 crossbar can be constructed using one of five different configurations, three of which are shown in Figure 5.11. It is obvious that the configuration in Figure 5.11(a) provides the highest connectivity, but it uses the largest number of vertical levels (12 levels). In contrast, the configuration in Figure 5.11(c) offers the lowest connectivity, but it utilizes the smallest number of vertical levels (3 levels).

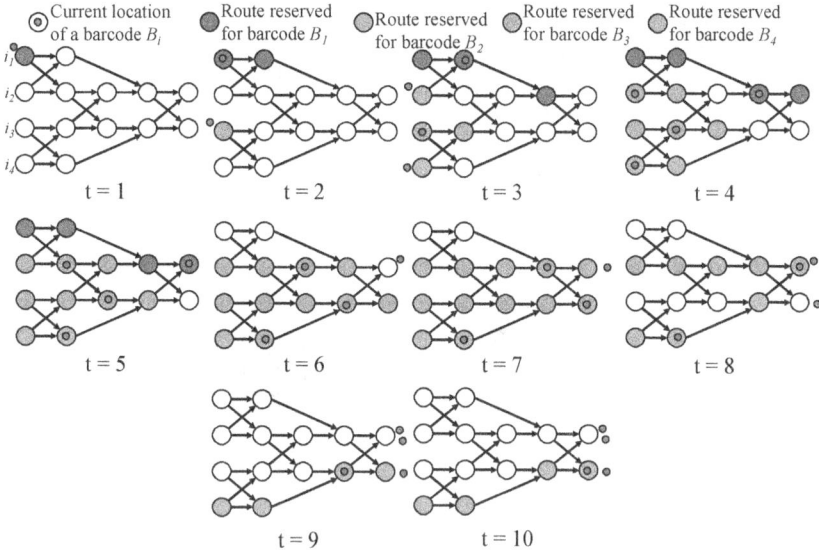

FIGURE 5.10
Incremental routing of 4 barcoding droplets (4-to-2 biochip).

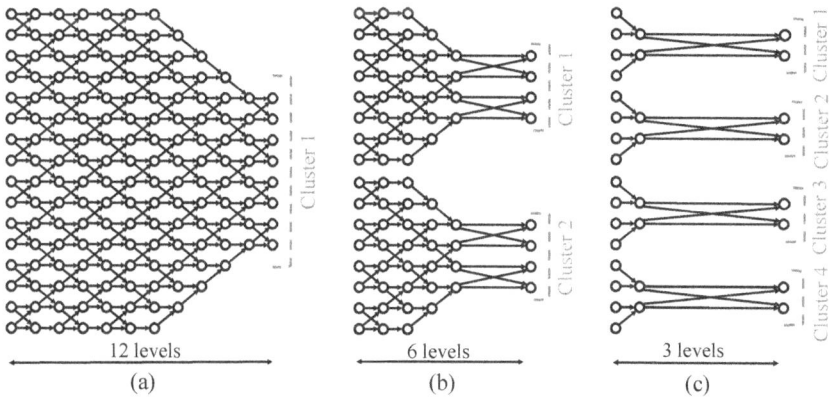

FIGURE 5.11
Architectural variants of a 16-to-8 crossbar: (a) $\mathcal{F}_{16\times 8}^{1}$; (b) $\mathcal{F}_{16\times 8}^{2}$; (c) $\mathcal{F}_{16\times 8}^{4}$.

We observe that a crossbar configuration can be constructed hierarchically using a set of fully connected crossbars. For instance, a 16-to-8 configuration can be constructed using a single 8-to-4 crossbar $\mathcal{F}_{8\times 4}$ and two 4-to-2 crossbars $\mathcal{F}_{4\times 2}$ (Figure 5.11(c)), or it can be designed using a single 16-to-8 fully connected crossbar $\mathcal{F}_{16\times 8}$ (Figure 5.11(a)). Clearly, the hierarchical design approach not only facilitates the control of the crossbar connectivity, but it also allows us to reuse the definitions and routing methods proposed earlier for fully connected fabrics. In other words, we can automatically generate the graph models of such crossbar variants using the algorithm described in Appendix B; only a minor modification is required to systematically combine the graph models associated with the constituting fully connected crossbars into a unified model.

A formal discussion of a crossbar configuration can be established by partitioning the crossbar outputs into clusters. The output ports within a single cluster are accessible to the same set of input ports, i.e., a cluster forms a fully connected crossbar. On the other hand, the output ports that belong to two different clusters cannot be accessed from the same input. By using this characterization mechanism, we can define a crossbar configuration based on the number of output clusters cl. For example, the configuration in Figure 5.11(a) has a single cluster ($cl = 1$), the configuration in Figure 5.11(b) has two clusters ($cl = 2$), and the configuration in Figure 5.11(c) has four clusters ($cl = 4$). Hence, we denote a crossbar configuration by $\mathcal{F}^{cl}_{n_b \times m_b}$, where n_b and m_b are the number of inputs and outputs, respectively. Also, note that we can use $\mathcal{F}^1_{n_b \times m_b}$ and $\mathcal{F}_{n_b \times m_b}$ interchangeably since $\mathcal{F}^1_{n_b \times m_b}$ is the only crossbar variant that is fully connected.

The proposed crossbar configurations can be employed to route droplets using our methods from the previous subsections. Our algorithms will automatically comply with the clustering constraints imposed by the new crossbar connectivity, since these constraints are captured by the graph model. In Section 5.6, we study the impact of crossbar configurations on single-cell performance via simulations.

5.5   Protocol Modeling Using Markov Chains

In this section, we present a probabilistic scheme to address the stochasticity of protocol conditions. The outcome of this step is used as an input to CoSyn.

Discrete-Time Markov Chains and Stochastic Systems

DTMCs constitute a formal method to model stochastic systems, such as in biology [81], that exhibit a discrete state space [18]. DTMCs are similar to finite-state machines, but enriched with probabilistic transitions to model random phenomena. A transition from state w_i to state w_j is associated with a one-step transition probability that must depend only on the two states and not on the previous transitions—this property is referred to as Markov property.

Formally, a DTMC $\mathcal{M}$ is specified as a tuple $\mathcal{M} = \langle \mathcal{W}, w_{init}, \mathcal{U}, \mathcal{J} \rangle$, where: (i) $\mathcal{W}$ is a finite set of states (state space) and $w_{init} \in \mathcal{W}$ is the initial state; (ii) $\mathcal{U} : \mathcal{W} \times \mathcal{W} \to [0,1]$ is a transition probability matrix such that

$\sum_{w_j \in \mathcal{W}} \mathcal{U}(w_i, w_j) = 1 \; \forall w_i \in \mathcal{W}$; (iii) $\mathcal{J} : \mathcal{W} \to 2^{AP}$ is a labeling function that assigns each state to a set of atomic propositions from a denumerable proposition set AP [18]. These labels can be utilized by a probabilistic model-checking tool to express DTMC model properties with the help of temporal logic.

Properties of a DTMC model can be described using Probabilistic Computation Tree Logic (PCTL) [139]. PCTL formulae are interpreted over the DTMC states. For instance, consider a predicate ν that is always satisfied in state w_i (i.e., $w_i \models \nu$), we can use the following PCTL formulae to describe path or numerical queries:

1. $Qy_1 := P_{\geq 0.95}[\lozenge \nu]$, which inquires if the modeled system can eventually reach a state that satisfies ν with a probability that is greater than or equal to 0.95. The operator $\lozenge$ means "eventually."

2. $Qy_2 := P_{max=?}[\lozenge^{<10}\nu]$, which seeks the maximum probability that the system will eventually reach a state w satisfying ν in less than 10 steps.

The query Qy_1 is considered to be a reachability query, whereas Qy_2 is a numerical query. These queries and others can be used to investigate the evolution of a stochastic system that is modeled using DTMC. More details about DTMC model checking and PCTL syntax can be found in [139].

Probabilistic Modeling Approach

While the graph model described in Section 5.2 can effectively support type-driven single-cell execution, it suffers from the following drawbacks: (1) it assumes that the protocol conditions (e.g., fixation time and incubation temperature) are insensitive to cell types and consequently they do not have an impact on the protocol efficiency; (2) it also assumes that the optimal settings of these conditions can be uniquely defined. These assumptions however may not be valid in many real-life scenarios, particularly due to the inherent stochasticity in cellular interactions.

To analyze the stochastic behavior of single-cell protocols, we collected experimental data for a population of yeast cells after conducting several benchtop implementations of the ChIP protocol. The collected data (Table 5.3) shows the normalized immunoprecipitation (IP) value, i.e., protocol efficiency[2], while varying some experimental conditions such as the number of washing steps after IP; this table is referred to as *protocol-condition space*. By analyzing the obtained data, we observe that changing the protocol settings leads to different IP outcomes; the settings shown in Case #6 provide the highest efficiency. In addition, despite the use of biological replicates in each case, the results obtained by these replicates are not identical and they follow a normal distribution $\mathcal{N}$. Hence, according to this analysis, it is necessary to redesign the protocol model to address the above challenges.

---

[2]The higher the IP value, the more "enriched" the DNA is, leading to higher efficiency.

TABLE 5.3

Experimental data describing protocol-condition space for ChIP using yeast cells.

Case number	# Replicates	Fixation time (min)	Incubation Temp. (°C)	# Washes	IP (mean, var.)
1	32	10	25	4	(16.4, 3.4)
2	6	10	18	4	(18.2, 5.1)
3	5	10	32	4	(15.8, 3.7)
4	4	8	25	4	(13.2, 2.9)
5	2	8	18	4	(16.9, 6.8)
6	3	10	32	4	(22.1, 5.6)
7	4	15	25	4	(18.7, 4.3)
8	3	15	18	4	(19.1, 3.6)
9	3	15	32	4	(16.8, 6.2)
10	5	10	25	8	(16.7, 4.1)
11	3	5	25	4	(12.2, 3.1)
12	3	30	25	4	(15.4, 4.7)

An effective solution to this problem is based on a probabilistic approach. The steps of this approach are given below:

(**1**) For each cell type, we collect data from previous experiments and construct the associated protocol-condition space, similar to the example in Table 5.3.

(**2**) We specify an arbitrary lower-bound threshold BT for the protocol efficiency. For example, we choose $BT = 16.5$ for the example in Table 5.3. Based on this threshold, we classify cell population (replicates) into two sets: a set of cells that contributes to high efficiency, denoted by ($\mathcal{HE}$), and another set of cells that contributes to low efficiency, denoted by ($\mathcal{LE}$).

(**3**) Let A_e be an event that an arbitrary input cell belongs to $\mathcal{HE}$. We compute the probability that the input cell contributes to high efficiency, i.e., belongs to $\mathcal{HE}$, as follows: $P(A_e) = \frac{|\mathcal{HE}|}{|\mathcal{HE}|+|\mathcal{LE}|}$. Based on Table 5.3, $P(A_e) = \frac{26}{73} = 0.36$.

(**4**) Consider a certain protocol condition such as the fixation time. We extract all possible settings S_{ei} of this condition from the table. According to Table 5.3 and based on the fixation time, S_{e1}: 5-min, S_{e2}: 8-min, S_{e3}: 10-min, S_{e4}:15-min, and S_{e5}: 30-min.

(**5**) Let B_{ei} be an event that an arbitrary cell is processed using the setting S_{ei}. We compute the probability that an arbitrary cell will be processed using the setting S_{ei} as follows: $P(B_{ei}) = \frac{|S_{ei}|}{\sum_i |S_{ei}|}$, where $|S_{ei}|$ represents the number of replicates processed using S_{ei}. By considering (S_{e3}: 10-min) from Table 5.3, we obtain $P(B_{e3}) = \frac{51}{73} = 0.7$.

(**6**) We next compute the conditional probability $P(B_{ei}|A_e) = \frac{P(B_{ei} \cap A_e)}{P(A_e)}$, which represents the probability that an arbitrary cell will be processed using S_{ei} given that this cell belongs to $\mathcal{HE}$. The probability $P(B_{ei} \cap A_e)$ represents the percentage of cell population that satisfy the following conditions:

(i) the cells are processed using the setting S_{ei}; *(ii)* the cell belong to $\mathcal{HE}$. In the example shown in Table 5.3, $P(B_{e3} \cap A_e) = \frac{14}{73} = 0.2$, and therefore $P(B_{e3}|A_e) = 0.53$.

(7) We apply Bayesian inference to estimate the posterior probability $P(A_e|B_{ei})$, which indicates the probability that an arbitrary cell contributes to high performance given that this cell is processed using the setting S_{ei}—Bayes' theorem mathematically states that $P(A_e|B_{ei}) = \frac{P(B_{ei}|A_e) \cdot P(A_e)}{P(B_{ei})}$. According to the example in Table 5.3, $P(A_e|B_{e3}) = \frac{0.53 \times 0.36}{0.7} = 0.27$. We can also compute $P(A_e|B_{e1})$, $P(A_e|B_{e3})$, and the other posterior probabilities similarly.

(8) We normalize $P(A_e|B_{ei})$ using the following relation: $\hat{P}(A_e|B_{ei}) = \frac{P(A_e|B_{ei})}{\sum_i (A_e|B_{ei})}$.

By adopting the above systematic approach, we are able to capture the inherent stochasticity in the protocol environment. This approach therefore allows us to redefine the protocol interactions via a Markov-chain model.

Modeling Using Discrete-Time Markov Chains

Consider a given sequence of cells, where each cell belongs to one of n_c cell types. We consider that cell-type selection follows a uniform distribution $P(X)$; X is a random variable associated with the percentage of the cell-population count for each type. Based on the probabilistic approach discussed earlier, each cell type is associated with specific experimental steps for single-cell analysis and unique probabilistic distributions for selecting protocol settings in each step. We also take into account cost limitations of the overall experimental environment such as the maximum quantities of master mixes and washing liquids. Such cost limitations (also known as experiment budget) play a key role in tuning the overall performance of the protocol.

Based on the above characteristics, our objective is defined as follows: *Given a certain experiment budget, find the protocol strategy for a given sequence of cells that maximizes the probability that the cells contribute to high efficiency.* To achieve this goal, we model the execution of single-cell analysis as a DTMC and solve the strategy synthesis problem via probabilistic model checking. Strategy synthesis is a problem that is concerned with finding a strategy, in our case a sequence of protocol steps, which satisfies a property or optimizes a long-term objective such as the probability of protocol success [141]. Strategy synthesis has widely been used in many applications including planning of robots motion under uncertainty [144], security analysis via synthesis of malicious strategies [242], and dynamic power management using optimal control strategies [21].

The outcome of strategy synthesis can then be fed into CoSyn for resource allocation. In the future, we plan to integrate both strategy synthesis and resource allocation (CoSyn) into a single unified framework.

Figure 5.12 shows a DTMC model for single-cell analysis for $n_c = 3$. The model constitutes a finite set of states $\mathcal{W}$, which represents a sequence of biochemical operations (e.g., *mRNA Prep*) or analysis procedures (e.g.,

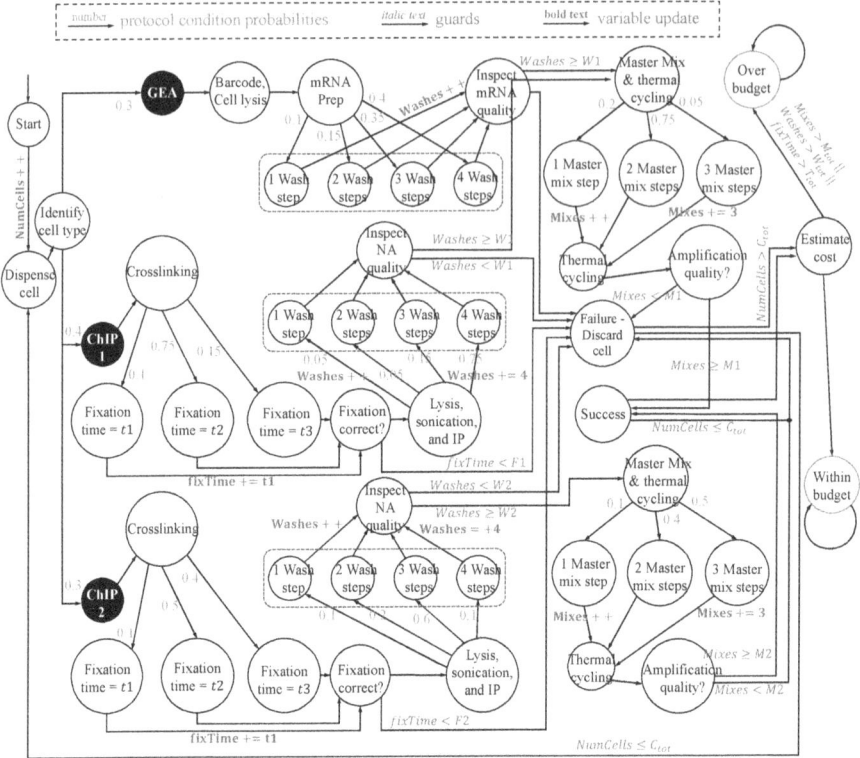

FIGURE 5.12
A DTMC model for a single-cell protocol that considers three cell types.

Estimate cost). In addition to the set $\mathcal{W}$, we define a transition probability matrix $\mathcal{U} : \mathcal{W} \times \mathcal{W} \rightarrow [0,1]$ that represents decisions related to the protocol conditions and the associated probabilities; these probabilities are shown in Figure 5.12. Recall that we compute these probabilities using Bayesian inference as discussed earlier.

To model the experiment budget constraints (and other constraints), we augment the state transitions of the proposed DTMC model with a set of *guards*, which ensure the fulfillment of these constraints and therefore transition the system towards success (*Within budget*) or failure (*Over budget*). For example, in Figure 5.12, the transition from state "*Success*" to state "*Dispense cell*" is possible only if the number of processed cells is still below the total population count C_{tot}. Clearly, these guards, which are tuned based on the experiment budget, play a key role in computing the optimal protocol strategy. In Section 5.6, we examine the budget's role in controlling the protocol efficiency using PCTL.

Note that the above discussion considers a static DTMC model, which is computed only once based on the protocol-condition space. However, since stochasticity is ubiquitous in single-cell applications, protocol conditions can significantly change over time, and therefore there is a need to update the transition probabilities in the DTMC model. To cope with this dynamic behavior, a closed-loop system can be constructed where multiple iterations can be performed. In this closed-loop system, the results obtained by the hybrid microfluidic system are added to the protocol-condition space, allowing a new set of transition probabilities to be computed; the new DTMC model is then used in the next iteration of strategy synthesis and single-cell analysis. The number of cells per iteration can be specified by the user. By following this approach, the DTMC model can be changed to adapt to protocol dynamics.

5.6 Simulation Results

We implemented CoSyn using C++. All evaluations were carried out using a 2.7 GHz Intel i5 CPU with 8 GB RAM. The set of bioassays constituting the single-cell analysis protocol (Section 5.1) were used as a benchmark. Cell types were assigned to the cells using a uniform distribution function.

Since this is the first work on synthesis for hybrid microfluidic platforms, we have developed two baseline frameworks: (1) architectural baseline (ArcSyn), wherein the barcoding fabric is valveless and it utilizes a one-to-one mapping between syringe pumps and DMFB ports as in Figure 5.3(b); (2) algorithmic baseline (ReSyn), in which resource allocation is initially performed for each microfluidic domain separately. However, we must ensure that the system behavior at the boundary between the two domains is deterministic—the synthesis tool for the DMFB must be aware of the order of the barcoding droplets generated from the valve-based crossbar. The only way to meet this constraint is to disallow pipelining in the valve-based system; thus an upper bound on the number of barcoding droplets that can be processed simultaneously is equal to m_b. Since we consider a large number of cells, we divide the cells into batches, each of a maximum size of m_b cells, such that ReSyn executes them iteratively.

In the rest of this section, we assume a microfluidic platform that contains a fully connected crossbar, except Section 5.6, in which we investigate variants of the crossbar architecture.

Comparison with Baselines

We evaluate the performance of CoSyn, ArcSyn, and ReSyn in terms of the total completion time for the protocol, measured in minutes (we assume $T_b = T_f = 0.2$ s). The assumption of $T_f = 0.2$ s is based on the fact that a DMFB can flexibly function at any rate in the range from 1 to 100 Hz, as described in Section 3.3. Therefore, a rate of 5 Hz is a valid choice. On the other side, $T_b = 0.2$ s is associated with a previous flow-based design that was operated using a pneumatic transporter with a pumping rate of 50 Hz. The reported flow speed in this design is in the range of 10 to 20 mm/s [150]. In our example, a transposer with a length of 4 mm with the proper channel configurations (PDMS material and diameter) can lead to a flow speed of 4 mm/0.2 s = 20 mm/s [119].

For valve-based routing, we consider the graph theoretic scheme (i.e., Method 1 from Section 5.4). We fix the number of input cells to 100, and we consider 20 and 40 barcoding inputs (or cell types). To ensure that this evaluation is independent of the platform architecture, the results were obtained using a DMFB with no resource constraints. The selected number of cells is based on a design that was introduced earlier in [212]. Usually, a cell suspension (an experiment session) using a microfluidic prototype includes a number of cells in the range of 100 to 1000 cells. Larger number of cells per suspension requires more effort during cell isolation and sorting.

Figure 5.13 compares the three synthesis frameworks in terms of completion times. ReSyn leads to the highest completion times due to the loose coordination between the DMFB and the valve-based crossbar. The completion time of CoSyn is close to the lower bound, which is obtained using ArcSyn. ArcSyn uses the maximum number of barcoding outputs due to the one-to-one mapping between the barcoding inputs and outputs. Hence, these results indicate that pipelined valve-based routing and the coordination between the components of CoSyn play a key role in increasing cell-analysis throughput.

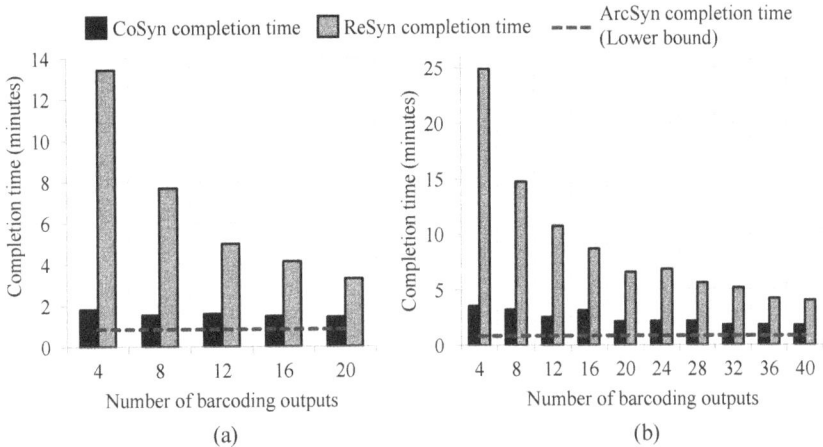

FIGURE 5.13
Comparison between CoSyn, ArcSyn, and ReSyn in terms of completion time: (a) using 20 barcoding inputs, (b) using 40 barcoding inputs.

Tradeoffs of Valve-Based Routing Schemes

Next, we evaluate the performance and the computation time of the valve-based routing schemes by using CoSyn simulations. We refer to CoSyn that utilizes the graph-theoretic method (Method 1) and the incremental method (Method 2) as CoSyn-Graph and CoSyn-Inc, respectively. We fix the number of input cells to 1000 and we consider a DMFB with no resource constraints. Figure 5.14(a) and Figure 5.14(b) compare CoSyn-Graph and CoSyn-Inc in terms of completion times and overall computation times, respectively, while varying the number of inputs n_b and outputs m_b in the crossbar.

As shown in Figure 5.14(a), we observe that the completion times of both methods exhibit a parabolic behavior with respect to the crossbar size $n_b \times m_b$. First, the completion times decrease when we increase the crossbar size from 20×8 to 50×20; the increase in the crossbar size within this range allows more barcoding droplets to be consumed by the crossbar, which in turn overshadows the negative impact of adding more vertical levels. However, when the size increases beyond 50×20, the impact of having additional vertical levels appears to be more prominent, leading to an increase in the completion times for both methods. Finally, we also observe that the completion time obtained by CoSyn-Inc is at least the completion time obtained by CoSyn-Graph, and the difference in the completion time between CoSyn-Graph and CoSyn-Inc becomes significant when the crossbar size is increased.

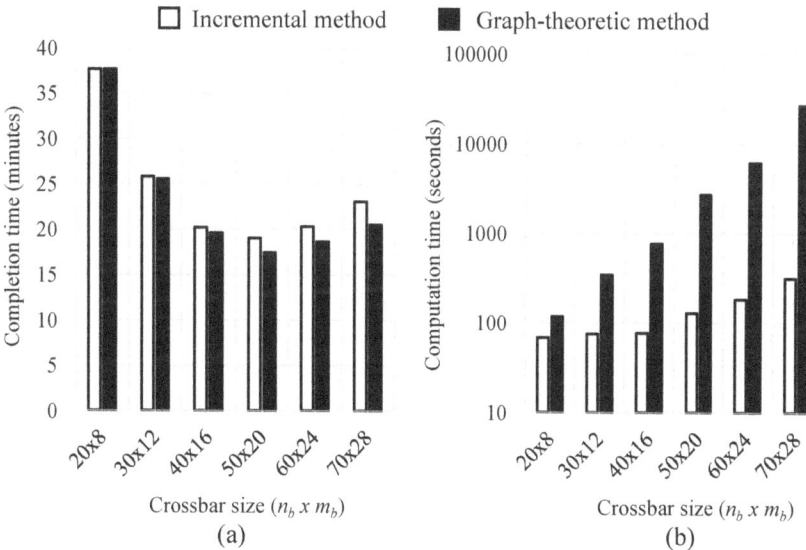

FIGURE 5.14
Comparison between CoSyn-Graph and CoSyn-Inc: (a) completion time, (b) computation time (logarithmic scale).

Despite the high performance gained by using CoSyn-Graph, Figure 5.14(b) shows that the computation time of CoSyn-Graph dramatically increases when larger crossbar designs are used. Therefore, it is obvious that CoSyn-Inc needs to be used to dynamically synthesize large-scale single-cell analysis.

Design-Quality Assessment

We also evaluate the quality of the designs generated by CoSyn-Graph in terms of the number of single-cell experiments that can be completed for a given time limit and the given number of DMFB resources. In addition, we also quantify the fraction of input cells that can be processed simultaneously by the given set of DMFB resources. Our objective here is to investigate the conditions under which CoSyn-Graph is effective. Therefore, we introduce the following terms:

Cell-analysis density: The number of cells (samples) that completed analysis during a specific window of time (cell throughput), using a given array of electrodes. The time window is set to be a minute and the size of the array is equal to 100 electrodes.

DMFB capacity: A real number $z_c \in [0,1]$ that provides the fraction of input cells that can be processed simultaneously using DMFB resources. For example, a capacity of 1 indicates that there are sufficient resources to process all the cells simultaneously. On the other hand, a capacity of 0.5 means that the existing resources are sufficient for simultaneously processing only half of the cells.

We investigate the design-quality for valve-based crossbars by evaluating the cell-analysis density of CoSyn-Graph and ArcSyn. We simulate the execution of 50 cells using 4 barcoding outputs (Figure 5.15(a)), 8 barcoding outputs (Figure 5.15(b)), 12 barcoding outputs (Figure 5.15(c)), and 20 barcoding outputs (Figure 5.15(d)). The density values are computed while the capacity is varied. By comparing the density values for CoSyn-Graph and ArcSyn, we observe two regimes: (1) Regime I in which the cell-analysis density of CoSyn-Graph is higher, i.e., it is more effective; (2) Regime II in which the density of CoSyn-Graph is less than or equal to the density of ArcSyn. Regime I highlights the fact that CoSyn-Graph efficiently exploits valve-based barcoding, and the power of valve-based pipelining is evident when the DMFB resources are limited. On the other hand, the overprovisioning of resources leads to Regime II, where a lower cell-analysis density is reported. Finally, we note that Regime I shrinks as we increase the number of barcoding outputs; this is expected since CoSyn-Graph is more effective in the realistic case of a limited number of barcoding interfaces.

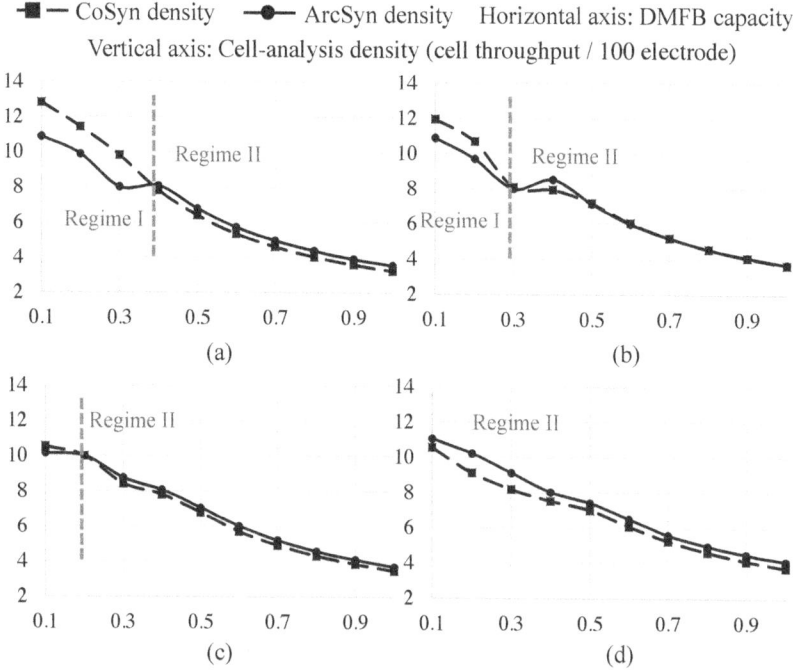

FIGURE 5.15
Comparison between CoSyn-Graph and ArcSyn in terms of cell-analysis density using (a) 4 barcoding outputs, (b) 8 barcoding outputs, (c) 12 barcoding outputs, and (d) 20 barcoding outputs.

Crossbar Connectivity: Performance vs. Flexibility

We also study the impact of crossbar connectivity on the system performance and the routing flexibility. We use architecture variants $\mathcal{F}^{cl}_{n_b \times m_b}$ of an n_b-to-m_b crossbar; the parameters n_b, m_b, and cl denote the number of crossbar inputs, the number of crossbar outputs, and the number of output clusters, respectively (Section 5.4). For evaluation purposes, we simply measure the flexibility, denoted by $fx(m_b, cl)$, of a crossbar variant $\mathcal{F}^{cl}_{n_b \times m_b}$ using the following relation: $fx(m_b, cl) = \frac{m_b}{cl}$, which represents the effective number of output ports reachable from an input port. Also, we simulate the execution of 1000 cells using CoSyn-Inc, and we consider an 80-to-32 crossbar ($n_b = 80$ and $m_b = 32$). Moreover, to ensure that this evaluation applies to any DMFB architecture, the results were obtained using three DMFBs that have the following resource capacities z_c: 0.6, 0.8, and 1.

As shown in Table. 5.4, as the number of output clusters increases, the flexibility fx of the crossbar design decreases. On the other hand, as shown in Figure 5.16, we observe that the total completion time decreases when the

TABLE 5.4
Flexibility of crossbar variants (fx).

1 Cluster	2 Clusters	4 Clusters	8 Clusters
32	16	8	4

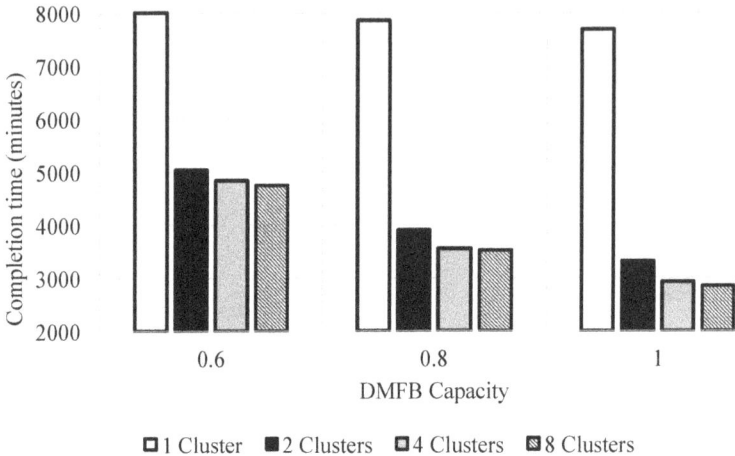

☐ 1 Cluster   ■ 2 Clusters   ☐ 4 Clusters   ▨ 8 Clusters

FIGURE 5.16
Comparison between crossbar variants using protocol completion time.

number of clusters increases—this result is due to the reduction in the number of vertical levels q and thus the routing distance of the barcoding droplets.

Analysis of Markov Model

To understand the impact of protocol conditions and experiment budget on the protocol efficiency, we run model checking on the DTMC model described in Figure 5.12, where the number of cells is equal to 6. We investigate the impact of three experimental factors: the number of washing steps (W_{Tot}), the number of fixation time steps $(fixTime_{Tot})$, and the number of master mixes $(Mixes_{Tot})$. For this purpose, we use PRISM [140] to implement the DTMC model along with the following PCTL query that expresses the model objective:

$$Q_y := P_{max=?}[(\Box(!Failure))\ U\ (\Diamond WithinBudget)]$$

The query Q_y seeks the maximum probability that the protocol will globally avoid the state *Failure* "until" eventually reaching the success state, i.e., the state *WithinBudget*. The operator $\Box$ means "globally," the operator U means "until," and the operator $\Diamond$ means "eventually." The maximum probability can be computed using *value iteration* algorithm [48]. By running this

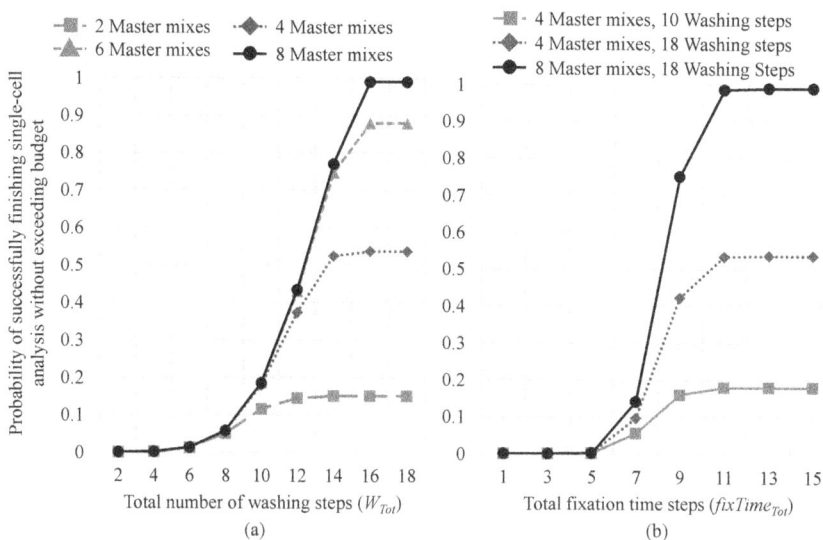

FIGURE 5.17
Impact of experiment budget on the probability of successful completion of single-cell analysis under the optimal protocol strategy: (a) with varying number of washing steps; (b) with varying fixation time.

query using PRISM, we can obtain the optimal protocol strategy that maximizes the probability of satisfying the target predicate, i.e., the processed cells contribute to high efficiency and the experiment budget is not exceeded.

Figure 5.17 shows the obtained results, which illustrate the impact of experiment budget on the probability that single-cell analysis completes successfully under the optimal protocol strategy. It is obvious that the success probability increases when we increase the experiment budget. We also observe that the success probability is close to 1 only if there are sufficiently large quantities of washing liquid and master mixes and if we allow more time for fixation.

These results highlight the need to incorporate stochastic behavior of a single-cell protocol into any synthesis methodology since it has a significant impact on the protocol efficiency.

5.7 Chapter Summary

In this chapter, we have introduced the first automated design method for single-cell analysis using a cyber-physical microfluidic platform. This design coordinates the control of diverse microfluidic components and concurrently processes a large number of sample pathways. We have also presented a probabilistic model of the single-cell analysis protocol based on DTMCs. This model captures experimental settings where protocol conditions are specified in a probabilistic manner. The proposed platform has been evaluated on the basis of the time needed for analysis and the biochip size needed for realistic test cases.

A limitation of CoSyn is that it assumes direct-addressing of valves using pressure ports; this design approach is prohibitively expensive especially with large reconfigurable fabrics. In Chapter 6, we address this limitation by using a pin-constrained design. We also study the impact of this design on the performance of a single-cell application; this study is guided by a system model that is analyzed using computational fluid dynamics simulations.

6

Timing-Driven Synthesis with Pin Constraints: Single-Cell Screening

Reconfigurable flow-based microfluidic biochips (RFBs), such as the valve-based fabrics introduced in Chapter 5, enable single-cell analysis by providing unprecedented fluid-handling capabilities for nanoliter droplets transported in a flow medium [134]. RFBs can be efficiently used for screening a stream of heterogeneous cells in a high-throughput manner, whereby every cell can be individually encapsulated into an aqueous droplet, incubated with a secreted enzyme-reporter molecule, sorted using a fluorescence sensing [19], and barcoded for single-cell analysis [300].

While Chapter 5 addresses design-automation challenges associated with droplet routing in RFBs (as part of a hybrid microfluidic system), the described method is cost-prohibitive and it cannot tackle the complexities of RFBs designed for high-throughput single-cell screening because of the following challenges:

(1) Scalable valve control: A large number of valves are needed to sort and barcode several cell types. These valves also need to be controlled in real-time in response to identified cell types; the control actions for these valves are not known *a priori*. In Chapter 5, an RFB solution was introduced for scalable barcoding using a fully reconfigurable valve-based crossbar, where an n_b-to-m_b crossbar can route a barcoding droplet from any of the n_b inputs to any of the limited m_b outputs. However, it can be shown that a simple 40-to-4 crossbar requires at least 1344 valves, thus direct-addressing of valves using pressure ports is prohibitively expensive.

(2) Impact of chip parameters on performance: Parameters such as channel diameter and elasticity of the deformable membranes significantly influence the performance of the microfluidic application [225, 133, 65]. For example, consider a microfluidic channel as depicted in Figure 6.1(a). We carried out computational fluid dynamics (CFD) simulations using COMSOL [2] to characterize laminar flow at the channel output when a pressure step function is applied to the input. As shown in Figure 6.1(b), the steady-state pressure at the output decreases as the input pressure and the channel diameter Dr decrease. Next, we assume that this channel is connected to a 90-kPa pressure source and it is used to actuate a valve that requires at least a 2-kPa pulse to deform the membrane (highlighted using a horizontal thick line in Figure 6.1(b)). We find that reliable valve actuation can be achieved only if

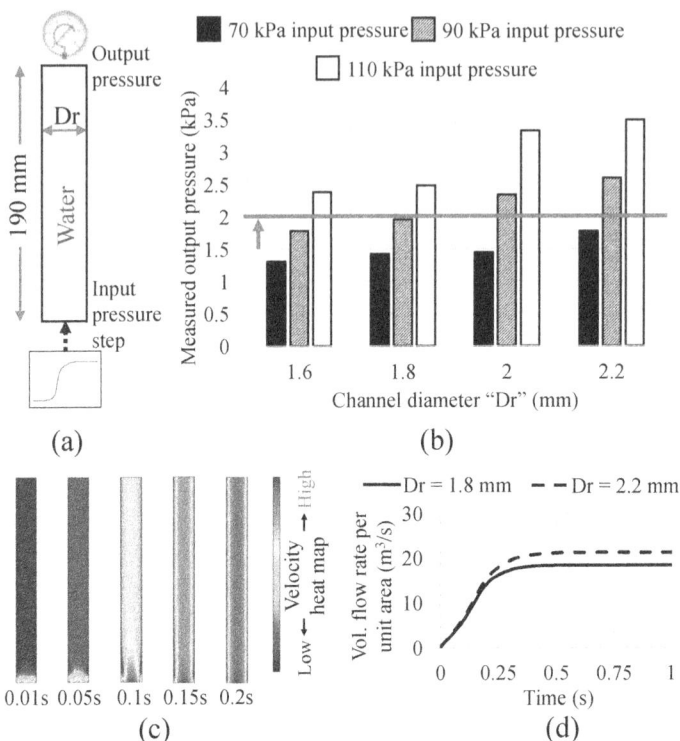

FIGURE 6.1
The correlation between the parameters of a microfluidic channel and system performance: (a) geometric specifications, (b) evaluation of output pressure, (c) velocity profile of pressure-driven mechanisms, (d) dependence of flow rate on Dr.

Dr is larger than 2 mm; see Figure 6.1(b). Similarly, by analyzing the flow rate of this pressure-driven mechanism (Figure 6.1(c)), we find that the flow rate decreases as Dr decreases (Figure 6.1(d)).

Therefore, neglecting the correlation between chip parameters and single-cell screening efficiency will invariably lead to unacceptable screening rates or even unexpected behavior. It is important to investigate the delay associated with pressure-driven fluid transport to precisely predict the timing characteristics of a screening biochip. The result of this analysis can be utilized as a *latency constraint* to synthesize a biochip that is free of timing violations.

In this chapter, we address the above challenges by investigating the design of a pin-constrained RFB for scalable single-cell screening [119, 120]. We utilize the delay model of pressure-driven transport to design a cost-effective RFB with high throughput. The contributions of this chapter are as follows:

- We present a formal definition for the architecture of an RFB that is used for high-throughput single-cell screening.

- We describe a graph-theoretic algorithm that generates all possible screening paths within an RFB. This algorithm is used as a pre-processing step, and it is a key component in streamlining the use of our synthesis methodology.

- We describe a pin-sharing scheme that offers real-time multiplexed control of valves. An RC model is derived to describe the delay of the multiplexed control path and used to characterize chip performance. This model is validated using CFD simulations.

- We describe and evaluate a timing-driven synthesis solution for the control of RFBs. The solution (referred to as Sortex: cell <u>sor</u>ter using multipl<u>ex</u>ed control) reduces the number of pins subject to performance constraints associated with single-cell screening.

The rest of the chapter is organized as follows. Section 6.1 describes prior work on pin-constrained design and an RFB architecture for single-cell screening. In Section 6.2, we introduce multiplexed control and the RC model. Next, Section 6.3 presents the timing-driven synthesis technique. Experimental evaluation is presented in Section 6.4 and conclusions are drawn in Section 6.5.

6.1 Preliminaries

In this section, we review pin-constrained design techniques for flow-based biochips and describe an RFB for single-cell screening.

Pin-Constrained Design of Flow-Based Biochips

A number of synthesis solutions were proposed with the aim of minimizing the number of control pins [101, 173, 275]. These methods however are inadequate for high-throughput screening due to the following reasons:

1. For control-pin minimization, [173] relies on activation-based compatibility; valve-actuation patterns of a protocol are mapped to a pin-count minimization strategy. A more flexible approach in [101] explores compatibility among basic control actions of individual fluidic operations. However, both techniques assume that "compatible" valves can be simultaneously addressed using the same pressure source—this assumption is valid only with a small number of compatible valves due to fan-out limits. Moreover, these approaches do not enable independent actuation of valves, thereby preventing the reconfigurability that is necessary for single-cell screening.

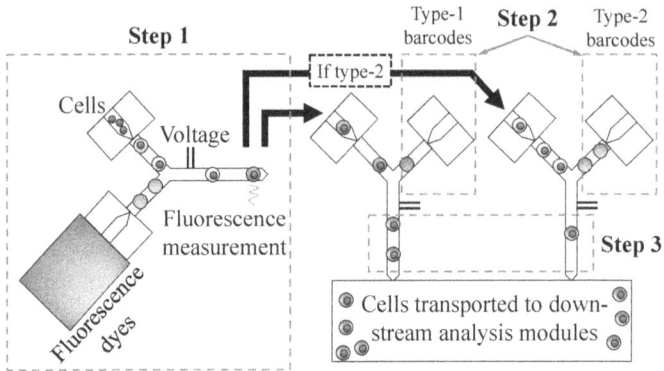

FIGURE 6.2
Steps of single-cell screening, considering two cell types.

2. The work in [101] considered pressure-propagation delay within fluidic-channel routing using a *path-length* model. However, this model does not capture the correlation between fluid dynamics and biochip parameters (e.g., channel width and elasticity). Furthermore, [101] uses only the longest pressure-propagation delay of a pin-sharing valve group to assess performance. This approach cannot be used with multiplexed control of independently addressable valves.

3. A control-layer multiplexer structure was adopted in [275] to minimize the number of control pins. Unlike the above-mentioned methods, this mechanism allows biochip valves to be flexibly addressed, and it is similar to our proposed pin-constrained method. However, [275] considers only reliability issues, e.g., pressure degradation, associated with pin-switching activities, and it does not take into consideration fluid dynamics and its impact on the application throughput.

An RFB for Single-Cell Screening

Single-cell screening is a biochemical process that aims to classify a stream of droplet-encapsulated cells based on their biological behavior and label them for down-stream analysis [170]. Hence, this biochemical process consists of three main steps, as shown in Figure 6.2: (i) Step 1: identification of cell types by observing certain fluorescence biomarkers; (ii) Step 2: suspension of suitable molecular labels (barcodes); (iii) Step 3: mixing each cell with a suitable barcode depending on the cell's type. These steps need to be applied in a high-throughput manner to allow the processing of any number of cells. The outcome of this process is a collection of labeled cells that can be steadily transferred to other modules where further down-stream analysis, e.g., based on polymerase chain reaction, can be performed (Chapter 5).

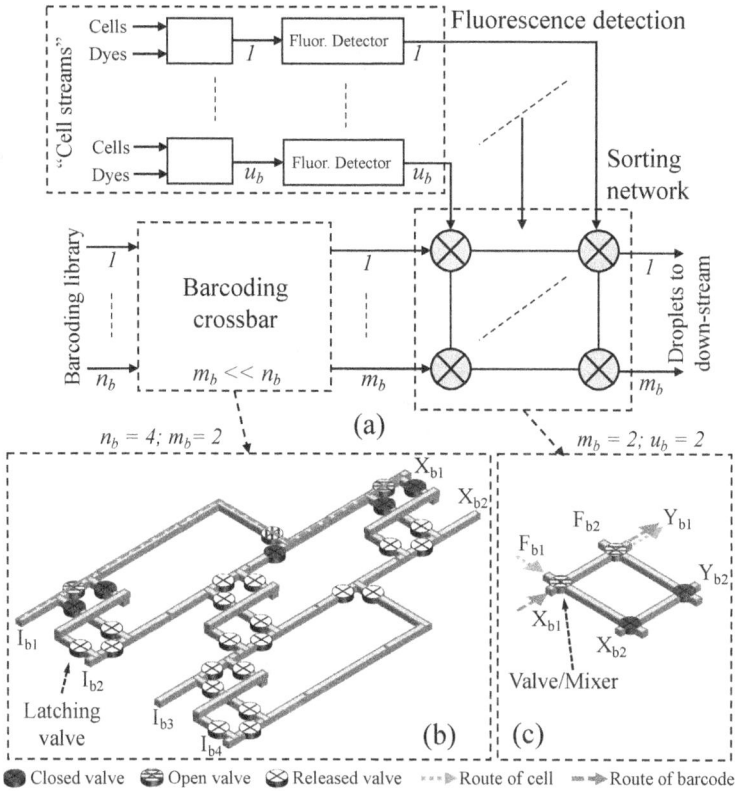

FIGURE 6.3
An RFB for single-cell screening: (a) platform modules, (b) a 4-to-2 barcoding
module, (c) a 2-by-2 sorting network [134].

The above specifications can be realized using flow-based biochips, which
can be configured to perform high-throughput operations. More specifically,
RFBs can be used to systematically process any number of cells in real time
while allowing biochip resources to be shared among different cells at different
times. Figure 6.3(a) shows an RFB architecture that screens u_b streams
of cells, classifies cells into n_b types, and barcodes the cells via m_b ports
($m_b \ll n_b$). To execute the steps of single-cell screening (Figure 6.2), this
platform consists of three modules: valve-less fluorescence detection [19], barcoding
crossbar [232], and sorting network [134]. Adaptation is achieved via
the detection of samples in the fluorescence-detection module, whereas reconfiguration
is carried out *in response* at both the barcoding crossbar and the
sorting network.

A valve-based crossbar is employed to barcode droplets and route them
towards the sorting network. A barcoding droplet is routed from an input port

$I_{bk} \in \{I_{b1}, I_{b2}, ..., I_{bn_b}\}$ to an output port $X_{bi} \in \{X_{b1}, X_{b2}, ..., X_{bm_b}\}$ through channels, and the routing path is configured online using a set of valves $\mathcal{L}^p = \{L_1, L_2, ..., L_{p_v}\}$; p_v is the number of crossbar valves. Figure 6.3(b) shows a 4-to-2 crossbar that routes two different barcoding droplets concurrently.

Connected to the crossbar is a sorting module that mixes a sample droplet from a stream $F_{bj} \in \{F_{b1}, F_{b2}, ..., F_{bu_b}\}$ with a barcoding droplet generated from X_{bi}. This function needs to be implemented by a programmable microfluidic platform that performs biochemical operations *on-the-fly*. For this purpose, we exploit the programmable microfluidic network proposed in [134], where an u_b-by-m_b network can dynamically process any pair of $u_b \times m_b$ input droplets. A set of valves $\mathcal{L}^q = \{L_1, L_2, ..., L_{q_v}\}$, where q_v is the number of sorting-network flow valves, are used to mix barcodes with cells and route mixed droplets to output ports in the set $\{Y_{b1}, Y_{b2}, ..., Y_{bm_b}\}$. Fig 6.3(c) illustrates a 2-by-2 network.

While this RFB design offers reconfigurability, the efficiency of single-cell screening depends on the topology of the flow channels and how fast the flow valves in the set $\mathcal{L} = \mathcal{L}^p \cup \mathcal{L}^q$ can be actuated. Since we consider a fixed topology for the flow channels, we focus on delays associated with valve actuation (pressurization or de-pressurization). We demonstrate in the next section that an effective pin-constrained control methodology for an RFB must consider the timing overhead of valve actuation. Therefore, we consider valve-based modules (the barcoding crossbar and the sorting network) to investigate the action and delay of valve controls required to route a barcode from I_{bk} : $\{k \in \mathbb{N}, k \leq n_b\}$ to X_{bi} : $\{i \in \mathbb{N}, i \leq m_b\}$ and to route the barcoded cell from F_{bj} : $\{j \in \mathbb{N}, j \leq u_b\}$ to Y_{bi} : $\{i \in \mathbb{N}, i \leq m_b\}$. This action is referred to as a *control vector* and is denoted by ϑ_i : $\{\vartheta_i \in \{0, 1\}^{x_v}, i \in \mathbb{N}, i \leq CV\}$, where x_v is the number of chip valves ($x_v = p_v + q_v$) and CV is the total number of possible control vectors. The parameter CV can also be interpreted as the number of possible flow paths that can be utilized by a single-cell sample of any type. In addition, we use $\theta_{sc}(x_v, y_v)$ to denote a screening biochip that contains x_v valves and is actuated using y_v control pins.

Our goal in this chapter is to provide an architectural-level synthesis scheme that allows the control of x_v valves using y_v pins ($y_v \ll x_v$), while the latency of every control vector ϑ_i is less than a threshold. A solution to this problem leads to a biochip that provides a desired screening throughput.

Motivating Example

Consider an experimental setting that involves screening of a sequence of n_e cells and classifying them into 4 groups (or types) of populations. All cells are initially tagged with two types of dyes (green-fluorescent GF and yellow-fluorescent YF dyes), which can be used to classify the cells via a fluorescence-measurement utility. The type of a cell is identified based on the emission rate of both GF and YF observed from the cell. For example, assume that the intensity of a fluorescence signal is normalized to the range from 0 to 1, where

0 corresponds to the "absence of fluorescence emission" and 1 corresponds to the "highest fluorescence emission". A cell is assigned to the first population if the detected fluorescence from both GF and YF is below 0.5, whereas it is assigned to the second population if the detected signal from GF is below 0.5 but the detected signal from YF is equal to or above 0.5. A similar argument applies to the third and fourth cell populations.

Our objective is to design an RFB that efficiently screens the above cells. The biochip design needs to be cost-effective; that is, it should be controlled using the minimum number of control pins that allows a fluorescence-identified cell to be appropriately barcoded and routed to an output port within a duration of D. Moreover, this biochip is designed to be connected to a digital-microfluidic biochip—which performs down-stream analysis—via two capillary interfaces, as described in Section 5.1.

The above requirements can be fulfilled using the RFB shown in Figure 6.3(a), where the number of barcoding ports n_b is 4 (Figure 6.3(b)) and the number of output ports m_b is 2 (Figure 6.3(c)). Since the processing time of fluorescence measurement tends to be non-negligible especially when more dyes are used, screening performance can be enhanced by increasing the number of cell streams—we consider two parallel cell streams ($u_b = 2$), as shown in Figure 6.3(c). Clearly, the above reconfigurable architecture provides the following advantages:

- It decouples single-cell specifications (cell counts and types) from the interfacing constraints (output ports).

- It is a scalable architecture that can process any number of cells regardless of the cell ordering or cell-type distribution.

For the sake of simplicity, suppose that the proposed design is used to screen a single cell ($n_e = 1$), which is found to belong to the first type. Based on this finding, a barcode needs to be dispensed from I_{b1} and routed through the barcoding module. Next, the barcode is mixed with the cell droplet in a reaction chamber that belongs to the sorting network. The dashed and dotted lines in Figure 6.3(b-c) show the routes of the barcoding droplet and the cell droplet, respectively.

Obviously, the routing of these two related droplets cannot be executed unless some valves (shown as diagonal-patterned valves in Figure 6.3(b-c)) are opened while others (shown as black-colored valves in Figure 6.3(b-c)) are closed. The speed of this action, and therefore the duration of cell screening, is determined based on how fast these valves can be actuated. If these valves are directly addressable using dedicated control pins, then each pin can be placed adjacent to its associated flow valve, leading to the highest speed of valve actuation and the lowest duration of single-cell screening. However, by adopting a pin-constrained design approach, the duration of single-cell screening may exceed the threshold D, therefore more control pins must to be included. The proposed pin-constrained design and the trade-offs between pin-count reduction and system performance are described in Section 6.2.

6.2   Multiplexed Control and Delay

In this section, we explain the pin-constrained design methodology and the associated delay model.

Multiplexed Control

To reduce the number of control pins, we allow several valves in a screening biochip to share a few pins using *time-division multiplexing* (TDM) [87]. Figure 6.4(b) shows an example of multiplexed control for an 8-valve flow channel (shown in Figure 6.4(a)); i.e., $CV = 1$. As shown in Figure 6.4(b), the dark-colored circles represent the control pins of the biochip whereas the white-colored circles represent the valves.

To implement TDM, two types of control pins are needed: (1) A set of primary pins $\mathcal{C}^a$ (represented by circles surrounding $\mathcal{C}^a$), used to provide the pressure (or vacuum) through primary control channels to actuate valves in the flow channel; (2) a set of demultiplexing pins $\mathcal{C}^g$ (represented by circles surrounding $\mathcal{C}^g$), used to direct the pressure-driven flow from a primary pin to a particular valve in the flow channel through demultiplexing control channels, thereby allowing flow valves to be independently addressable. Each pin in $\mathcal{C}^g$ is connected to an off-chip two-position solenoid valve, which can pressurize and de-pressurize a set of valves, referred to as control valves $\widetilde{\mathcal{L}}$ (Figure 6.4(b)) to avoid confusion with the flow valves $\mathcal{L}$ in Figure 6.4(a).

For example, to actuate flow valve L_2 in Figure 6.4(a), demultiplexing pins C_1^g, C_2^g, and C_3^g (shown in Figure 6.4(b)) are first activated to switch the control valves l_{13}, l_9, and l_2, respectively[1], as shown in Step 1 in Figure 6.4(c) By switching these control valves, a control path is now opened between the primary pin C_1^a and the desired flow valve L_2. Next, C_1^a is actuated to open or close L_2 (Step 2 in Figure 6.4(c))), which in turn is a latching valve that can maintain its open or closed state while disconnected from the controller [87].

Latching behavior can be imparted to flow valves by using a modified valve structure that can trap vacuum (pressure) to hold the valve open (closed). In this work, we use the structure shown in Figure 6.5(a) [87]. The pressure, vacuum, and latching valves are normally closed (Figure 1.6). When a pulse of vacuum (pressure) is applied to the "Input," the vacuum (pressure) valve opens and the pressure (vacuum) valve remains closed. As a result, the connection between the vacuum valve and the latching valve is depressurized (pressurized) forcing the latching valve to open (close). Figure 6.5(b-c) show the mechanisms of opening and closing the "latching" flow valve, respectively.

---

[1]Recall that C_2^g is a two-position selonoid valve. Therefore, when C_2^g is activated to open control valve l_9, control valve l_{11} is also opened, whereas both l_{10} and l_{12} are closed automatically.

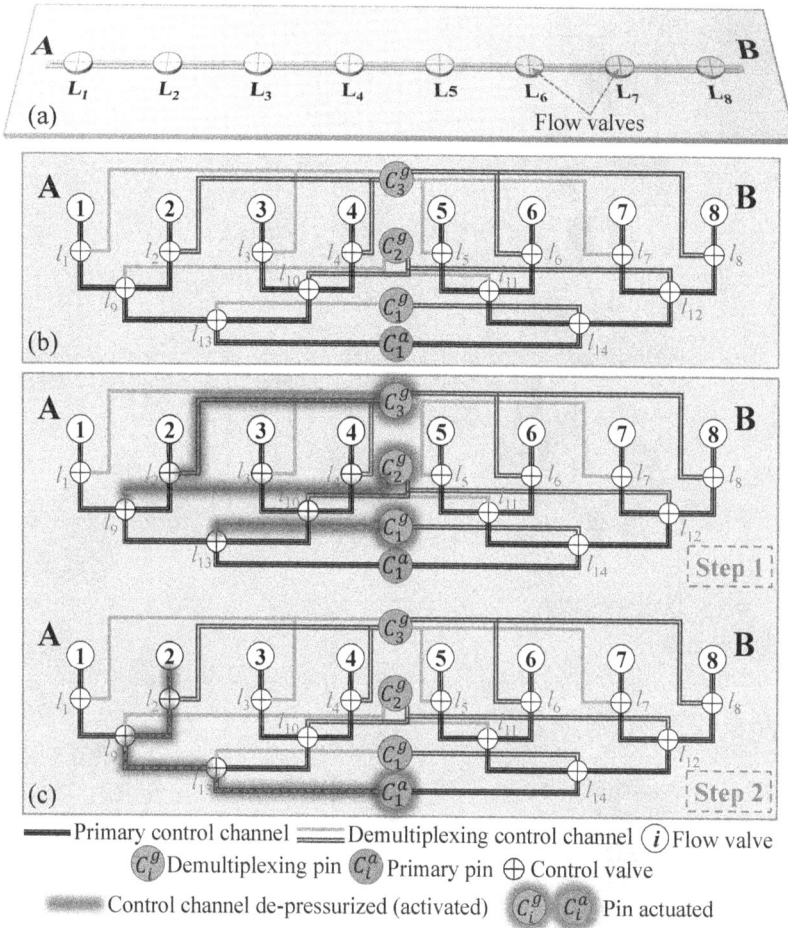

FIGURE 6.4
(a) An 8-valve channel; (b) Multiplexed control of the channel using 4 control pins. (c) Steps of multiplexed control to actuate L_2.

Note that we can route a fluid through the flow channel in Figure 6.4(a) from A to B or vice versa only after all the 8 valves are actuated. In addition, the flow circuitry in Figure 6.4(a) and the multiplexed control circuitry in Figure 6.4(b) are located on different layers.

By using multiplexed control and by considering the valve-control signal as a binary signal, x_v flow valves can be independently actuated using only a single primary pin and $\lceil \log_2 x_v \rceil$ demultiplexing pins; i.e., $y_v = \lceil \log_2 x_v \rceil + 1$ provides a lower bound on the number of control pins needed to actuate X valves. For example, the 8 valves in Figure 6.4(a) can be addressed using only 4 control pins. Similarly, the valves of an 1024-valve biochip can be addressed using only 11 pins.

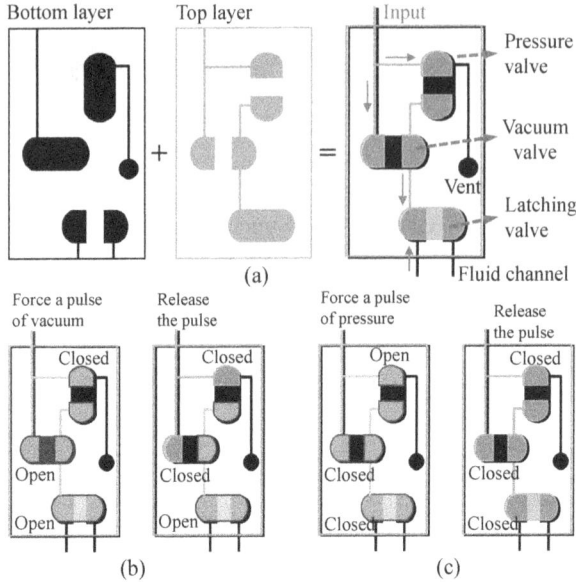

FIGURE 6.5

Illustration of (a) assembling, (b) opening, and (c) closing a latching flow valve.

Note however that the addressing of a single flow valve requires a sequence of pin actuations, which imposes timing overhead. For the 8-valve channel in Figure 6.4(a), Figure 6.6(a-b) illustrates multiplexed control for actuating all the channel valves using 4 pins and 6 pins, respectively. Although Figure 6.6 does not show the delays of control pulses through the control circuitry (the delays are expected to be larger for the 4-pin design), it is evident that the multiplexed control procedure is less complicated with 6 control pins. For example, to actuate L_2 using a 6-pin design, only three control pins—C_1^g, C_2^g, and C_1^a—need to be actuated, following the two-step procedure described earlier in Figure 6.4(c). Moreover, unlike the design in Figure 6.6(a), the multiplexed control in Figure 6.6(b) allows L_2 and another flow valve such as L_5 to be controlled simultaneously since they are not actuated using the same control pins. Hence, there is a tradeoff between pin-count reduction and the complexity of multiplexed control, and therefore the performance of single-cell screening.

Consider a screening biochip $\theta_{sc}(x_v, y_v)$ with multiplexed control. Also, consider a variable d_{xy}^i that specifies the delay contributed by a pin C_y^i of type i to actuate a valve L_x. Let λ_{xy}^a be a connectivity function such that $\lambda_{xy}^a = d_{xy}^a$ if a primary pin $C_y^a \in \mathcal{C}^a$ is used to actuate a flow valve $L_x \in \mathcal{L}$ with actuation delay d_{xy}^a, and $\lambda_{xy}^a = 0$ otherwise. Likewise, $\lambda_{xy}^g = d_{xy}^g$ indicates

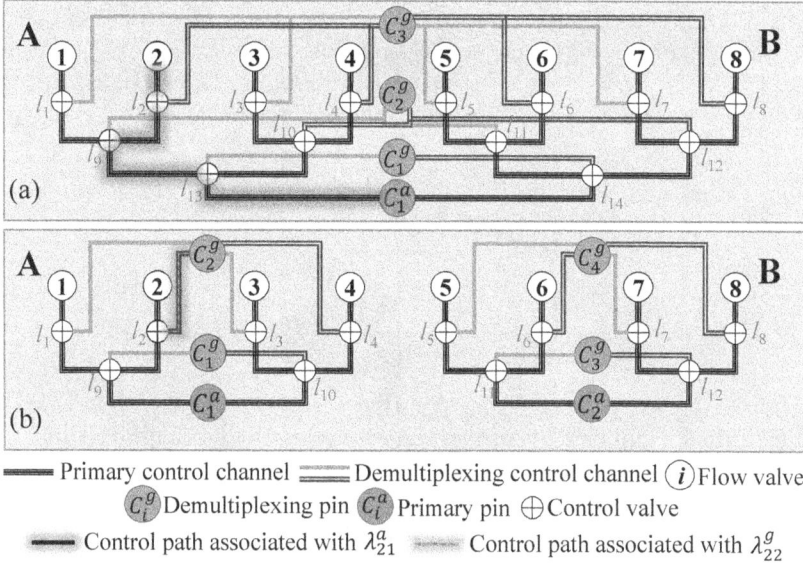

FIGURE 6.6
Multiplexed control of an 8-valve channel using (a) 4 control pins and (b) 6 control pins.

that a demultiplexing pin $C_y^g \in \mathcal{C}^g$ is used in the actuation of L_x with delay d_{xy}^g.

Definition 6.1 *The connectivity vector $\phi_x \in \mathbb{R}^{y_v}$ of a valve L_x is defined as $\phi_x = [\lambda_{x1}^a \ \lambda_{x2}^a \ ... \ \lambda_{xa_v}^a | \lambda_{x1}^g \ \lambda_{x2}^g \ ... \ \lambda_{xg_v}^g] = [\phi_x^a | \phi_x^g]$, where a_v and g_v are the number of primary pins and demultiplexing pins, respectively ($y_v = a_v + g_v$), ϕ_x^a is the connectivity associated with the primary pins, and ϕ_x^g is the connectivity associated with the demultiplexing pins. In addition, the connectivity matrix that describes the connectivity of all flow valves in $\theta_{sc}(x_v, y_v)$ is defined as*

$$\Phi = [\Phi^a | \Phi^g] = \begin{bmatrix} \phi_1 \\ \phi_2 \\ ... \\ \phi_{x_v} \end{bmatrix}$$

where Φ^a and Φ^g are the connectivity matrices associated with the primary pins and demultiplexing pins, respectively. □

We use the multiplexed control designs in Figure 6.6(a-b) for explanation. In Figure 6.6(a), the actuation of any flow valve $L_x \in \{L_1, L_2, ..., L_8\}$ is accomplished using the same primary pin C_1^a and the same three demultiplexing pins $\{C_1^g, C_2^g, C_3^g\}$. As a result, $\phi_x = [\lambda_{x1}^a | \lambda_{x1}^g \ \lambda_{x2}^g \ \lambda_{x3}^g], 1 \le x \le 8$. Figure 6.6(a) depicts the control path associated with λ_{21}^a. On the other hand, in Figure 6.6(b), the actuation of valves $\{L_1, L_2, L_3, L_4\}$ is accomplished using primary pin C_1^a and demultiplexing pins $\{C_1^g, C_2^g\}$, whereas valves $\{L_5, L_6, L_7, L_8\}$ are addressed using primary pin C_2^a and demultiplexing pins $\{C_3^g, C_4^g\}$. Therefore, ϕ_x and Φ are defined in this case as follows:

$$\phi_x = \begin{cases} [\lambda_{x1}^a \quad 0 \quad |\lambda_{x1}^g \quad \lambda_{x2}^g \quad 0 \quad 0 \quad] & \text{if } 1 \leq x < 5 \\ [0 \quad \lambda_{x2}^a| \quad 0 \quad 0 \quad \lambda_{x3}^g \quad \lambda_{x4}^g] & \text{if } 5 \leq x \leq 8 \end{cases}$$

$$\Phi = \begin{bmatrix} \phi_1 \\ \phi_2 \\ \cdots \\ \phi_8 \end{bmatrix}$$

Figure 6.6(b) depicts the control path associated with λ_{22}^g. Note that only a single primary pin can be used in the actuation of a certain flow valve (i.e., L_x cannot be addressed by both C_1^a and C_2^a). Hence, flow valves are divided into groups; each group is associated with a particular primary pin. To prevent control interference between groups, each demultiplexing pin can only be used in the actuation of flow valves that belong to the same group. By this grouping method, different groups of flow valves can be actuated concurrently.

While the flow valves $\{L_1, L_2, ..., L_8\}$ construct only a single control vector (i.e., $CV = 1$), the following definition considers the generic case where $CV > 1$.

Definition 6.2 *A control vector* $\vartheta_i : \{i \in \mathbb{N}, i \leq CV\}$ *is a vector of length* x_v *such that* $\vartheta_i[x] = 1$ *indicates that a flow valve* L_x *must be actuated as a part of the control vector, and* $\vartheta_i[x] = 0$ *otherwise. In addition, the control (screening) matrix of* $\theta_{sc}(x_v, y_v)$ *is defined as*

$$\Gamma = \begin{bmatrix} \vartheta_1 \\ \vartheta_2 \\ \cdots \\ \vartheta_{CV} \end{bmatrix}$$

$\square$

The simplest form of an RFB that contains more than a single control vector is a 2-by-2 crossbar (also known as a transposer [113]); see Figure 6.7. Such a transposer consists of six valves $\{L_1, L_2, ..., L_6\}$ that can be configured to route a biochemical sample through four different paths ($CV = 4$): I_{b1} to X_{b1}, I_{b1} to X_{b2}, I_{b2} to X_{b1}, and I_{b2} to X_{b2}. To activate the second path (i.e., I_{b1} to X_{b2}), valves $\{L_2, L_4\}$ must be opened and valves $\{L_1, L_6\}$ must be closed. Hence the control vector ϑ_2 is defined as follows: $\vartheta_2 = [1\ 1\ 0\ 1\ 0\ 1]$. Note that a pressure pulse for closing a valve and a vacuum pulse for opening the same valve are considered to have the same delay characteristics [87].

Next, we compute the latency associated with a control vector ϑ_i. Recall that Φ is an augmented matrix; $\Phi = (\Phi^a|\Phi^g)$, where Φ^a and Φ^g are the connectivity matrices associated with the primary pins and demultiplexing pins,

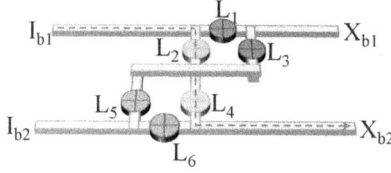

FIGURE 6.7
A transposer design that exhibits 4 different flow paths (i.e., 4 control vectors).

respectively (Definition 6.1). By examining the multiplexed control procedure for each flow valve, we observe that the demultiplexing pins are used first to actuate the control valves, which in turn select the control path within the multiplexed control circuitry. Next, the primary pin is activated to actuate the target flow valve. With this procedure, demultiplexing pins can be activated concurrently, whereas a primary pin can be activated only after a control path is available. This observation leads us to the following lemma—the proof can be found in Appendix C.

Lemma 6.1 *The control delay of a flow valve L_x is a value $b_x = \max\{\phi_x^a\} + \max\{\phi_x^g\}$, and the control delay vector for all the valves in $\theta_{sc}(x_v, y_v)$ is $\Psi = [b_1 \; b_2 \; ... \; b_{x_v}]^\top$.*

We use the multiplexed control example in Figure 6.6(b) to illustrate the above lemma. Recall that $\phi_2 = [\phi_2^a | \phi_2^g] = [\lambda_{21}^a \; 0 | \lambda_{21}^g \; \lambda_{22}^g \; 0 \; 0]$. Since the demultiplexing pins $\{C_1^g, C_2^g\}$ are actuated concurrently, then the control delay is associated with the worst-case (i.e., maximum) delay among $\{\lambda_{21}^g, \lambda_{22}^g\}$, which is defined as $\max\{\lambda_{21}^g, \lambda_{22}^g, 0, 0\} = \max\{\phi_2^g\}$. Since the primary pin C_1^a is actuated after the control path is opened, then its associated delay is defined as $\lambda_{21}^a = \max\{\phi_2^a\}$, and the effective control delay associated with L_2 is $b_2 = \max\{\phi_2^a\} + \max\{\phi_2^g\} = \lambda_{21}^a + \max\{\lambda_{21}^g, \lambda_{22}^g\}$.

Based on this key result, we can compute the latency of a control vector ϑ_i, while taking into consideration the grouping of flow valves according to the assigned primary pins. The control valves that belong to the same group cannot be actuated concurrently since they share the same primary pin, but different valve groups can be actuated concurrently. The concurrent actuation of different valve groups (even those which belong to the same control vector; see Figure 6.6(b)) can be captured by considering only the worst-case delay among all the groups. For this purpose, we define the connectivity of a primary pin C_y^a as $\rho_y = [\lambda_{1y}^a \; \lambda_{2y}^a \; ... \; \lambda_{x_v y}^a]^\top \in \mathbb{R}^{x_v}$, $y \in [1, a_v]$. We use a binary vector $sign(\rho_y)$ to specify the set of flow valves that belong to primary pin group C_y^a. In other words, the vector $sign(\rho_y)$ contains 1's if certain flow valves are actuated using C_y^a, and 0's otherwise.

Consider the vector operator $\circ$ that provides the element-wise product of two vectors, i.e., $E \circ B = [e_1 \; e_2 \; ... \; e_n] \circ [b_1 \; b_2 \; ... \; b_n] = [e_1 b_1 \; e_2 b_2 \; ... \; e_n b_n]$. Hence, the vector $sign(\rho_y) \circ \Psi$ specifies the control delays associated with the flow valves in group C_y^a. A subset of the flow valves in C_y^a belongs to a control vector ϑ_i and therefore they need to be actuated serially if ϑ_i is selected for single-cell screening. As a result, the scalar quantity $\vartheta_i \cdot (sign(\rho_y) \circ \Psi)$ computes the cumulative control delay associated with a set of flow valves that is characterized as follows: (1) the flow valves belong to group C_y^a; (2) the actuation of these valves is needed for activating ϑ_i. By computing the cumulative control delays associated with all groups of flow valves within ϑ_i, we obtain the worst-case control delay among all groups as follows: $\max\limits_{1 \leq y \leq a_v} \{\vartheta_i \cdot (sign(\rho_y) \circ \Psi)\}$.

Figure 6.7 is used to illustrate the above discussion. Recall that the second control vector is $\vartheta_2 = [1\ 1\ 0\ 1\ 0\ 1]$. First, assume that all six valves belong to the same group, i.e., they are actuated using the same primary pin C_1^a. In this case, $sign(\rho_1) = [1\ 1\ ...\ 1]$. To activate ϑ_2, L_1, L_2, L_4, L_6 need to be actuated serially; therefore the cumulative control latency, denoted by α_2, is the sum of the control delays associated with these valves. In other words, $\alpha_2 = \vartheta_2 \cdot (sign(\rho_1) \circ \Psi) = [1\ 1\ 0\ 1\ 0\ 1] \cdot [b_1\ b_2\ ...]^\intercal = b_1 + b_2 + b_4 + b_6$. Second, if two primary pins $\{C_1^a, C_2^a\}$ are used, where $sign(\rho_1) = [1\ 1\ 1\ 0\ 0\ 0]$ and $sign(\rho_2) = [0\ 0\ 0\ 1\ 1\ 1]$, the cumulative control latency α_2 is computed as $\alpha_2 = \max\limits_{1 \leq y \leq 2} \{[1\ 1\ 0\ 1\ 0\ 1] \cdot (sign(\rho_y) \circ \Psi)\} = \max\{b_1 + b_2, b_4 + b_6\}$.

Based on this observation, we have derived the following theorem (proven in Appendix D), which forms the basis for the Sortex synthesis procedure (Section 6.3).

Theorem 6.1 *If α_i, where $\alpha_i \in \mathbb{R}, i \in \mathbb{N}, i \leq CV$, is the cumulative control latency value associated with a control vector ϑ_i in a chip $\theta_{sc}(x_v, y_v)$, then $\alpha_i = \max\limits_{1 \leq y \leq a_v} \{\vartheta_i \cdot (sign(\rho_y) \circ \Psi)\}$, where $\circ$ is the element-wise product. In addition, the cumulative control latency vector for all control vectors in $\theta_{sc}(x_v, y_v)$ is $\Theta = \max\limits_{1 \leq y \leq a_v} \{\Gamma \cdot (sign(\rho_y) \circ \Psi)\}$.*

The above discussion is focused on the latency of multiplexed control in an RFB. Since the topology of flow channels in the proposed RFB is fixed, we can easily estimate the flow latency of samples through these channels and therefore obtain an accurate estimate of biochip throughput. Note that a sample can be routed through a flow path only after the associated control vector is activated. Therefore, if the flow latency vector associated with the biochip control vectors is defined as $\Theta_f \in \mathbb{R}^{CV}$, then the effective latency vector of $\theta_{sc}(x_v, y_v)$ is $\Theta + \Theta_f$ and the worst-case latency is $ly = \max\{\Theta + \Theta_f\}$. To optimize the throughput of single-cell screening, our synthesis method in Section 6.3 optimizes the multiplexed control scheme in an RFB such that y_v is minimized and $ly \leq ly_{max}$, where ly_{max} is a predefined value.

Delay of Pressure-Driven Fluid Transport

The study of fluidic delay can ideally be performed based on a closed-form analytical solution, which can be derived from the fundamental Navier-Stokes and continuity equations [252]. However, the derivation of such an analytical solution or an approximation is extremely complicated, even for basic channel topologies. The power of CFD software tools such as COMSOL [2] is that they provide a solution for a conventional fluidic/mechanical property, e.g., pressure, velocity, or turbulence.

On the other hand, interconnect delay in VLSI circuits has been studied thoroughly over the past three decades. Delay models provide a closed-form approximated solution of signal delay based on well-defined circuit models such as resistors and capacitors. Circuit models are therefore used to compute interconnect delay, and they represent the basis for all static-timing analysis (STA) tools used nowadays in electronic circuit design [56].

Hence, inspired by the success achieved by such models in electronic circuits, we aim to approximate the fluidic delay also based on well-defined fluidic components whose behaviors are similar to their counterparts in electronic circuits. By leveraging previous delay models, we can reach an acceptable formulation of fluidic delay.

In electrical circuits, the delay of an electrical signal through a wire can be characterized using a delay model [56]; a widely used delay model, especially with wires characterized as RC trees or ladders, is the Elmore delay model [56]. In analogy with electrical circuits, the delay of laminar flow through a long elastic channel can be approximated using an equivalent Elmore delay model, which is a practical alternative to complex CFD simulations. For this purpose, there is a need to define the model components, i.e., the hydraulic resistance R and the hydraulic compliance M [135].

A laminar flow of a fluid through a long channel can be described using the Hagen-Poiseuille equation: $QH = \frac{\pi \cdot Dr^4 \cdot \Delta Pr}{128 \mu_v \cdot Nr}$, where QH is the flow rate, Dr is the channel diameter, $\Delta Pr = Pr_{in} - Pr_{out}$ is the pressure drop across the channel, μ_v is the dynamic viscosity of the fluid, and Nr is the length of the channel. The analog of this law in electrical circuits is Ohm's law [189]. We use this analogy to estimate the *hydraulic resistance R*, which is defined as $R = \frac{\Delta Pr}{QH} = \frac{128 \mu_v \cdot Nr}{\pi \cdot Dr^4}$.

The above model makes the assumption that a microfluidic channel is rigid. However, RFBs are fabricated using elastic material (e.g., PDMS) hence pressure can cause the cross-sectional area of a channel to change [91]; see Figure 6.8(a). The capability of an elastic channel to store fluid when pressurized is known as hydraulic compliance (similar to capacitance in electrical circuits). To verify this behavior, we conducted transient CFD simulation using COMSOL for laminar flow in a PDMS channel. Figure 6.8(b) shows channel deformation at different time steps when a pressure step Pr_1 is applied to the input. This deformation can be interpreted as the potential of a channel to store fluid when pressurized; this is referred to as *hydraulic compliance M*.

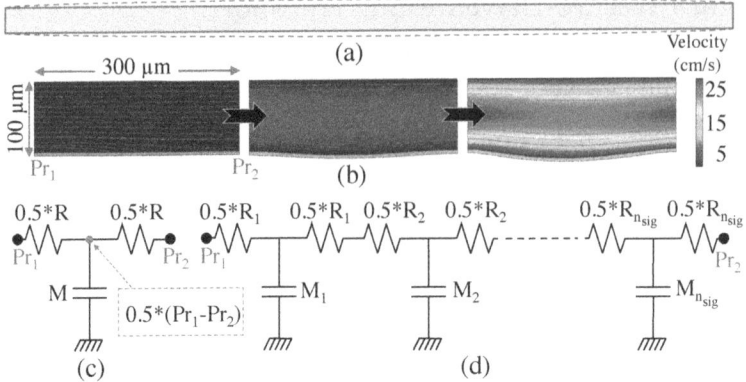

FIGURE 6.8
(a) A PDMS microfluidic channel expands when pressurized. (b) Result of transient CFD simulation for an elastic channel under pressure. (c) Lumped-RC model. (d) Distributed-RC model.

Hence, the following equation can be used to compute the hydraulic compliance: $M = \frac{dVr}{dPr}$, where $\frac{dVr}{dPr}$ is the rate by which channel volume Vr changes when pressure Pr is applied. To provide a closed-form formula, we re-write the above relation as $M = Vr \cdot \frac{1}{Vr} \cdot \frac{dVr}{dPr} \approx Vr \cdot \gamma$, where γ is the channel dilatability (an elasticity property). This formula is aligned with an observation from the CFD simulation that the deformation distance and therefore M are correlated with the material elasticity. By considering an elastic circular channel, M is computed as follows: $M = \frac{\pi}{4} \cdot Dr^2 \cdot \gamma \cdot Nr$, where γ is the channel dilatability, Dr is the channel diameter, and Nr is the channel length.

The closed-form equations for R and M can be used to estimate fluidic delay. Furthermore, since the values of R and M are specified based on the channel geometry and material elasticity, the fluidic delay, and therefore the screening throughput, can be tuned based on these parameters. According to [252], a straightforward approach for modeling an elastic channel is by using a lumped-RC model, as shown in Figure 6.8(c). For this model, the fluidic delay of the channel is simply $R \cdot M$. However, this lumped model does not take into account the significant change in pressure across the channel. To capture the variation in pressure and its impact on fluidic delay, we use a distributed-RC ladder, as shown in Figure 6.8(d). This model offers three advantages: (1) increasing the number of model segments n_{sig} enhances model accuracy since every segment exhibits an infinitesimal change in pressure; (2) similar to the models of electrical interconnects [56], modeling various segments of a channel provides the opportunity to design a width-varying channel—this design method can be employed to minimize both the fluidic delay and the channel area; (3) the delay associated with a distributed-RC ladder (containing n_{sig} segments) can be estimated using the Elmore model [56]:

$d_{el} = \sum_{i=1}^{n_{sig}} \left(M_i \cdot \left(\sum_{j=1}^{i} R_j \right) \right)$, where d_{el} is the Elmore delay, R_j and M_j are the resistance and compliance of segment j, respectively. Hence, we adopt this model in our synthesis framework to estimate the delay associated with flow and multiplexed-control paths.

Advantages of Using the Distributed Model

We demonstrate the advantages of using the distributed model through the following example. We consider designing a flow-based channel that has the following specifications: (1) the input terminal of the channel is connected to a pressure source through a pin of diameter 150 μm, and the pressure source is configured to pump water at a rate of 33.5 mm^3/s with an input pressure of 700 kPa; (2) the channel length needs to be 2 cm; (3) the output terminal of the channel is used to control a valve, and the valve membrane can safely be actuated using a pressure in the range of 50 and 100 kPa; (4) the delay of fluidic transport across the channel needs to be less than or equal to 4 seconds.

Based on the above specifications, we provide two designs of the flow channel. The first design, shown in Figure 6.9(a), is a uniform channel that has a constant diameter along its length, whereas the second design (Figure 6.9(b)) is a non-uniform channel that has a varying diameter along it length. We analyze the fluidic delay associated with both designs using the lumped-RC model, denoted by RC$_{lump}$, and the distributed-RC model, denoted by RC$_{dist}$. Moreover, we compute the pressure levels across the channels in order to ensure that the given specifications are fulfilled; see Figure 6.9. To compute the pressure when using RC$_{dist}$ for delay estimation, we adopt a simplified method where we divide the channel into segments and compute the pressure drop across each segment separately using Hagen-Poiseuille equation[2]. We use Pr_{dist} to denote this distributed pressure model and Pr_{lump} to denote the lumped model.

Figure 6.9(a) shows the pressure and delay associated with the uniform channel. The computed pressure within the channel is similar for both Pr_{lump} and Pr_{dist}, and both lead to an output pressure (652 kPa) that is larger than 100 kPa, i.e., it does not fulfill the above specifications. However, by using RC$_{dist}$ (i.e., Elmore model) to estimate the delay, we observe that this design satisfies the delay constraint ($d_{con} \leq 4$ seconds); this is opposite to the result we obtained using RC$_{lump}$. In other words, the pessimistic nature of the lumped model may lead to incorrect assumptions about the reliability of the channel design.

Next, we use a non-uniform channel that has a varying diameter along its length, as shown in Figure 6.9(b). In this case, the model Pr_{lump} relies on the average diameter $\left(\frac{Dr_{in} + Dr_{out}}{2} \right)$ to compute the pressure, whereas Pr_{dist}

---

[2]Unlike electrical circuits, modeling fluid flow in a channel relies on the continuum concept [252], meaning that we cannot compute the pressure drop across a channel using conventional AC steady-state analysis methodologies that are used in electrical circuits, or more specifically, RC-ladder circuits.

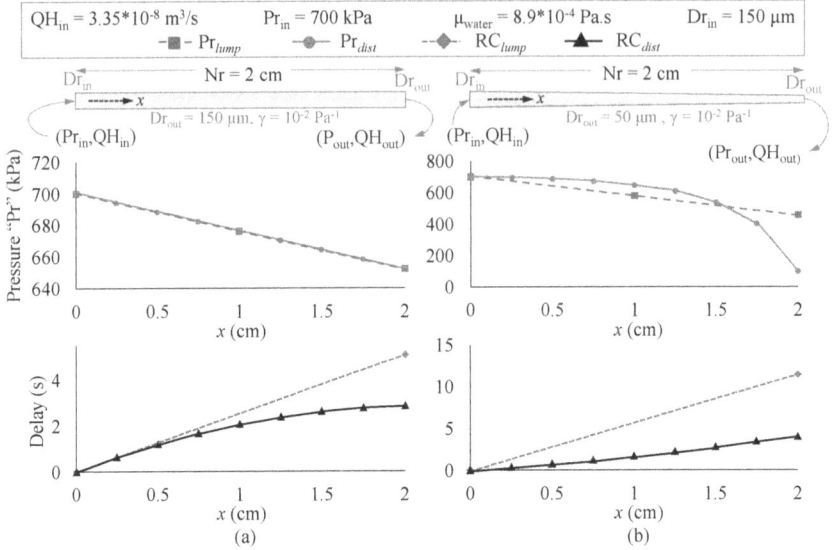

FIGURE 6.9
Comparison in terms of pressure and fluidic delay between the distributed model and the lumped model using (a) a uniform flow channel, and (b) a non-uniform flow channel.

uses the average diameter within every segment. We notice that the pressure curve obtained by Pr_{lump} is highly inaccurate. Furthermore, unlike Pr_{dist}, the model Pr_{lump} leads to an incorrect assumption that the proposed channel design violates the pressure constraint.

The above example corroborates the fact that using a distributed model provides more accurate results compared to lumped models. In this work, we focus only on delay analysis using RC_{dist}—pressure-drop analysis is left for future research.

6.3   Sortex: Synthesis Solution

In this section, we discuss the problem formulation and the proposed synthesis framework (Sortex). The notation used in this chapter is summarized in Table 6.1.

TABLE 6.1

Notation used in Chapter 6.

$\mathcal{L}$	The set of flow valves $\{L_1, L_2, ..., L_{x_v}\}$
$\mathcal{I}$	The adjacency matrix showing the connectivity of $\mathcal{L}$
G_l	a direct graph that models $\mathcal{L}$ and the connections $\mathcal{I}$
$\widetilde{\mathcal{L}}$	The set of control valves $\{l_1, l_2, ...\}$
$\mathcal{C}^a$	The set of primary pins $\{C_1^a, C_1^a, ...\}$
$\mathcal{C}^g$	The set of demultiplexing pins $\{C_1^g, C_1^g, ...\}$
d_{xy}^i	The delay contributed by a pin C_y^i of type i to actuate a valve L_x
λ_{xy}^i	The connectivity function relating a pin C_y^i and a valve L_x
ϕ_x	The connectivity vector of a valve L_x
Φ	The connectivity matrix of a biochip $\theta_{sc}(x_v, y_v)$
ρ_y	The connectivity of a primary pin C_y^a
Ψ	The control delay vector for all valves in $\theta_{sc}(x_v, y_v)$
ϑ_i	A control vector; $i \leq CV$; CV: The number of control vectors
Γ	The control screening matrix of $\theta_{sc}(x_v, y_v)$
α_i	The cumulative control latency associated with ϑ_i
Θ	The cumulative control latency vector for all ϑ_i in $\theta_{sc}(x_v, y_v)$
ly	The worst-case latency in $\theta_{sc}(x_v, y_v)$
ly_{max}	The maximum latency threshold
$\mathcal{UC}$	The lengths of the control channels; $\mathcal{UC} = \{uc_1, uc_2, ...\}$
$\mathcal{FC}$	The widths of the control channels; $\mathcal{FC} = \{fc_1, fc_2, ...\}$

Problem Formulation

We consider the following problem formulation in this work.

Input: (1) The specifications of a screening biochip $\theta_{sc}(x_v, y_v)$, which includes: *(i)* the locations of all the valves $\mathcal{L}$ and $\widetilde{\mathcal{L}}$; *(ii)* the channels connecting the flow valves $\mathcal{L}$ using an adjacency matrix form $\mathcal{I}$; *(iii)* the maximum number of control pins ($\mathcal{C}^a$ and $\mathcal{C}^g$) and their locations; *(iv)* possible values of the channel width $\{Dr_1, Dr_2, ...\}$.

(2) A fluidic delay model based on a distributed RC ladder.

(3) A threshold $ly_{max} \in \mathbb{R}$ that represents the maximum latency.

(4) An upper bound Ar_{max} on the channel area.

Output: (1) The screening matrix Γ. (2) The connectivity matrix Φ. (3) The worst-case latency ly. (4) The dimensions of the control channels: lengths $\mathcal{UC} = \{uc_1, uc_2, ...\}$ and widths $\mathcal{FC} = \{fc_1, fc_2, ...\}$.

Constraints: (1) Screening latency constraint ($ly \leq ly_{max}$). (2) Area constraint ($\sum_i (uc_i \times fc_i) \leq Ar_{max}$).

Objective: Minimize the number of control pins y_v used for multiplexed control of biochip valves.

Note that efficient channel routing is beyond the scope of this work. Sortex calculates control delays by using Manhattan distances between different entities.

This is motivated by the fact that today's flow-based designs such as the described multiplexed-control circuitry can be fabricated as a multi-layered structure using 3-D printers [93]. Similar to their counterparts in electronic circuits, advances in the fabrication technology of flow-based systems have made an enormous impact in increasing the density of integrated valves in a cost-effective manner while complex channel routing is overcome by using 3-D valve integration. Figure 6.10 shows an example of a multi-layered RFB [134]. Nonetheless, even if the design is constrained with a limited number of layers, Sortex can be improved to enable basic routing functionality. This basic improvement can allow Sortex to reject a control-channel connection if a certain routing feature is violated. Sortex can also be extended further by incorporating routing as in [101]. The design of comprehensive routing techniques to accompany Sortex is left for future work.

Computation of the Screening Matrix Γ

By using the sets $\mathcal{L}$ and $\mathcal{I}$, we represent the set of flow valves and their channel connections as a directed acyclic graph (DAG) $G_l = (V_l, E_l)$, where a vertex $v_{li} \in V_l$ models a flow valve $L_i \in \mathcal{L}$, and an edge $e_{li} \in E_l$ represents a flow channel connecting two valves. The orientation of an edge e_{li} is specified based on the direction of fluid transport inside the associated channel. Note that a flow channel is allowed to transport fluid only in one direction at all times to avoid collisions. To simplify the computation, we also map the inlets and outlets of the flow network into *virtual* vertices, which are not considered in the final outcome of Γ. Figure 6.11 depicts the graph model of a 60-valve screening biochip.

Our goal is to leverage the graph model G_l to enumerate all possible screening paths in the biochip. To achieve this goal, we adopt depth-first search (DFS) to traverse all possible paths between the inlet and outlet virtual vertices. Note that we need to traverse the whole graph, considering all inlet and outlet virtual vertices, only once in order to print all paths (time complexity $O(|V_l| + |E_l|)$). However, recall that a control vector ϑ_i includes all flow valves that need to be actuated to transport a droplet within the associated screening path. In other words, we need to consider not only the valves given by the

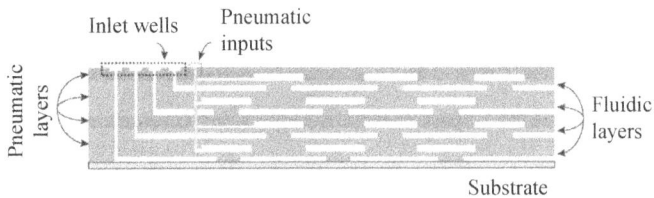

FIGURE 6.10
Cross-sectional view of a 3-D RFB [134].

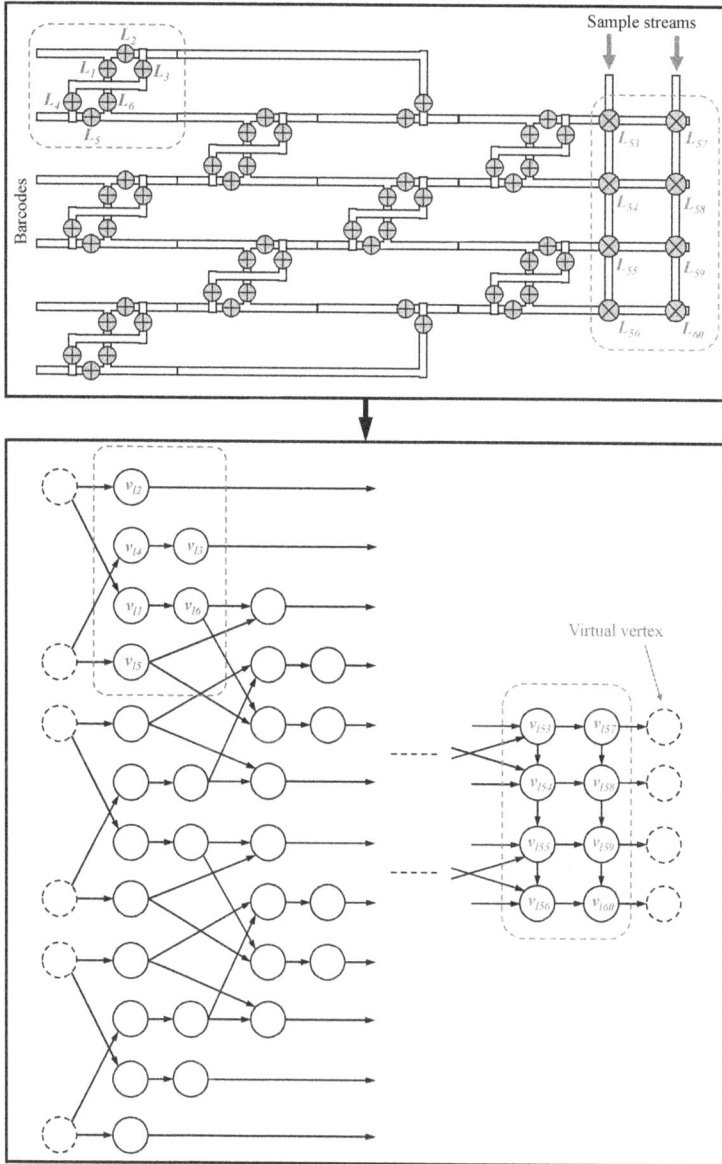

FIGURE 6.11
The graph model of a 60-valve RFB.

DFS procedure, but also the neighboring valves that need to be actuated in order to generate Γ. For this purpose, after running DFS, we loop over the valves to specify their neighboring valves (using $\mathcal{I}$) and add those neighbors to the screening path. As a result, the time complexity of this algorithm is

$O(CV \cdot x_v^2)$, where CV is the number of control vectors and x_v is the number of flow valves.

Sortex Algorithm

Recall that flow valves are assigned to groups; each group is actuated using a dedicated primary pin. To reduce the number of control pins, synthesis for multiplexed control must increase the number of flow valves assigned to a group.

We use a heuristic method based on divide-and-conquer (Algorithm 6.1). Initially, the proposed method selects a primary pin (Line 13) and iterates over the flow valves in a pairwise manner (Line 10) trying to connect every pair to the current primary pin (Line 16). When the multiplexed-control scheme is expanded, timing analysis is performed using the Elmore delay model from Section 6.2 and the relation from Theorem 6.1 to ensure that the latency constraint is not violated (Line 17). If no constraint violation is detected, the expanded multiplexed control is accepted and the associated valve/pin variables are updated (Lines 18-21). However, if a violation is detected, the proposed connection to the current primary pin is declined and a new primary pin is selected for connection (Lines 23-24). In this case, the flow valves are divided into two groups; one group combines the valves that have already been connected to the first primary pin and the second group combines the remaining valves (including the current pair)—we call this scheme *divide-and-conquer*. When the remaining flow valves are connected to a second primary pin, we further divide the set of valves into two groups according to the latency constraint. This process continues until all the flow valves are addressed.

The order of selection of the valves impacts computational performance. Random selection of valves may lead to high CPU time because distant valves can be combined in a single group, causing divide-and-conquer to be invoked frequently; i.e., unnecessarily increasing the number of control pins. To address this issue, we create a priority queue of the unaddressed flow valves (Line 2). The priority of valve selection is decided according to the following policy: (1) select a valve L_f randomly and place it at the head of the queue; (2) select a set of flow valves that does not belong to the same control vectors as L_f and sort these valves in an increasing order according to their Manhattan distance from L_f; (3) sort the remaining valves according to their Manhattan distance from L_f. This priority scheme (worst-case time complexity $O(CV \cdot x_v + x_v \cdot \log x_v)$) is invoked whenever a divide-and-conquer action takes place (Line 25).

We next connect a new pair of flow valves to a primary pin (i.e., TestMC function in Algorithm 6.1). To expand the multiplexed control, control paths are designed to manage the actuation of the new flow valves. The synthesis of a control path is governed by two aspects: (1) the selection of control valves $\widetilde{\mathcal{L}} = \{l_1, l_2, ...\}$ through which the control path is designed; (2) the selection of a new demultiplexing pin (if needed) in order to actuate the control valve. Consider the design of the control paths for the RFB in Figure 6.12(a). To

Algorithm 6.1 Sortex Procedure

1: $\Gamma \leftarrow \text{ConstructScreeningMatrix}(\mathcal{L}, \mathcal{I})$;
2: $\mathcal{VQ} \leftarrow \text{ConstructValvesPriorityQueue}(\mathcal{L})$;
3: $\text{AssignControlValvesToLevels}(\widetilde{\mathcal{L}})$;
4: $\text{SortDecrChannelWidthsRange}()$;
5: $j \leftarrow 0$;   ▷ Iterator
6: **repeat**
7:     $\mathcal{CC}^a \leftarrow \emptyset; \mathcal{CC}^g \leftarrow \emptyset$;   ▷ Connected pins so far
8:     $\mathcal{UC} \leftarrow \emptyset; \mathcal{FC} \leftarrow \emptyset; \Phi \leftarrow [0]; ly \leftarrow 0$;
9:     $Dr_j \leftarrow \text{ConsiderChannelWidth}()$;
10:     **for** $(L_i, L_{i+1}) \in \mathcal{VQ}$ **do**
11:         $\mathcal{XC} \leftarrow \emptyset; \mathcal{YC} \leftarrow \emptyset$;   ▷ Connected pins to the pair
12:         **if** $\mathcal{CC}^a = \emptyset$ **then**
13:             $\mathcal{XC} \leftarrow \text{SelectNearestPrimaryPinLocation}()$;
14:             $\mathcal{YC} \leftarrow \text{BuildMC}(L_i, L_{i+1}, \mathcal{XC}, \text{"2-Pin"})$;
15:         **else**
16:             $\{\Phi, \text{Conf}\} \leftarrow \text{TestMC}(L_i, L_{i+1})$;
17:             $ly \leftarrow \text{CalculateLatencyUsingElmore}(\Phi, Dr_j)$;
18:             **if** $ly \le ly_{max}$ and $\text{Conf} = \text{"1-Pin"}$ **then**
19:                 $\mathcal{YC} \leftarrow \text{BuildMC}(L_i, L_{i+1}, \text{"1-Pin"})$;
20:             **else if** $ly \le ly_{max}$ and $\text{Conf} = \text{"0-Pin"}$ **then**
21:                 $\text{BuildMC}(L_i, L_{i+1}, \text{"0-Pin"})$;
22:             **else**
23:                 $\mathcal{XC} \leftarrow \text{SelectNearestPrimaryPinLocation}()$;
24:                 $\{\mathcal{XC}, \mathcal{YC}\} \leftarrow \text{BuildMC}(L_i, L_{i+1}, \mathcal{XC}, \text{"2-Pin"})$;
25:                 $\text{UpdatePriorityQueue}(\mathcal{VQ})$;
26:             $\text{UpdateVariables}(\mathcal{UC}, \mathcal{FC}, \Phi, ly)$;
27:             $\mathcal{CC}^a \leftarrow \mathcal{CC}^a \cup \mathcal{XC}; \mathcal{CC}^g \leftarrow \mathcal{CC}^g \cup \mathcal{YC}$;
28:             $Ar \leftarrow \text{CalculateChannelsArea}(\mathcal{UC}, \mathcal{FC})$;
29:             **if** $Ar > Ar_{max}$ **then** break;
30:     $j \leftarrow j + 1$;
31: **until** $Ar \le Ar_{max}$
32: **return** $\{\mathcal{UC}, \mathcal{FC}, \mathcal{CC}^a, \mathcal{CC}^g, ly, \Phi\}$;

simplify the selection of control valves at each iteration, we initially split $\widetilde{\mathcal{L}}$ into logical levels (Line 3), where the first level contains control valves that will be directly connected to the flow valves, e.g., l_1 in Figure 6.12(b) and $\{l_1, l_2\}$ in Figure 6.12(c). Such an organization is performed in advance and it enforces two conditions: (1) two control valves cannot be connected using primary control channels if they belong to the same logical level; (2) a single demultiplexing pin can only be used to actuate the control valves that belong to the same level and are used in the actuation of a single group of flow valves.

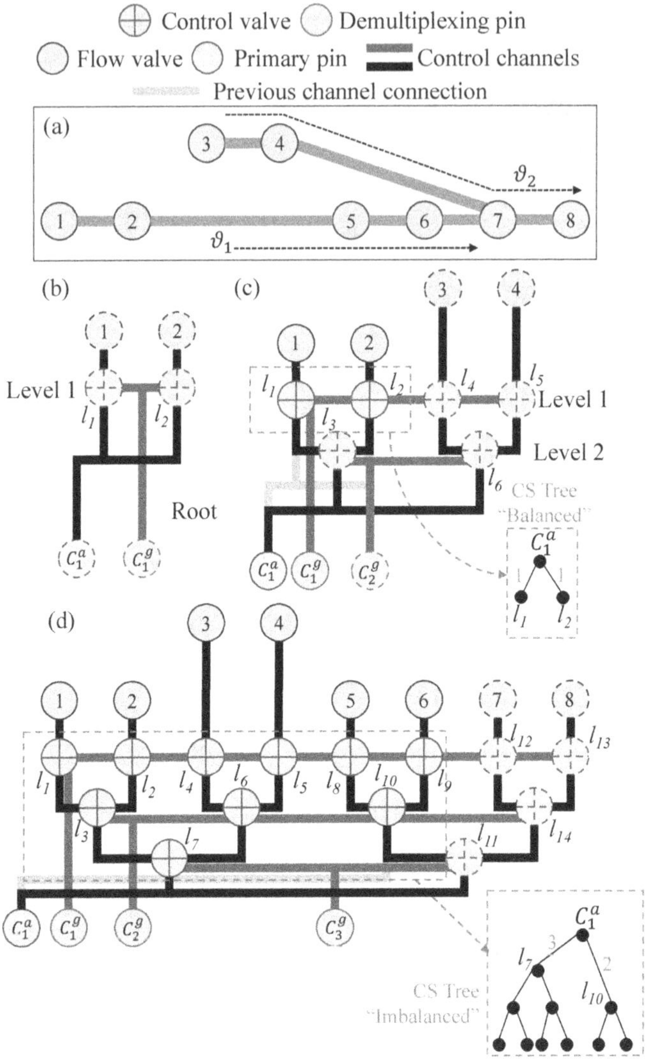

FIGURE 6.12
Multiplexed control configurations used to connect a pair of flow valves. (a)
An 8-valve RFB with two control vectors. (b) 2-Pin configuration. (c) 1-Pin
configuration. (d) 0-Pin configuration.

For example, in Figure 6.12(c), $\{l_1, l_2\}$ can be actuated using only C_1^g, which
cannot be used to actuate l_3.

Based on the above hierarchy of control valves, the synthesis of control
paths associated with a new pair of flow valves can be performed based on
three different configurations (associated with the variable Conf in Line 16),

as shown in Figure 6.12(b-d). The first (2-Pin) configuration is applied when a new primary pin is needed. This configuration is adopted either at the first iteration (Line 12-14) or when there is a need for a new primary pin in order to prevent violation of the screening-latency constraint (Lines 23-24). In addition, a new demultiplexing pin is also needed. For example, in Figure 6.12(b), a new primary pin C_1^a is connected to $\{L_1, L_2\}$ using control valves l_1 and l_2 that belong to Level 1. Since l_1 and l_2 are the first valves to be used in Level 1, a new demultiplexing pin C_1^g is connected.

The second and third configurations address the case where a new primary pin is not needed. These configurations, however, differ in how control valves are selected, and therefore, whether a demultiplexing pin is needed. To demonstrate the difference between the two configurations, we map the control valves and their connectivity into a binary tree, referred to as control-valves (CS) tree. Each node represents a connected control valve, and the root represents the primary pin that is connected to these control valves; see Figure 6.12(c-d). Each node has two interfaces (left and right) which are connected to two sub-trees. Each interface is characterized by the maximum height of the associated sub-tree. A CS tree is balanced only if the interfaces of every node in the tree exhibit equal heights.

The CS tree evolves at every iteration whenever a new pair of flow valves is selected; the tree evolves in a top-down fashion; new leaves are added when new control valves are connected. The second (1-Pin) configuration is selected if the CS tree is currently balanced. Figure 6.12(c) shows an example for 1-Pin. Before the multiplexed-control scheme is expanded for connecting the pair $\{L_3, L_4\}$, we observe that the CS tree contains only two nodes $\{l_1, l_2\}$ and it is balanced. For connecting $\{L_3, L_4\}$ to the same primary pin C_1^a, four new control valves $\{l_3, l_4, l_5, l_6\}$ are connected. Since l_3 and l_6 are the first valves to be connected at Level 2, a new demultiplexing pin C_2^g is selected and the CS tree is expanded.

On the other hand, the third (0-Pin) configuration is applied if either the CS tree or any sub-tree within the CS tree is imbalanced. For example, in Figure 6.12(d), we investigate the multiplexed-control paths when a new pair $\{L_7, L_8\}$ is selected. The CS tree is imbalanced, because the difference between the heights of the root is 1; this difference is called the imbalance factor. Valve l_{10} is therefore called the source of imbalance and its level (Level 2) is called the level of imbalance. To connect $\{L_7, L_8\}$ to the primary pin C_1^a, four control valves $\{l_{11}, l_{12}, l_{13}, l_{14}\}$ are connected. The valves $\{l_{12}, l_{13}\}$ connect the pair $\{L_7, L_8\}$ and therefore they are located at Level 1. The control valve l_{14} connects both $\{l_{11}, l_{12}\}$ and it is located at Level 2. However, l_{11} is located at Level zl which is determined as follows: zl = level of imbalance + 1. The control valve l_{11} is used to connect the source of imbalance l_{10}, the new control valve l_{14}, and the primary pin C_1^a. This configuration does not allocate new control pins. By using this balancing scheme, we reduce the height of the tree and hence decrease the number of demultiplexing pins.

The worst-case time complexity of this algorithm is $O(x_v \cdot |\widetilde{\mathcal{L}}| \cdot a_v \cdot g_v)$.

6.4   Experimental Results

We implemented Sortex in a software simulation environment. All evaluations were carried out using a 3.4 GHz Intel i7 CPU with 12 GB RAM. The architecture of single-cell screening biochip (Section 6.1) was used as a benchmark, and four biochip configurations of different complexity were adopted; see Table 6.2 ($|\mathcal{L}|$: number of flow valves; $|\tilde{\mathcal{L}}|$: number of control valves). During simulation, we consider a single cell stream ($u_b = 1$), and we set an upper bound on the number of control entities (i.e., control pins and valves); locations of these entities were specified in advance. We opt to fix the locations of the control entities since considering these entities as variable would cause the problem complexity to become untractable. Also, the channel width is always fixed and it is assumed to be 3 μm, unless stated otherwise.

We evaluate the performance of Sortex using two metrics: (1) the worst-case screening latency, measured in seconds (s); (2) the number of used control pins. In all evaluations, we consider air as a channel medium ($\mu_v = 1.983 * 10^{-5} Pa \cdot s$), and the channel dilatability is set to $\gamma = 10^{-5} Pa^{-1}$ (PDMS).

Comparison with Direct-Addressing

We compare Sortex with the direct-addressing (**DA**) method, in which every valve is addressed by a dedicated control pin. We evaluate **DA** and Sortex in terms of the screening latency and we study the convergence of latency for Sortex by varying the number of control pins. Figure 6.13(a-d) show the screening-latency results for CNF_1, CNF_2, CNF_3 and CNF_4, respectively.

Based on Figure 6.13, the general trend of the screening latency is that the latency decreases when the number of control pins is increased. With more control pins, the complexity of multiplexed control decreases, therefore the latency decreases. However, we also observe that there is still a gap between the latency for Sortex and for **DA**. Such a gap may still exist if we increase the number of control pins even further due to the complex connectivity of the multiplexed control compared to the **DA** control.

Although our simulation results are associated with a single-cell stream ($u_b = 1$), we expect that increasing u_b will have a similar impact as increasing

TABLE 6.2

Biochip configurations used in evaluation.

Conf.	(n_b, m_b)	$|\mathcal{L}|$	$|\tilde{\mathcal{L}}|$	Chip Complexity
CNF_1	(6,4)	52	72	Low
CNF_2	(8,6)	118	456	Low-Medium
CNF_3	(10,8)	208	2,784	High-Medium
CNF_4	(12,8)	272	6,132	High

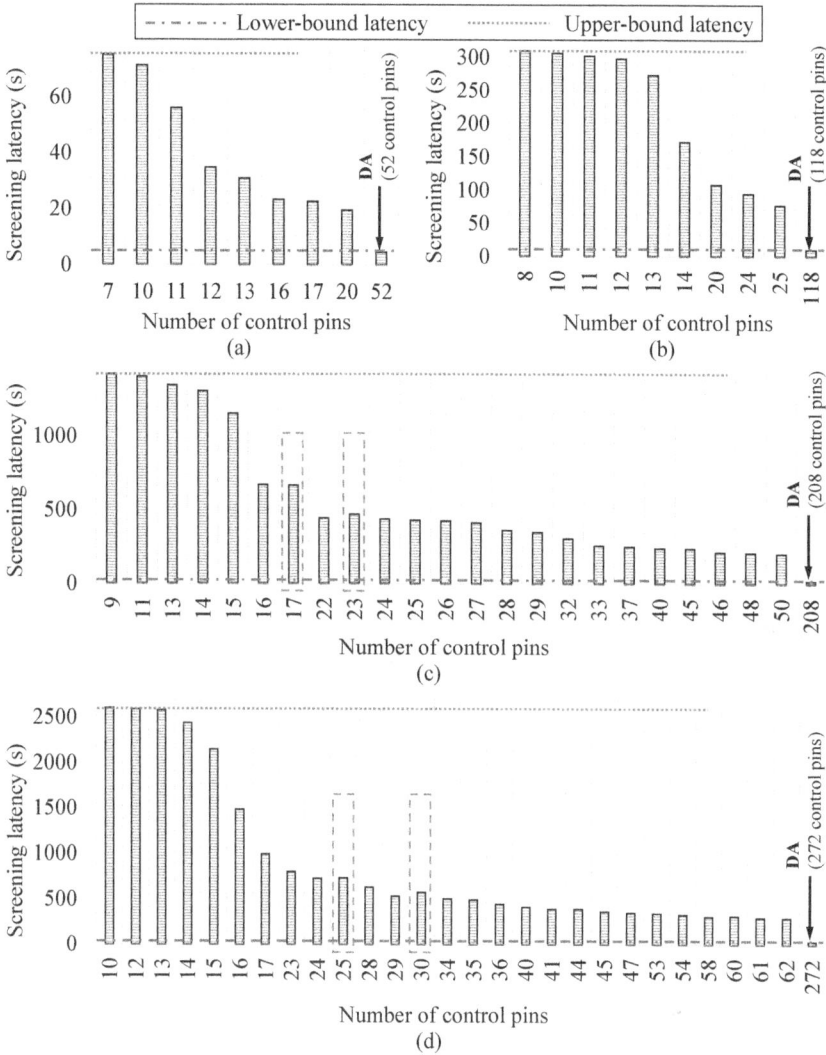

FIGURE 6.13
Comparison between Sortex and **DA**: (a) screening latency (ly) for CNF_1, (b) ly for CNF_2, (c) ly for CNF_3, (d) ly for CNF_4.

n_b and m_b on the trade-off between the number of control pins and the screening latency. As shown in Figure 6.13, an increase in n_b or m_b or both is inherently accompanied by an increase in the number of intermediate flow channels and valves; thus, making valve control based on a limited number of pins more costly, i.e., leads to a higher screening latency. We observe such an impact by comparing the results of Figure 6.13(a), Figure 6.13(b), Figure 6.13(c), and

Figure 6.13(d). Similarly, based on the sorting-network design in Figure 6.3(c), we also predict that an increase in u_b requires additional flow channels and valves to be incorporated, leading to a parallel increase in the complexity of valve control. As a result, we believe that increasing u_b will also lead to the same results as in Figure 6.13, but with a higher latency profile.

By examining the latency results in Figure 6.13(c-d), we observe some cases in which using more control pins can lead to higher latency values. Such counter-intuitive results, highlighted with dotted boxes, can be seen, for example, in Figure 6.13(c) when using 23 instead of 22 control pins, and in Figure 6.13(d) when using 25 instead of 24 control pins. We interpret this outcome as follows. First, by fixing the total number of control pins, multiplexed control of complex biochip designs can be developed with different configurations of primary pins, demultiplexing pins, and control valves, entailing different performance characteristics. To investigate this finding, we run Sortex hundreds of times on CNF_4 while scanning several values of the latency constraint ly_{max}; the relationship between the resulting latency ly and the associated control pins is shown in Figure 6.14. We observe that with a specific pin-count, e.g., 16 control pins, Sortex can provide different latency characteristics, depending on the adopted pin-valve configuration during simulation.

Second, achieving optimal results not only requires "local" balancing of CS trees (i.e., for each primary pin independently), but it also necessitates global balancing across all primary and demultiplexing pins—this requirement cannot be met using our greedy method, which seeks to connect primary pins and balance their associated CS trees one at a time. As a result, optimal combinations of control pins and control valves may not be realized in Sortex.

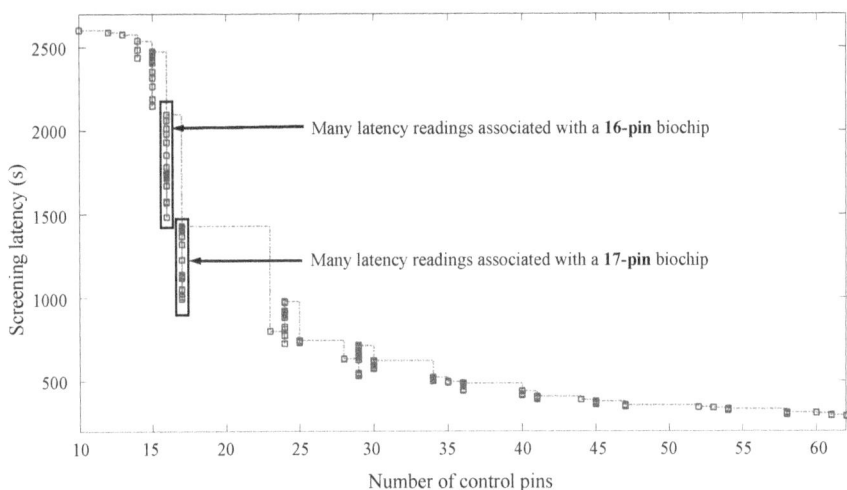

FIGURE 6.14
Detailed analysis of the screening latency for CNF_4.

While pin-constrained design has been studied earlier [101, 269, 275], the trade-off analysis presented in this section cannot be applied to these prior methods; therefore a meaningful comparison is not applicable in this case. First, recall that our objective is to develop a pin-constrained design that also supports dynamic reconfiguration, which is a key requirement in today's biochemical frameworks. A design that does not support dynamic reconfiguration cannot execute single-cell screening. Therefore, a major requirement of the control design is that each valve must be independently addressable. Pin-sharing methods, which connect multiple valves to the same pin, do not fulfill this requirement, thereby they limit system reconfigurability.

Second, as described above, the proposed Sortex algorithm is not optimal; Sortex may not provide the optimal number of control pins nor the optimal configuration of multiplexed control (the connectivity of control pins) that are suitable for a certain screening latency. Using prior pin-sharing methods is not possible unless we decide in advance the sequence of valve actuations in the reconfigurable chip. Even if we can predict the sequence of valve actuations before the start of single-cell screening, an optimization that uses pin sharing may lead to a higher pin-count reduction, but it may not lead to efficient utilization of biochip channels during the execution of single-cell screening due to pin-sharing constraints. Also, the higher the pin-count reduction, the lower the biochip utilization. Hence, we realize that the goals of pin-count reduction are contradictory with the goals of design reconfigurability and utilization unless new design paradigms such as multiplexed control are used.

Finally, even though Sortex in not optimal, it can be used to analyze the relationship between screening latency and the number of control pins, unlike previous methods. As shown in Figure 6.14, despite the fact that Sortex may not provide the lowest number of control pins in some cases, the trend of this relationship shows that screening latency decreases when the number of control pins is increased. An optimal version of Sortex may help in eliminating the irregularities in this relationship, thus providing a "smooth" curve that replaces the one in Figure 6.14.

Analysis of the Distributed-RC Model

We investigate the sensitivity of the distributed-RC model and its impact on the synthesis performance. To perform this study, we analyze the impact of changing the number of segments in the channel delay model, denoted by n_{sig} in Section 6.2, on the worst-case screening latency and the number of utilized pins. Figure 6.15(a-d) shows the screening-latency results for CNF_1, CNF_2, CNF_3, and CNF_4, respectively, while varying the number of segments (n_{sig}) and the number of pins.

As shown in Figure 6.15, it is obvious that reducing n_{sig} from 80 segments down to only 2 segments triggers significant degradation in synthesis performance (screening latency) and the performance gap increases as the number of control pins decreases. This result corroborates our findings in Section 6.2.

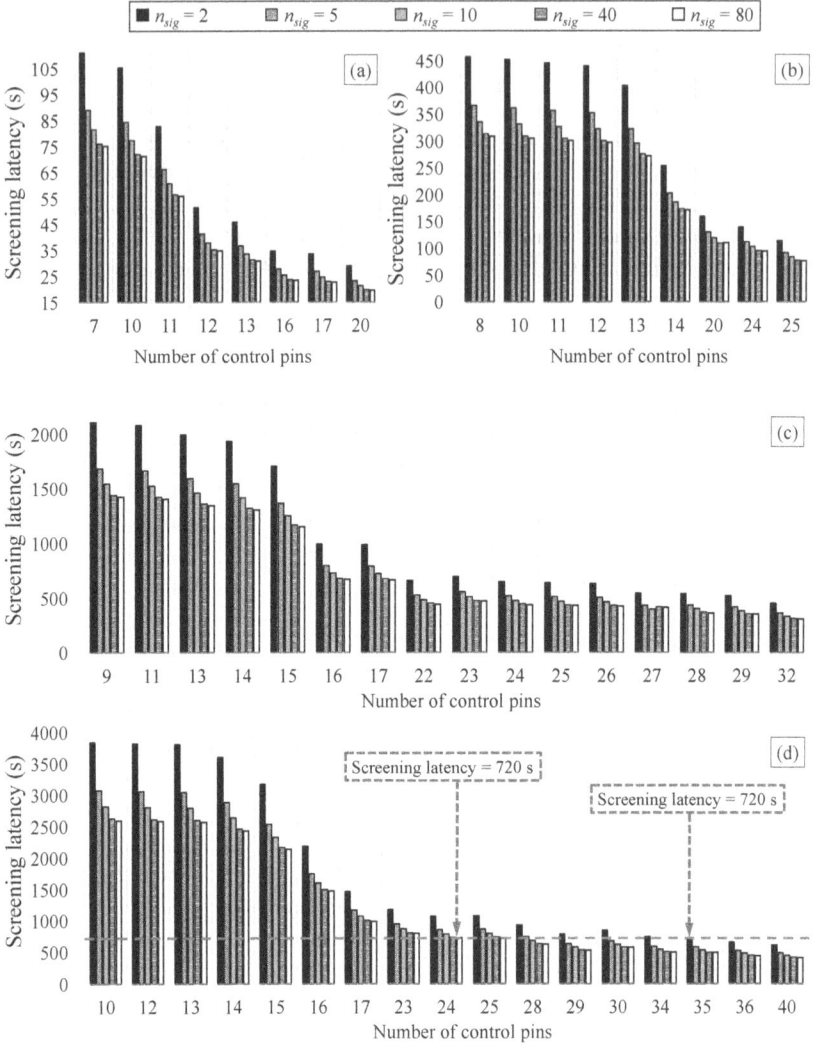

FIGURE 6.15
Impact of the number of model segments n_{sig} on the performance using (a) CNF_1, (b) CNF_2, (c) CNF_3, and (d) CNF_4.

In fact, we can identify cases where adopting fewer segments can force Sortex to use more control pins to satisfy the latency constraint. For example, by adopting a 2-segment delay model in CNF_4 (Figure 6.15(d)), we can satisfy a screening-latency constraint of 720 seconds (highlighted in a dotted line) only using 35 control pins, whereas the same latency constraint is achievable in a 40-segment or an 80-segment model using only 24 control pins. Furthermore, it is also noticeable from Figure 6.15(a-d) that insignificant performance gain is achieved by increasing n_{sig} beyond 40 segments.

The performance gain we obtain by increasing n_{sig} may be impeded by an increasing computational cost. To investigate this cost, we simulated synthesis for n_{sig} =1, 2, 5, 10, 20, 40, 60, 80, 100, 120, 140, 160, 180, and 200 using all the four biochip configurations. Next, we consider the result for n_{sig} =1 as a reference and we use it to normalize the results for the rest of n_{sig} values. By using the normalized values, we then compute the average computation time across all the four configurations. Figure 6.16 shows the computation-time results, referred to as normalized computational cost, based on the above procedure.

As expected, increasing n_{sig} is accompanied by a significant increase in the computational cost; this may impact our method's capacity to support larger biochip designs. As a result, when choosing n_{sig}, it is necessary to carefully weigh the benefits of getting higher performance (by increasing n_{sig}) against the risks of losing support for highly complex designs.

Impact of Channel Width and Latency Constraint

We next evaluate the impact of chip parameters and design constraints on screening performance. We focus here on the channel-width parameter and the worst-case latency.

For each biochip configuration, we carried out two sets of synthesis simulations. In each set, we consider a specific latency constraint and report the number of control pins and the worst-case screening latency while varying the channel width. The latency constraints considered for CNF_1 are 35 s and 70

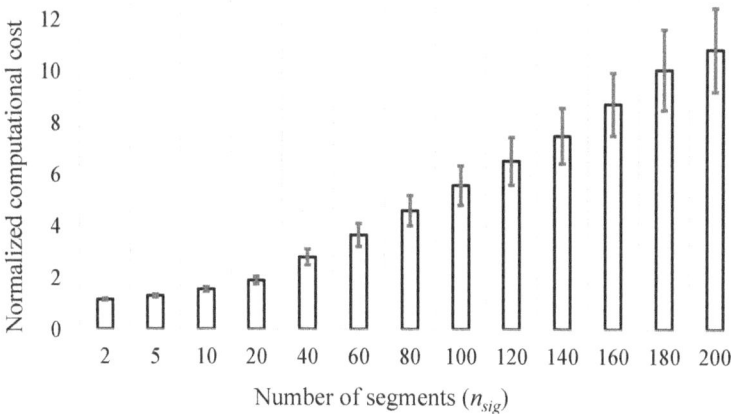

FIGURE 6.16

Impact of the number of model segments n_{sig} on the computational cost.

s, and their associated results are shown in Figure 6.17(a) and Figure 6.17(b), respectively. The latency constraints for CNF_2 are 170 s and 340 s, and their associated results are depicted in Figure 6.17(c) and Figure 6.17(d), respectively. Also, the latency constraints for CNF_3 are 294 s and 588 s, and their associated results are shown in Figure 6.18(a) and Figure 6.18(b), respectively. Similarly, for CNF_4, the latency constraints are 294 s and 588 s, and their associated results are shown in Figure 6.18(c) and Figure 6.18(d), respectively.

We first investigate the number of control pins in Figures 6.17–6.18. In all eight cases, we observe that fewer control pins are needed (to satisfy the latency constraint) when the channel width is increased. This result is intuitive because using a wider channel causes the Elmore delay for fluid transport to

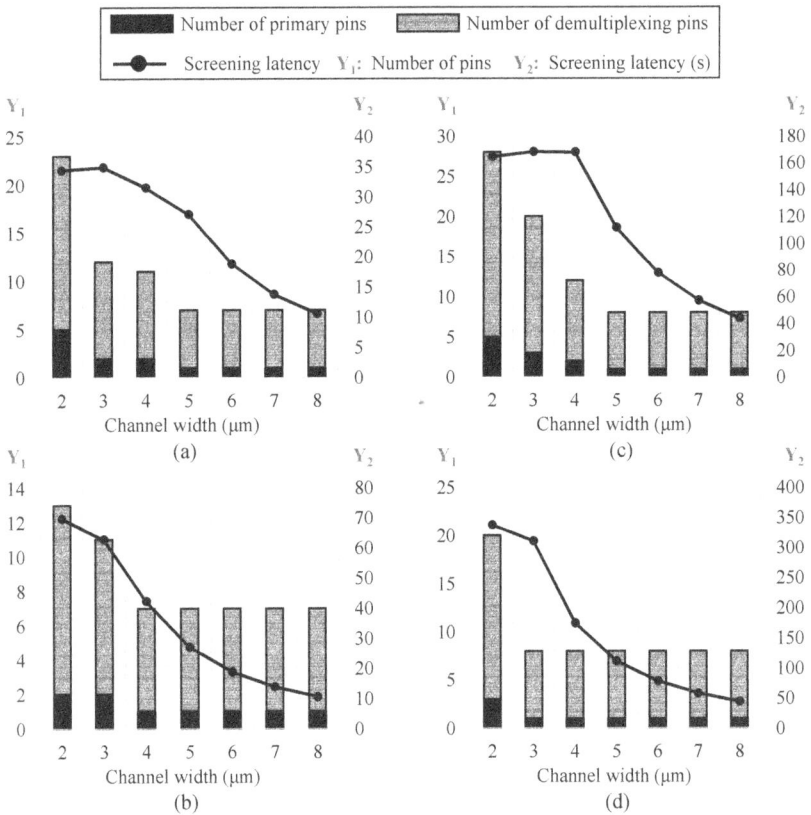

FIGURE 6.17
Impact of channel width and latency constraint ly_{max} on the performance: (a) CNF_1, $ly_{max} = 35$ s; (b) CNF_1, $ly_{max} = 70$ s; (c) CNF_2, $ly_{max} = 170$ s; (d) CNF_2, $ly_{max} = 340$ s.

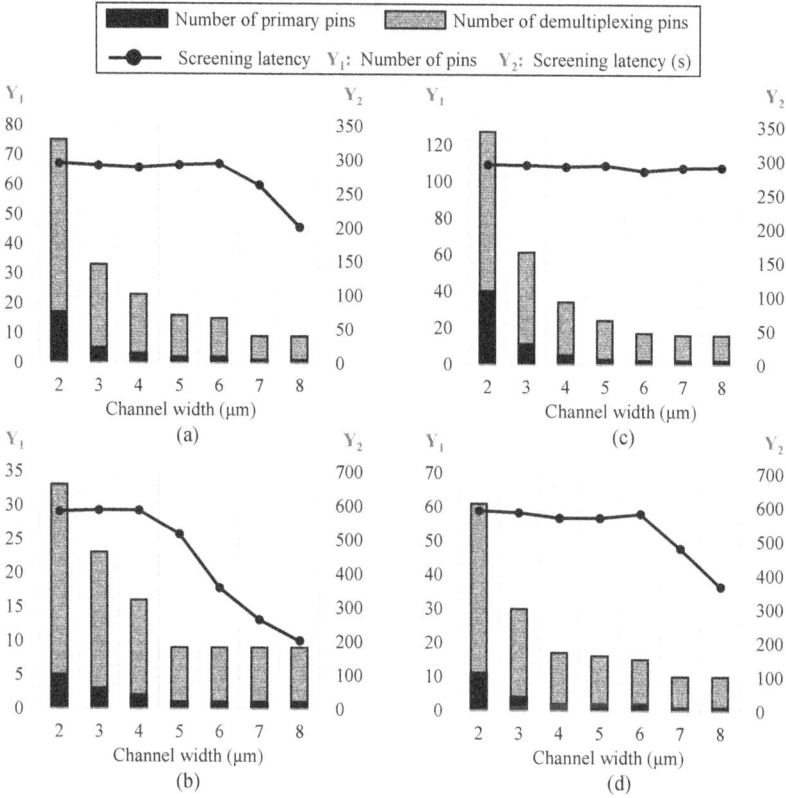

FIGURE 6.18
Impact of channel width and latency constraint ly_{max} on the performance:
(a) CNF_3, $ly_{max} = 294$ s; (b) CNF_3, $ly_{max} = 588$ s; (c) CNF_4, $ly_{max} = 294$
s; (d) CNF_4, $ly_{max} = 588$ s.

be minimized, thus reducing the number divide-and-conquer procedures in
Sortex (Section 6.3).

Second, we investigate the screening latency reported in Figures 6.17-6.18.
We observe that in Figure 6.18(b), for example, that the latency is nearly
constant when the channel width is increased from 2 μm to 4 μm. Since the
increase in the channel width leads to pin-count reduction, the impact of
pin-count reduction on increasing the latency is equivalent to the impact of
increasing the channel width on reducing the latency. We also observe that
the latency is decreased in Figure 6.17(b) when the channel width is larger
than 4 μm as no further pin-count reduction can be achieved, i.e., the minimum number of control pins is reached ($\lceil \log_2 x_v \rceil + 1$). The same argument
also applies to all other cases in Figures 6.17-6.18. These results show the

impact of the channel width, particularly narrow channels, on the screening performance.

We finally investigate the impact of changing the latency constraint on the number of control pins. As expected, by relaxing the latency constraint (i.e., increasing its value), the number of pins is reduced. For example, by comparing Figure 6.18(c) with Figure 6.18(d), the number of pins is decreased from 117 to 61 for a design with a 2 μm-channel width.

6.5   Chapter Summary

We have introduced a timing-driven design method for a pin-constrained RFB that performs single-cell screening. The proposed design synthesizes multiplexed control that actuates biochip valves independently while using shared control pins; thus minimizing the number of control pins. We have also described our methodology for supporting high-throughput single-cell screening, which is based on satisfying a screening-latency constraint. According to this methodology, we have first analyzed parameters of fluid dynamics in biochip micro-channels using CFD simulation. Next, we have employed these parameters, namely hydraulic resistance and compliance, to construct a delay model of pressure-driven transport; such a model is adopted for targeting high-throughput single-cell screening. The proposed method has been evaluated based on the screening delay and the pin count.

Part III

Parameter-Space Exploration and Error Recovery

7

Synthesis for Parameter-Space Exploration: Synthetic Biocircuits

In previous chapters, we presented optimization techniques for on-chip execution of gene-expression analysis. We specifically focused on design problems that arise at the second and third stages of the biomolecular workflow (Section 1.6)—we developed cyber-physical microfluidic solutions that dynamically process a collection of heterogeneous samples and enable the quantification of gene expression among these samples. In this chapter, we move a step further to address important design problems arising at the first stage of the biomolecular workflow. At this stage, the realization of gene-expression analysis is no longer the main goal; instead, our main goal now is to provide a microfluidic solution that tunes complex gene-regulatory parameters and thereby enables parameter-space exploration of various genomic organisms. This capability is a true enabler for optimizing synthetic biosystems.

Biological systems are inherently cyber-physical because they can sense and react to the surrounding environment. To exploit this phenomenon for practical applications, synthetic biology has emerged over the past decade with the goal of engineering biological parts that can perform new and useful functions. Applications of synthetic biology include environmental monitoring, production of therapeutics, creation of new material, etc. [278]. Such applications can be assembled via pathways (biocircuits) that function like integrated circuits.

Recent advances in microfluidic technology offer attractive platforms that efficiently emulate complex biocircuits on a lab-on-chip. Biocircuits are governed by a number of gene-regulatory parameters, and a fundamental challenge in synthesizing them is the lack of automated design tools that are capable of exploring the large parameter space efficiently, while optimizing synthesis time and reagent cost. A microfluidic assay, referred to as biocircuit-regulatory scanning (BRS), can be used to produce representative reagent-mixtures needed for modulating the parameter space.

In this chapter, we introduce an optimization flow for systematic exploration of the parameter space of a biocircuit [106]. The proposed flow, named BioScan, includes:

- a statistical approach to select suitable volumetric ratios of reagents in each mixture droplet;

- a high-level synthesis method that generates the specifications of a BRS-assay;

- a physical-level synthesis technique that implements the BRS-assay on a micro-electrode dot-array (MEDA) biochip; and

- a Dirichlet-regressor to control the accuracy of the parameter space.

The rest of the chapter is organized as follows. Section 7.1 presents an overview of the PSE problem and prior synthesis methods. Section 7.2 describes the proposed flow for PSE. A statistical method for space sampling is presented in Section 7.3. Next, Section 7.4 describes the proposed synthesis framework used to implement mixture production on MEDA biochips. Subsequently, details of the high-level synthesis and the physical-level synthesis are introduced in Section 7.5 and Section 7.6, respectively. Experimental evaluation is presented in Section 7.7 and conclusions are drawn in Section 7.8.

7.1　Background

Parameter-Space Exploration

A synthetic biocircuit consists of a regulatory network of genetic parameters that interact through gene-expression activities. For example, consider a simple biocircuit that has the following parameters (Fig. 7.1(a)) [94]: (1) PX_1, a gene that expresses "AraC" (a transcription activator); (2) PX_2, a gene that expresses "TetR" (a transcription repressor); (3) PX_3, a gene that expresses GFP (the fluorescent protein output of the circuit). As shown in Fig. 7.1(a), the activity of "AraC" regulates (through activation) both PX_2 and PX_3, but the activity of "TetR" in PX_2 regulates (through repression) PX_3. Such interactions are demonstrated in the regulatory network diagram in Fig. 7.1(b). Clearly, there are two incoherent regulations of GFP expression at PX_3. As a result, careful modulation of PX_1, PX_2, and PX_3 is necessary to achieve the desired function.

Hence, an essential step toward reliable performance of synthetic biocircuits is to maintain the expression of circuit parameters in an optimal range. Such an optimization can be realized using a technique known as *parameter-space exploration* (PSE) [94], which aims to scan all possible combinations of circuit parameters. In practice, PSE can be implemented by generating numerous droplets (a droplet is referred to as a *mixture MX_i*) of equal volume with varying volumetric ratios of biochemical reagents. A volumetric ratio of a reagent j in MX_i is referred to as a *concentration factor* (CF) and is denoted by $cx_{(i,j)}$, which can variably stimulates gene transcription and translation of biocircuit parameters. Such an assay is referred to as biocircuit-regulatory scanning (BRS). For example, to study the parameter space of the circuit

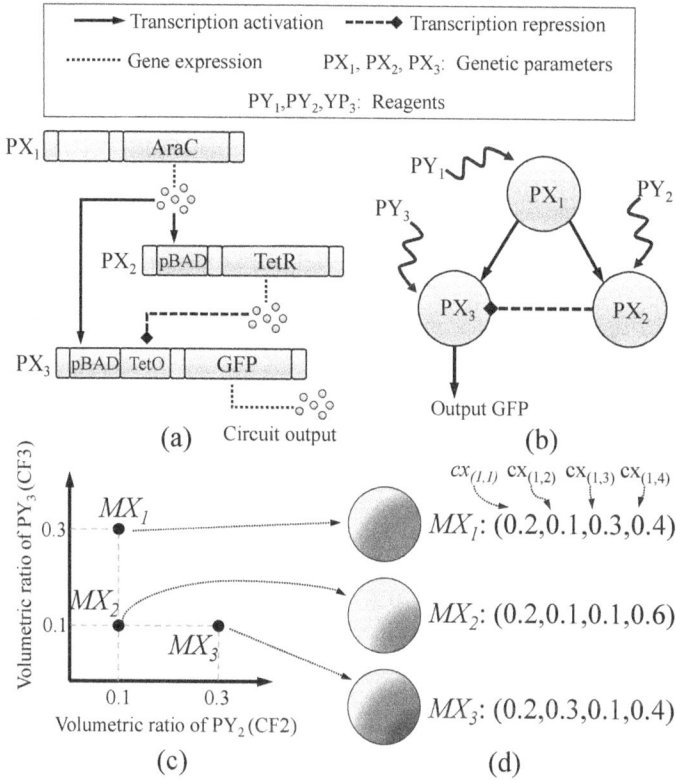

FIGURE 7.1
Study of a 3-parameter biocircuit [94]: (a) Schematic of the biocircuit; (b) a gene-expression regulatory network representation of the biocircuit; (c) CF space associated with the biocircuit (reduced to a 2-D space); (d) CF profiles of three mixtures used for biocircuit PSE.

in Fig. 7.1(a), we can implement a BRS assay that generates three mixtures MX_1, MX_2, and MX_3, as shown in Fig. 7.1(c). Each mixture constitutes four types of reagents: (1) PY_1, which modulates parameter PX_1; (2) PY_2, which modulates parameter PX_2; (3) PY_3, which modulates parameter PX_3; (4) distilled water PY_4, which is used to ensure the droplet volume remains constant. Also, each mixture contains a unique combination of CFs, referred to as *CF profile*, as shown in Fig. 7.1(d). For simplicity, we allow PY_1 to have a fixed CF value ($cx_{(i,1)} = 0.2$) among all mixtures, whereas the CF values of PY_2, PY_3, and PY_4 are changed. The CF profiles of the three mixtures are selected such that their representation in the *CF space* can fill as much space as possible; see Fig. 7.1(c). These mixtures are incubated and then analyzed using fluorescence detectors to record GFP expression.

Ideally, by generating numerous droplets, the range of any constituent CF can span the full concentration range between 0% and 100% of a droplet volume. This process, however, is cost-prohibitive, especially with the exponential growth in the number of parameters. Hence, a technical barrier facing PSE is the development of a systematic methodology that enables dense scanning of a CF space. This methodology, moreover, requires an experimental framework that offers fine-grained mixing capabilities to enable biochemical composition of CFs. Recently, a framework based on a flow-based biochip supported with dynamic control of reagent-flow rates has enabled PSE for the circuit of Fig. 7.1(a) [94]. Despite the novelty of this design, it suffers from the following drawbacks:

(1) Scalability Limitation: The flow-based solution performs passive mixing, which relies only on molecular diffusion of fluids without any external energy. However, complete diffusion of several reagents can be accomplished only if a long channel is used [126]—channel length is proportional to the number of reagents. Besides, passive mixing in flow-based systems is too slow compared to other active mixing schemes.

(2) Manual Configuration: The above design utilizes sinusoidal flow rates of reagents with pre-computed function periods to scan a large volume of the CF space. However, different biocircuits may have different numbers of parameters and therefore the number of reagents may also change. As the number of reagents is varied, a new set of sinusoidal functions must be computed; such a process is time-consuming and it can be a significant challenge when a large number of reagents is involved.

(3) Abandoned CF Subspace: A fixed droplet volume cannot be maintained unless an additional input flow of distilled water (DW) is introduced. Note that the flow-rate function of DW is computed based on the other sinusoidal functions. This technique allows all mixtures to have equal volume; however, they contain large amounts of DW and therefore this approach causes a large CF subspace to be abandoned. For example, by applying the above technique to a 3-CF setting, the value of each CF can vary only between 0% and 100%/3=33% (Fig. 7.1(c)).

To overcome the above limitations, we need a PSE microfluidic framework that offers reconfigurable mixing and droplet-actuation capabilities, thus enabling flexible composition of several biochemical reagents. This requirement can be met using a digital microfluidic biochip (DMFB), which allows the manipulation of nanoliter droplets using an array of electrodes (Section 1.1). We consider a specific architecture of digital microfluidics referred to as the micro-electrode-dot-array (MEDA) architecture, which can support mixing of various reagents with considerably higher resolution, as described in Section 1.1.

Prior Sample-Preparation Techniques

Early research on sample preparation focused on optimizing the dilution process for a single sample with the goal of minimizing the amount of waste droplets [213, 61, 201]. In [148], the process of sample dilution has been optimized using the $(m_x : n_x)$ mixing model offered by MEDA. However, these methods are not capable of handling mixtures that contain three or more reagents.

To support dilution gradients in quantitative analysis, multi-target sample-preparation techniques have been introduced [175, 200, 98]. These techniques generate multiple droplets of the same sample, but with different concentration levels. Each droplet therefore contains only a sample and a buffer solution. However, these methods are limited to (1 : 1) mixing and they cannot support the preparation of multiple mixtures that constitute a large number of reagents.

For producing a desired multi-reagent mixture, synthesis methods have been developed to generate a bottom-up mixing tree that encodes the successive composition of reagent volumetric ratios [215, 24]. These methods can be easily adapted to our space-exploration problem by running multiple iterations of the algorithm; every iteration specifies the mixing of an individual mixture. This approach, however, may lead to a significant increase in the amount of waste droplets and in the protocol completion time.

To make DMFBs useful for dense PSE in synthetic biology, we need a new top-down synthesis methodology that allows concurrent production of several mixtures with maximum precision, especially in the presence of reagent-usage constraints.

7.2 PSE Based on MEDA Biochips

Recall that the current PSE technique for synthetic biology suffers from scalability limitation, complex manual configuration, and abandonment of a large CF subspace. Our goal is to develop a design flow that addresses the above drawbacks; thus enabling systematic PSE for any biocircuit model. Our methodology for achieving this goal is based on four key attributes; Table 7.1 describes these attributes and explains how they address the above drawbacks. The assessment of these attributes is described in Section 7.7.

To realize the above attributes, we propose the design flow shown in Fig. 7.2. First, to enable systematic scanning of any k-dimensional CF space, a statistical sampling approach is used to generate tm points in the CF space (Step 2), where tm is initially specified by the user. By using this sampling scheme, we ensure that the scanning process covers all regions of the CF space with equal probability (i.e., maximizing space filling). Second, having computed the CF profiles of the target tm mixtures, an integer linear programming (ILP)-based synthesis method is presented to implement a BRS assay, which specifies the mixture-production strategy for all the samples on a MEDA biochip (Steps 3). This method can be optimized to reduce the usage

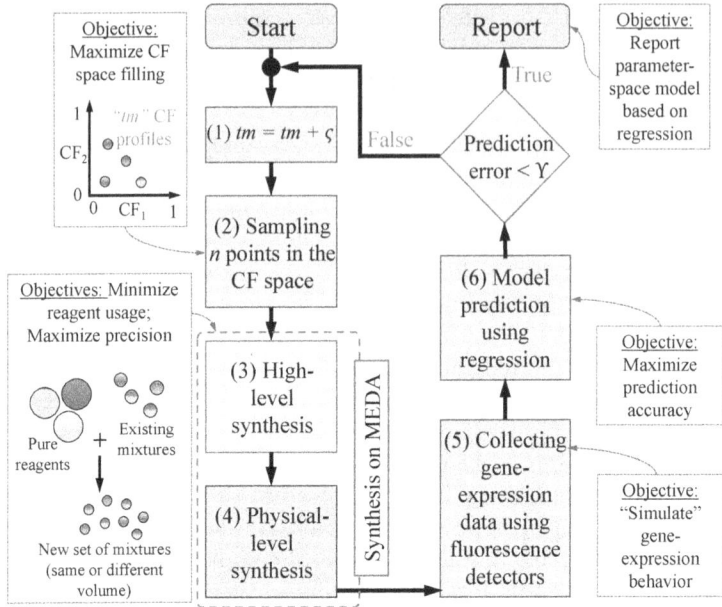

FIGURE 7.2
The proposed design flow.

of reagents and to maximize the precision of the mixture-production process. Third, the mixture-production strategy is implemented on MEDA using physical-level synthesis (Step 4) and data related to gene-expression analysis for all tm samples is collected using on-chip sensors (Step 5)—from a design-automation perspective, it suffices to "simulate" the behavior of gene expression to generate appropriate gene-expression labels. Finally, the obtained data is processed using regression analysis to derive the model that represents the parameter space (Step 6).

TABLE 7.1
Attributes of the proposed flow.

Attributes	Benefits
MEDA-driven	Reconfigurable mixing and droplet-actuation capabilities; fine-grained composition of several biochemical reagents.
Selective	Scanning any k-dimensional CF space.
Iterative	Support for CF spaces with various complexities.
Statistical	Ensures that space scanning covers all regions of the CF space; evaluates model-prediction accuracy.
Co-optimized	Co-optimization for different settings of cost constraints (e.g., reagent usage) and precision requirements.

A major challenge with the above sequence is the selection of a reasonable number of sample points tm in Step 1. Note that the value of tm depends on the CF space complexity, which cannot be predicted until the PSE completes the above sequence at least once. If a small value of tm is selected, model overfitting and poor prediction may occur. On the other hand, if a large value of tm is selected, a large amount of reagents may be unnecessarily consumed. To address the above problem, an iterative approach is employed, allowing additional mixtures to be prepared (Step 1) if model prediction needs to be enhanced.

A potential drawback of this iterative approach is that repetitive generation of new sets of mixtures may lead to significant reagent usage; thus leading to an increased cost. A solution to this problem is to generate the new mixtures not only using reagents stored on chip, but also by exploiting mixtures generated during previous iterations of PSE. This approach may reduce reagent usage, which is useful especially in settings where reservoir storage is constrained. This scheme, however, is computationally challenging and therefore it needs to be designed carefully through proper modeling and synthesis. We discuss the details of statistical sampling (Step 2) in Section 7.3, the high-level synthesis (Steps 3) in Section 7.5, the physical-level synthesis (Steps 4) in Section 7.6, and the regression analysis (Step 6) in Section 7.7.

By comparing the above design flow with Table 7.1 , we observe that Steps 3 and 4 provide "MEDA-driven" and "Co-optimized" attributes of the framework, whereas "Selective" and "Statistical" attributes are captured by Steps 2 and 6.

7.3 Sampling of Concentration Factor Space

In this section, we present a method of stratified sampling and explain the mapping to the CF space.

Stratified Sampling: Latin Hypercubes

Consider a continuous space $\mathcal{X} \subset [0, 1)^{rg}$ constructed using input variables $RR_x = \{RR_{(x,1)}, RR_{(x,2)}, ..., RR_{(x,rg)}\} \in \mathcal{X}$. We seek a sampling technique that selects an ensemble of tm samples $x = \{x_1, x_2, ..., x_{tm}\} \subset \mathcal{X}$, where $x_i = \{x_{(i,1)}, x_{(i,2)}, ..., x_{(i,rg)}\}|\ x_{(i,j)} \in RR_{(x,j)}$, such that it provides enhanced space-filling properties, i.e., it must ensure that all the space regions are roughly evenly sampled. Theoretically, space-filling criterion can be assessed using the Euclidean maximin (E_m) distance, which is defined to be the smallest distance between a pair of points in the sampling space [54]. Let

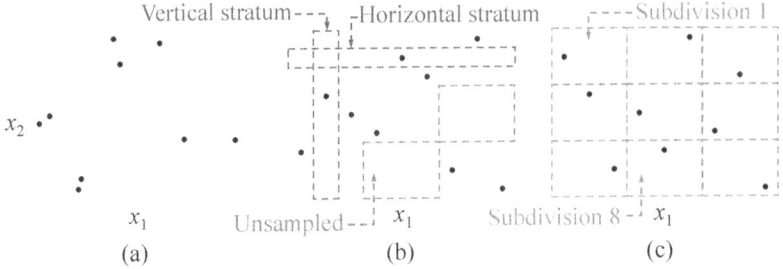

FIGURE 7.3
Sampling 9 points in a 2-D space using: (a) random sampling, (b) LHS, (c) OA-LHS ($ls = 9$).

$dd = [dd_{(1,2)} \; dd_{(1,3)} \; ... \; dd_{(tm-1,tm)}]$, where $dd_{(i,k)}$ is computed as follows:

$$dd_{(i,k)} = \sqrt{(x_{(i,1)} - x_{(k,1)})^2 + ... + (x_{(i,rg)} - x_{(k,rg)})^2}; \quad i < k$$

The E_m distance is the smallest $dd_{(i,k)} \in dd$. A larger value of E_m indicates that the distance between the closest points is big, indicating a better space-filling property.

It is known that pseudo-random sampling methods such as Monte Carlo sampling may result in poor space filling [186]. We overcome this limitation by using a classical *stratified* sampling technique known as *Latin Hypercube sampling* (LHS) [54], which divides the range of each input variable $RR_{(x,j)} \in RR_x$ into tm equally probable strata and samples once from each stratum. For each $RR_{(x,j)} \in RR_x$, the tm sampled input values are assigned at random to the tm strata, with all $tm!$ possible permutations being equally likely. This process is applied independently to each variable $RR_{(x,j)}$. LHS can be further enhanced by dividing the sampling space into ls equally probable subdivisions, and each subdivision is sampled with the same density; such an enhancement is known as *Orthogonal Array-based Latin Hypercube sampling* (OA-LHS) [253]. Figure 7.3 compares these methods using $tm = 9$ samples in a 2-D ($rg = 2$) space.

To apply OA-LHS to the CF space to generate a set of tm CF profiles $\{MX_1, MX_2, ..., MX_{tm}\}$, we need to take into consideration that the range of any input variable $RR_{(x,j)}$ (which represents the range of a CF) is defined as $RR_{(x,j)} \in [0,1)$. However, an additional requirement still applies; that is, ensuring that the sum of CFs in any CF profile (mixture) must equal 1. In other words, for any mixture MX_i, the following condition must be fulfilled: $\sum_{j=1}^{rg} x_{(i,j)} = 1$. Such a requirement cannot be fulfilled through direct application of OA-LHS to the CF space. We explain how to overcome this limitation in the next subsection.

In addition to OA-LHS, a number of statistical sampling techniques, known as *quasi-Monte Carlo* (QMC) techniques, can also be adopted for sampling the

CF space [186]. QMC methods such as Sobol' sequences apply a deterministic scheme to produce highly uniform samples of the unit hypercube; therefore, they offer enhanced space-filling properties. Details on QMC methods can be found in [169].

Mapping to CF Space

Consider a continuous space $\mathcal{Q} \subset [0, 1)^{rg}$ constructed using input variables $RR_q = \{RR_{(q,1)}, RR_{(q,2)}, ..., RR_{(q,rg)}\} \in \mathcal{Q}$. A point q in the space $\mathcal{Q}$ is defined as $q = \{q_1, q_2, ..., q_{tm}\} \subset \mathcal{Q}$, where $q_i = \{q_{(i,1)}, q_{(i,2)}, ..., q_{(i,rg)}\} | q_{(i,j)} \in RR_{(q,j)}$, and it must satisfy the following condition $\sum_{j=1}^{rg} q_{(i,j)} = 1; \forall i \in \{1, ..., tm\}$. The space $\mathcal{Q}$ can be graphically represented using a simplex that is formed using a barycentric coordinate system [222]. Figure 7.4(a) depicts the shapes of 1-simplex (2-D space) and 2-simplex (3-D space).

To adapt OA-LHS to the "simplex" CF space, we seek a mapping function f_{sim} that is defined as $f_{sim} : \mathcal{X} \to \mathcal{Q}$. A trivial implementation of the function f_{sim} is to uniformly sample points in the space $\mathcal{X}$ using OA-LHS then re-scale the points using the relation $q_{(i,j)} = \frac{x_{(i,j)}}{\sum_{j=1}^{rg} x_{(i,j)}}$; this method is referred to as *scaling-based mapping* . However, this approach severely degrades space filling

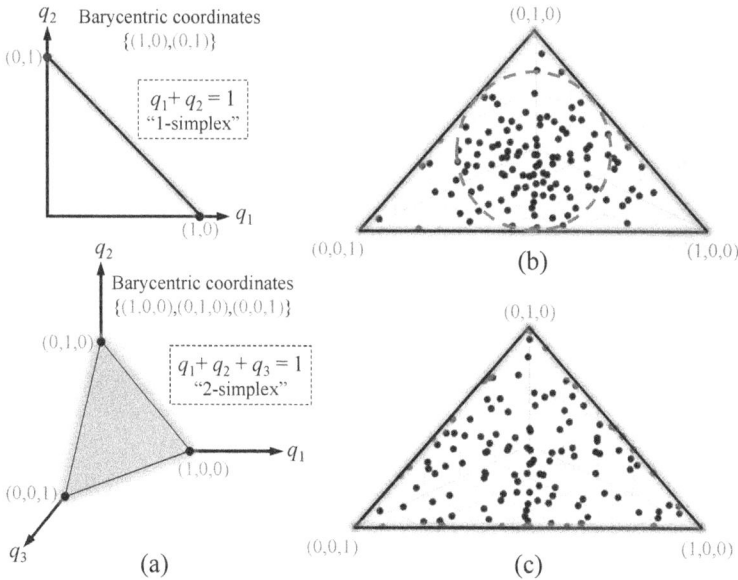

FIGURE 7.4
Mapping OA-LHS-sampled data to simplex CF space: (a) graphical representation of 1-simplex and 2-simplex; (b) scaling-based mapping leads to poor space filling; (c) Dirichlet-based mapping ($\tilde{\alpha} = 3$) preserves enhanced space filling.

since it tampers with the stratification property; see Figure 7.4(b). We develop an alternative implementation of f_{sim} that does not change the uniformity of the sampling by using the *Dirichlet distribution*, which is an exponential family distribution over a simplex, i.e., positive vectors that sum to one. Formally, if $x_{(i,j)} \in [0, 1)$ is sampled using OA-LHS based on a uniform distribution, then $f_{sim} : x_{(i,j)} \rightarrow q_{(i,j)}$ can be defined as

$$z_{(i,j)} = \frac{x_{(i,j)}^{\tilde{\alpha}-1} e^{-x_{(i,j)}}}{\Gamma(\tilde{\alpha})}; \quad q_{(i,j)} = \frac{z_{(i,j)}}{\sum_{j=1}^{rg} z_{(i,j)}}$$

Note that $\sum_{j=1}^{rg} q_{(i,j)} = 1$. A key property of this mapping, referred to as *Dirichlet-based mapping*, is that it applies an affine transformation that does not alter the space-filling properties in the simplex space [222]; see Figure 7.4(c).

The above sampling and mapping processes can be applied to any k-dimensional CF space, i.e., they are scalable. To further improve space filling while generating tm CF profiles, we run the above method several times and report the result associated with the largest E_m.

7.4   Synthesis Methodology

We next present a synthesis method that enables composition of mixtures on a MEDA biochip.

Definitions

We introduce the following definitions that are used in our synthesis method.
Target Mixtures: At any iteration of the flow in Figure 7.2, BioScan aims to compute the synthesis solution associated with tm new *target mixtures* $\{MX_1^t, MX_2^t, ..., MX_{tm}^t\}$. The CF profile of a mixture MX_i^t is defined as the set $\{cx_{(i,1)}^t, ..., cx_{(i,rg)}^t\} = \{q_{(i,1)}, ..., q_{(i,rg)}\}$; the values $q_{(i,j)}$ are obtained from the sampling step (Section 7.3). Also, each MX_i^t is characterized with a "reference" volume H_i^t that the synthesis framework aims to produce. Note that a potential solution for reducing reagent usage during mixture production is to reduce H_i^t.

For simplicity, we consider the same reference volume for all target mixtures, i.e., $\forall i : H_i^t = H^t$, and we also assume that a droplet volume is represented as a natural number that is multiple of the droplet volume on a single microelectrode. This abstraction is based on the fact that droplets are sandwiched between two plates; therefore volume expansion is only possible along the horizontal plane (Figure 1.5(a)). We refer to this measure of volume as a microelectrode (MC) unit.

Source Mixtures and Reagents: To generate target mixtures, a number of sm *source mixtures* $\{MX_1^s, MX_2^s, ..., MX_{sm}^s\}$, generated in a previous iteration, are used along with a number of rg reagent fluids $\{RG_1, RG_2, ..., RG_{rg}\}$. The CF profile of a source mixture MX_k^s is defined as the set $\{cx_{(k,1)}^s, ..., cx_{(k,rg)}^s\}$. Also, the CF profile of a reagent fluid G_j is defined as the set $\{cx_{(j,1)}^g, ..., cx_{(j,rg)}^g\}$; where $cx_{(j,a)}^g = 1$ only if $j = a$, and $cx_{(j,a)}^g = 0$ otherwise.

Aliquots: MEDA biochips enable the aliquoting of a source mixture MX_k^s of volume H_k^s or a reagent fluid RG_j of volume H_j^g into smaller droplets that can vary in volume. Hence, an aliquot of volume $\Lambda_{(i,k)} \leq H_k^s$ from a source mixture MX_k^s contributes to the generation of a target mixture MX_i^t. Similarly, an aliquot of volume $\Xi_{(i,j)} \leq H_j^g$ from a reagent RG_j contributes to the generation of the same mixture MX_i^t. Note that the lower bounds on $\Lambda_{(i,k)}$ and $\Xi_{(i,j)}$, denoted by $H_{k.min}^s$ and $H_{j.min}^g$, respectively, are controlled by the aliquoting constraints imposed by MEDA [298].

Degree of Concentration Accuracy: MEDA biochips discretize the CF space. We define parameter δ as the degree of concentration accuracy, whereby any concentration $cx_{(i,j)}^t$ can be expressed as $CX_{(i,j)}^t = \lceil cx_{(i,j)}^t / \frac{1}{\delta} \rceil$, where $CX_{(i,j)}^t$ is an integer. Similarly, $CX_{(i,j)}^s = \lceil cx_{(i,j)}^s / \frac{1}{\delta} \rceil$. For example, if $\delta = 128$, a concentration of 64% is expressed as $\lceil 0.64 / \frac{1}{128} \rceil = 82$. We use $CX_{(i,j)}^t$ or $CX_{(k,j)}^s$ to represent a CF in our synthesis flow.

Note that the largest number of target mixtures $\hat{tm}$ is impacted by δ— by using stars-and-bars combinatorics, $\hat{tm}$ can be computed as follows: $\hat{tm} = {}^{\delta-1}C_{rg-1}$, where ${}^{\delta-1}C_{rg-1}$ is the binomial coefficient $\binom{\delta-1}{rg-1} = \frac{(\delta-1)!}{(rg-1)!(\delta-rg)!}$. For example, with $\delta = 8$ and $rg = 4$, we find that $\hat{tm} = 35$.

BioScan Imprecision: A first objective of BioScan is to compute the values of $\Lambda_{(i,k)}$ and $\Xi_{(i,j)}$ that enable the production of target mixtures, considering the reference volume H^t. However, if H^t is small, the search space of all feasible values of $\Lambda_{(i,k)}$ and $\Xi_{(i,j)}$ becomes limited. As a result, the "actual" volume, denoted by $\hat{H}_i^t$, that is computed by BioScan may not be the same as H^t. The value of $\hat{H}_i^t$ can be computed as follows: $\hat{H}_i^t = \sum_k \Lambda_{(i,k)} + \sum_j \Xi_{(i,j)}$. Intuitively, the variation between $\hat{H}_i^t$ and H^t also leads to a variation between the target CF $CX_{(i,j)}^t$ and the "actual" CF, denoted by $\hat{CX}_{(i,j)}^t$. The actual CF is computed as follows: $\hat{CX}_{(i,j)}^t = \lceil \frac{\sum_k (CX_{(k,j)}^s \cdot \Lambda_{(i,k)}) + \delta \cdot \Xi_{(i,j)}}{\hat{H}_i^t} \rceil$.

To assess the impact of the variation between H^t and $\hat{H}_i^t$, we introduce a metric named *synthesis imprecision* Π, defined as

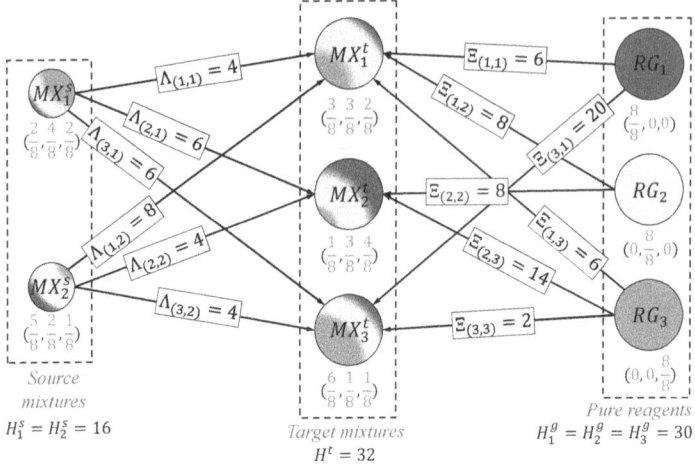

FIGURE 7.5
An example illustrating the preparation of three target mixtures using aliquots from two source mixtures and three reagent fluids ($\delta = 8$).

$$\Pi = \sum_{i,j} \Pi_{(i,j)} = \frac{1}{\delta} \sum_{i,j} \left| \hat{CX}^t_{(i,j)} \cdot \hat{H}^t_i - CX^t_{(i,j)} \cdot H^t \right|$$

$$= \frac{1}{\delta} \sum_{i,j} \left| \sum_k \left((CX^s_{(k,j)} \cdot \Lambda_{(i,k)}) + \delta \cdot \Xi_{(i,j)} - CX^t_{(i,j)} \cdot H^t \right) \right|$$

Hence, our goal is to minimize Π while also minimizing $\sum_j \Xi_{(i,j)}$ (reagent usage).

Figure 7.5 shows an example that illustrates the working principle of the synthesis method. In this example, three target mixtures of volume 32 MC units are generated using aliquots from two source mixtures of volume 16 MC units and three reagent fluids of volume 30 MC units. According to this solution, we obtain the lowest imprecision since $\Pi = 0$.

Problem Formulation

Based on the above discussion, we describe the optimization problem as follows.
Inputs: (i) The number of target mixtures tm and source mixtures sm. **(ii)** The number of reagents rg. **(iii)** The degree of accuracy δ. **(iv)** The source mixtures, each with concentration factors $(CX^s_{(k,1)}, CX^s_{(k,2)}, ..., CX^s_{(k,rg)})$ and volume H^s_k. **(v)** The target mixtures, each with reference concentration factors $(CX^t_{(i,1)}, CX^t_{(i,2)}, ..., CX^t_{(i,rg)})$ and volume H^t. **(vi)** Total liquid volume H^g_j of reagent j.

TABLE 7.2

Notation used in Section 7.5.

MX_i^t	A target mixture in the set $\{MX_1^t, ..., MX_{tm}^t\}$
MX_i^s	A source mixture in the set $\{MX_1^s, ..., MX_{sm}^s\}$
RG_i	A reagent in the set $\{RG_1, ..., RG_{rg}\}$
$cx_{(i,j)}$	The volumetric ratio (CF) of a reagent j in a mixture MX_i
H^t	The reference volume of MX_i^t
$\hat{H}_i^t$	The actual volume of MX_i^t
H_i^s	The volume of MX_i^s
H_i^g	The volume of RG_i
$\Lambda_{(i,k)}$	Volume of an aliquot generated from MX_k^s and used in MX_i^t
$\Xi_{(i,j)}$	Volume of an aliquot generated from RG_j and used in MX_i^t
Π	Synthesis imprecision
δ	Degree of concentration accuracy
E_m	Euclidean maximin distance
Υ_Ξ	Reagent usage threshold
Υ_Π	Synthesis imprecision threshold

Output: (i) Volume of aliquots $\Lambda_{(i,k)}$ and $\Xi_{(i,j)}$. (ii) Actual CF profile $\{\hat{CX}_{(i,1)}^t, ..., \hat{CX}_{(i,rg)}^t\}$ and actual volume $\hat{H}_i^t$ of every target mixture MX_i^t. (iii) BRS assay (Section 7.6).

Constraints: Aliquoting constraints of MEDA biochips (Section 7.6).

Objectives: (i) Minimize reagent usage. (ii) Minimize synthesis imprecision. (iii) Minimize the completion time of BRS.

Table 7.2 summarizes the notation described in the above problem. To solve this problem, we develop a two-stage synthesis framework. The first stage (Section 7.5) performs high-level synthesis, which is focused on computing the volume of aliquots $\Lambda_{(i,k)}$ and $\Xi_{(i,j)}$. In the second stage (Section 7.6), a physical-level synthesis algorithm is used to map the outcome of the high-level synthesis to a feasible MEDA-driven assay for mixture production. The objectives of the first stage are to minimize reagent usage and to minimize synthesis imprecision, whereas the objective of the second stage is to minimize the protocol completion time.

7.5 High-Level Synthesis

The high-level synthesis problem is harder than the generalized assignment problem (GAP), which is known to be NP-hard [192]. GAP focuses on assigning only discrete items to heterogeneous bins. Conversely, our problem allows volumetric distributions of input droplets to be assigned to target mixtures, and it also imposes budget/volume constraints on the input droplets. Nonetheless, high-level synthesis can be optimally solved by mapping the problem to an ILP model. The proposed model is developed to co-optimize synthesis imprecision and reagent usage. The model is described below.

$$\forall i \in \{1, ..., tm\}; j \in \{1, ..., rg\}; k \in \{1, ..., sm\}$$

$$\text{minimize} \quad \frac{\beta}{Z^*} \cdot \sum_{i,j} \Xi_{(i,j)} + \frac{1-\beta}{J^*} \cdot \sum_{i,j} \Pi_{(i,j)} \tag{7.1}$$

subject to:

$$H_{k.min}^s < \Lambda_{(i,k)} \leq H_k^s; \quad \sum_i \Lambda_{(i,k)} \leq H_k^s \tag{7.2}$$

$$H_{j.min}^g < \Xi_{(i,j)} \leq H_j^g; \quad \sum_i \Xi_{(i,j)} \leq H_j^g \tag{7.3}$$

$$\sum_k \Lambda_{(i,k)} + \sum_j \Xi_{(i,j)} - H^t \geq 0 \tag{7.4}$$

$$\Pi_{(i,j)} = \sum_k (CX_{(k,j)}^s \cdot \Lambda_{(i,k)}) + \delta \cdot \Xi_{(i,j)} - CX_{(i,j)}^t \cdot H^t \geq 0 \tag{7.5}$$

In (7.1), we describe the multi-objective function, which aims to minimize reagent usage (through minimizing $\sum_{i,j} \Xi_{(i,j)}$) and minimize synthesis imprecision (through minimizing $\sum_{i,j} \Pi_{(i,j)}$) simultaneously. The two objectives can be contradicting, thus proper modeling of the problem requires the two objectives to be normalized. For this purpose, we first optimize each of the objectives individually then divide the related objective term by the corresponding optimal value [167]. The parameters Z^* and J^* represent the optimal values of the first and second objectives, respectively.

The parameter $\beta \in [0, 1]$ is introduced to represent the weights for the problem objectives. If $\beta = 1$, then the objective is to minimize reagent usage regardless of the synthesis imprecision. On the other hand, if $\beta = 0$, then achieving the lowest imprecision becomes the main objective. Inequalities (7.2)-(7.3) specify volume constraints related to the aliquots. Inequalities (7.2) (Inequalities (7.3)) ensure that the volume of any source aliquot $\Lambda_{(i,k)}$ (reagent aliquot $\Xi_{(i,j)}$) is bounded. We also capture the variation between H^t and $\hat{H}_i^t$ using inequality (7.4) and the associated impact on the synthesis imprecision using inequality (7.5).

It is obvious that the value of H^t has a significant impact on the reagent usage and synthesis imprecision; therefore, it needs to be carefully selected. A small value of H^t may lead to reduction in the reagent usage, but it may also increase synthesis imprecision. On the other hand, a large value of H^t may lower synthesis imprecision, but it may also increase reagent usage, thus raising platform cost. In Algorithm 7.1, we provide a solution to this challenge; we define a sorted list $\boldsymbol{H}$ that includes multiple values of H^t and solve the

---

Algorithm 7.1 High-Level Synthesis

1: $\beta \leftarrow$ INITIALIZATION(); $Sol \leftarrow \emptyset$;
2: $\{H^s_{k.min}, H^g_{j.min}\} \leftarrow$ COMPUTELOWERBOUNDVOLUMES(H^s_k, H^g_j);
3: **for** $H^t \in \boldsymbol{H}$ **do** $\triangleright \boldsymbol{H}[1] < \boldsymbol{H}[2]$
4:    $\{\Lambda_{(i,k)}, \Xi_{(i,j)}, \Pi_{(i,j)}\} \leftarrow$ SOLVEILP($H^t, \beta, H^s_{k.min}, H^g_{j.min}$);
5:    **if** $\sum_{i,j} \Xi_{(i,j)} > \Upsilon_\Xi$ OR $\sum_{i,j} \Pi_{(i,j)} > \Upsilon_\Pi$ **then** break;
6:    **else** $Sol \leftarrow Sol \cup \{\Lambda_{(i,k)}, \Xi_{(i,j)}, \Pi_{(i,j)}\}$;
7: $BestSol \leftarrow$ SELECTLOWESTIMPRECISION(Sol); **return** $BestSol$;

---

above ILP problem once based on each value (Lines 3-4). The execution of the algorithm stops when a threshold of reagent usage, denoted by Υ_Ξ, or imprecision, denoted by Υ_Π, is exceeded (Line 5), and the solution with the lowest imprecision is selected (Line 7). We investigate the impact of H^t on the synthesis performance in Section 7.7.

7.6 Physical-Level Synthesis

In this section, we describe fluidic constraints related to MEDA and present a solution for the physical-level synthesis problem.

Droplet-Aliquoting Constraints

In principle, the synthesis of a BRS assay can be developed with the aim of executing the BRS assay on a conventional DMFB, which supports mixing and splitting operations only. MEDA biochips, on the other hand, support droplet aliquoting (Section 1.1), which allows a smaller target droplet, with principal radius of curvature W_2, to be aliquoted from another droplet, with principal radius of curvature W_1 ($W_1 > W_2$). This additional operation is a key enabler of fine-grained mixture production. Yet the main difficulty with droplet aliquoting, similar to droplet dispensing [209], is the control of the flow rate that leads to aliquot formation. Note that the original droplet, which is used before the aliquoting operation, has principal radius of curvature W_{src} that can be computed in terms of W_1 and W_2 as follows: $W_{src} \approx \sqrt{(W_1^2 + W_2^2)}$.

As shown in Figure 7.6, droplet aliquoting is performed in two steps: (1) finger formation, which initiates aliquoting by inducing a protrusion from the bigger droplet; (2) pinch-off, which breaks the protrusion to form an aliquot. According to [209], a protrusion can be successfully maintained only if the following condition is fulfilled:

$$0 < \frac{1}{W_2} - \frac{1}{W_1} < \frac{\epsilon_0 \cdot \epsilon_r \cdot (\mathcal{V}_e - \mathcal{V}_e^{th})^2}{2\gamma_{LM} \cdot d_e \cdot s_e} \tag{7.6}$$

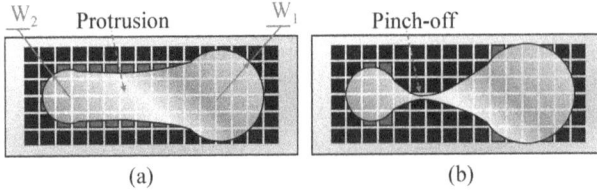

FIGURE 7.6
Steps of droplet aliquoting: (a) finger formation; (b) pinch-off.

where ϵ_0 and ϵ_r are the permittivity of free space and the relative permittivity of the insulator, respectively, $\mathcal{V}_e$ and $\mathcal{V}_e^{th}$ are the actuation voltage and the threshold voltage, respectively, γ_{LM} is the liquid-medium interfacial tension, d_e is the insulator thickness, and s_e is the spacing between the parallel plates. For simplicity, we consider $\varrho = \frac{\epsilon_0 \cdot \epsilon_r \cdot (\mathcal{V}_e - \mathcal{V}_e^{th})^2}{2\gamma_{LM} \cdot d_e \cdot s_e}$; thus we re-write inequality (7.6) as follows:

$$W_{1.max} = \frac{W_2}{1 - \varrho \cdot W_2} \tag{7.7}$$

$$W_{1.min} = W_2 \tag{7.8}$$

where $W_1 < W_{1.max}$ and $W_1 > W_{1.min}$. Note that $W_{1.max}$ must be positive, and therefore the following constraint must be satisfied:

$$W_{2.max} = \frac{1}{\varrho} \tag{7.9}$$

where $W_2 < W_{2.max}$. Finally, the smallest droplet that can be moved on MEDA covers at least a microelectrode, i.e., $W_{2.min} = \frac{L_e}{\sqrt{2}}$, where L_e is the microelectrode pitch and $W_2 > W_{2.min}$.

Figure 7.7 shows a graphical representation of the relation between W_1 and W_2 based on the above constraints[1]. For example, a target droplet with principal radius of curvature $W_2 = 140$ μm (i.e., of volume 11.5 nL) can be aliquoted from a droplet with principal radius of curvature W_1 only if 140 μm $< W_1 < 724$ μm. If $W_1 \geq 724$ μm, then the volume of the bigger droplet needs to be reduced until W_1 lies in the range (140, 724). The sequence of operations involved in this process is specified using the physical-level synthesis (described in the following subsection). Note that $W_{1.min}$ and $W_{2.min}$ can also be used to compute $H_{k.min}^s$ and $H_{j.min}^g$, which are required for the high-level synthesis.

Recall that the aliquot-volume solution of the high-level synthesis is presented using the MC unit; droplet volume measured in MC is referred to as H^{MC}. On the other hand, the radius of curvature and droplet volume in the

---

[1]The intuition behind the above constraints is that aliquoting requires an electrowetting force, specified by ϱ, that is sufficient to overcome the pressure gradient between the two droplets. A large gap in the pressure between the two droplets (due to the variation in radii of curvature) may prevent protrusion formation [209].

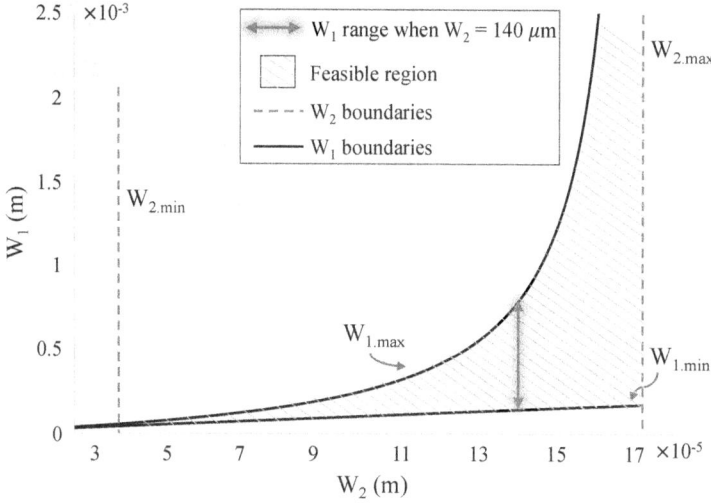

FIGURE 7.7
Illustration of aliquoting constraints ($\varrho = 5761.7$ m^{-1}; $s_e = 50$ μm).

aliquoting constraints are presented using metric units, e.g., meters (m) for radius and m^3 or liters for volume. Droplet volume measured in m^3 is referred to as H^{m^3}. Hence, conversion of volume measurement can be defined as follows[2]:

- $H^{m^3} = f_{ToM3} \cdot H^{MC}$; $f_{ToM3} = \frac{4\pi}{3} \cdot s_e \cdot L_e^2$
- $H^{MC} = f_{ToMC} \cdot H^{m^3}$; $f_{ToMC} = \lceil \frac{1}{f_{ToM3}} \rceil$

where L_e is the microelectrode pitch and s_e is the spacing between the parallel plates.

Formulation of Physical-Level Synthesis

The objective of physical-level synthesis is to map the high-level synthesis solution to a sequence of MEDA-enabled operations that implements a BRS assay with the lowest completion time. In addition, the constraints of droplet aliquoting must be satisfied.

Hence, a BRS assay consists of a sequence of MEDA-enabled operations (mixing, splitting, and droplet aliquoting), which can be modeled as a directed acyclic graph $G_a = (V_a, E_a)$, named a *composition graph*. A vertex $v_{ai} \in V_a$ represents a MEDA-enabled operation, which can be one of four types: (1) mixing; (2) splitting; (3) aliquoting; (4) null operation (it models a fluidic component at the start or at the end of a certain stage). Each operation type Δ_{op} is associated with a cost value $cost(\Delta_{op})$, where

---

[2]In the equilibrium state, the droplet liquid is shaped as an ellipsoid [156].

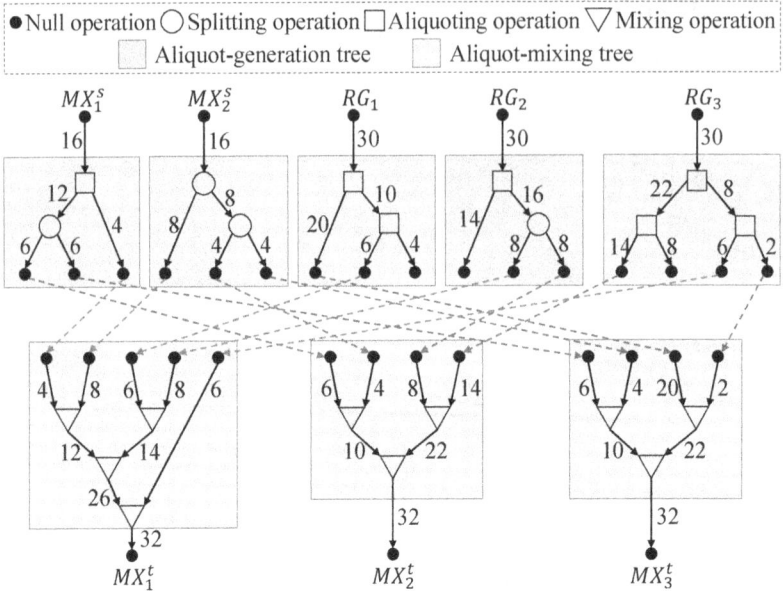

FIGURE 7.8

A composition graph G_a of a BRS assay that includes three target mixtures, two source mixtures, and three reagent fluids. Start and finish times of the operations are not shown in the figure.

$cost(\text{aliquoting}) > cost(\text{mixing}) > cost(\text{splitting}) > cost(\text{null}) = 0$—mixing and splitting are easier and quicker in execution compared to aliquoting. Hence, a vertex v_{ai} is defined as $v_{ai} = \{\psi_{ai}, ct_{ai}, st_{ai}, ft_{ai}\}$, where ψ_{ai} is the operation type, ct_{ai} represents the cost (i.e., time) of the fluidic operation implemented by v_{ai}, st_{ai} is the start time, and ft_{ai} is the finish time of the operation.

Also, an edge $e_{a(i,j)} = \{(v_{ai}, v_{aj}), u_{a(i,j)}\} \in E_a$ models the interdependency between a pair of operations v_{ai} and v_{aj}, and $e_{a(i,j)}$ is associated with a parameter $u_{a(i,j)}$ that captures the droplet volume resulting from v_{ai} and used by v_{aj}. Note that the maximum value for both the in-degree and out-degree of any vertex $v_{ai} \in V_a$ is 2. Also, similar to network-flow problems [58], the flow across any vertex v_{aj} must be consistent, meaning that the sum of input volumes must equal the sum of output volumes ($\sum_i u_{a(i,j)} = \sum_k u_{a(j,k)}$).

Figure 7.8 shows a composition graph G_a that is associated with the example in Figure 7.5. Note that G_a in Figure 7.8 is one of many composition graphs that can be used to implement the solution in Figure 7.5 on MEDA. For example, instead of generating the aliquots $\Lambda_{(i,2)}$ of MX_2^s using two splitting operations (see Figure 7.8), an alternative (but more costly) solution may involve two aliquoting operations to generate aliquots of volumes 4 MC, 4 MC, and 8 MC, respectively.

TABLE 7.3

Notation used in Section 7.6.

G_a	A composition graph
$\mathcal{Z}^g_{[o_1...o_N]}$	An aliquot-generation tree with N leaf nodes
$CT^g_{[o_1...o_N]}$	Overall cost of $\mathcal{Z}^g_{[o_1...o_N]}$
$\mathcal{Z}^g_{([o_1...o_N],j)}$	A node in $\mathcal{Z}^g_{[o_1...o_N]}$
$\psi^g_{([o_1...o_N],j)}$	Type of $\mathcal{Z}^g_{([o_1...o_N],j)}$
$ct^g_{([o_1...o_N],j)}$	Cost of $\mathcal{Z}^g_{([o_1...o_N],j)}$
$st^g_{([o_1...o_N],j)}$	Start time of $\mathcal{Z}^g_{([o_1...o_N],j)}$
$ft^g_{([o_1...o_N],j)}$	Finish time of $\mathcal{Z}^g_{([o_1...o_N],j)}$
$ht^g_{([o_1...o_N],j)}$	Height of $\mathcal{Z}^g_{([o_1...o_N],j)}$
$\mathcal{Z}^a_i$	An aliquot-mixing tree
$\mathcal{Z}^a_{(i,j)}$	A node in $\mathcal{Z}^a_i$
$st^a_{(i,j)}$	Start time of $\mathcal{Z}^a_{(i,j)}$
$ft^a_{(i,j)}$	Finish time of $\mathcal{Z}^a_{(i,j)}$

Hence, we describe the physical-level synthesis problem as follows:
Inputs: (i) Volume $\hat{H}^t_i$ of every target mixture MX^t_i. (ii) Volume H^s_k of every source mixture MX^s_k. (iii) Total liquid volume H^g_j of reagent RG_j. (iv) Volume of aliquots $\Lambda_{(i,k)}$ and $\Xi_{(i,j)}$ from the high-level synthesis.
Output: Composition graph G_a.
Constraints: Aliquoting constraints of MEDA biochips.
Objectives: Minimize the completion time ($\max_i ft_{ai}$).

Table 7.3 summarizes the notation used in the physical-level synthesis problem.

Physical-Level Synthesis Solution

An optimal composition graph is the one that has the lowest overall cost and the lowest completion time, and it can be found by enumerating all possible composition graphs and exhaustively searching for the optimal graph. Clearly, this is an impractical solution that entails a significant computation time. Therefore, we propose an alternative design methodology that divides the problem of computing a composition graph to two groups of subproblems: (1) computing aliquot-generation binary trees $\{\mathcal{Z}^g_1, \mathcal{Z}^g_2, ..., \mathcal{Z}^g_{sm+rg}\}$, and (2) computing aliquot-mixing binary trees $\{\mathcal{Z}^a_1, \mathcal{Z}^a_2, ..., \mathcal{Z}^a_{tm}\}$; see Figure 7.8. We observe that all the trees can be constructed separately before they are combined to form a composition graph G_a.

Construction of $\mathcal{Z}^g_i$: A tree $\mathcal{Z}^g_i$ represents a hierarchy of MEDA-enabled operations needed to generate aliquots from a source mixture MX^s_i or a reagent RG_i. The root of the tree represents the first MEDA-enabled operation applied to MX^s_i or RG_i and the leaf nodes represent the generated aliquots. We focus our discussion on the construction of $\mathcal{Z}_i^g$ that is associated with MX_i^s—a tree $\mathcal{Z}_i^g$ associated with RG_i can be constructed similarly. To simplify the construction of $\mathcal{Z}_i^g$, we consider a one-to-one mapping between the aliquot volumes $\Lambda_{(i,k)}$ and the leaf nodes; i.e., each aliquot volume is designated a certain leaf node in advance. Clearly, the selection of the mapping function may impact the performance of the synthesis because it partially decides the tree topology. Since our goal is to reduce the completion time (i.e., the cost), we consider a mapping function that likely reduces the number of aliquoting operations and increases the chances of satisfying the aliquoting constraints. This mapping function, denoted by f_{sort}, is simply implemented by sorting the aliquot volumes $\Lambda_{(i,k)}$ before assigning them to the leaf nodes, i.e., $f_{sort} : \{\Lambda_{(k,i)}\} \to [o_1, o_2, ..., o_N]$, where N is the number of aliquots and the vector $\boldsymbol{o}_i = [o_1, o_2, ..., o_N]$ contains pointers to the aliquots' volumes after sorting. Hence, the notations $\mathcal{Z}_i^g$, $\mathcal{Z}_{\boldsymbol{o}_i}^g$, and $\mathcal{Z}_{[o_1,o_2,...,o_N]}^g$ are used interchangeably.

Typically, the completion time of a protocol can be minimized by using a scheduling algorithm such as list scheduling [86]. However, a major advantage of modeling aliquots generation as a binary tree is that the scheduling problem can now be mapped to the problem of reducing the height of the tree, i.e., tree balancing, as well as the overall cost. For this purpose, we define the nodes of a tree $\mathcal{Z}_{\boldsymbol{o}_i}^g$ as $\mathcal{Z}_{(o_i,j)}^g = \{\psi_{(o_i,j)}^g, \; ct_{(o_i,j)}^g, \; st_{(o_i,j)}^g, \; ft_{(o_i,j)}^g, \; ht_{(o_i,j)}^g\}$, where $\psi_{(o_i,j)}^g$ is the operation type, $ct_{(o_i,j)}^g$ is the operation cost, $st_{(o_i,j)}^g$ is the start time, $ft_{(o_i,j)}^g$ is the finish time, and $ht_{(o_i,j)}^g$ is the height of the node. An objective function that captures the completion time, i.e., the tree height, and the overall cost can be defined as $CT_{\boldsymbol{o}_i}^g = CT_{[o_1...o_N]}^g = \sum_j ct_{(o_i,j)}^g \cdot ht_{(o_i,j)}^g$. Hence, our goal is to minimize $CT_{\boldsymbol{o}_i}^g$ for each tree $\mathcal{Z}_{\boldsymbol{o}_i}^g$.

Another relevant advantage of using binary trees is that binary search trees can be constructed optimally using recursive methods such as dynamic programming [58]. Similarly, we observe that the construction of $\mathcal{Z}_{\boldsymbol{o}_i}^g$ can be recursively decomposed into subproblems; each subproblem can be solved optimally, and the obtained solution can be reused to solve the original problem. For example, consider a source mixture of volume 16 nL that needs to be used to generate three aliquots A, B, and C, with volumes 8 nL, 4 nL, and 4 nL, respectively. To construct the tree $\mathcal{Z}_{[A \; B \; C]}^g$ shown in Figure 7.9(a), which is the tree with the lowest cost, we first construct two subtrees ($\mathcal{Z}_{[A \; B]}^g$ and $\mathcal{Z}_{[B \; C]}^g$) that represent two-aliquot problems (Step 1 in Figure 7.9(b)). In $\mathcal{Z}_{[A \; B]}^g$, the merging of A and B is modeled as an aliquoting operation, and the cost is therefore 2. On the other hand, in $\mathcal{Z}_{[B \; C]}^g$, the merging of B and C is modeled as a splitting operation, and the corresponding cost is therefore 1.

Next, these solutions are reused to compute (or "overlapped with ") the trees for a 3-aliquot problem (Step 2 in Figure 7.9(b)), and the solution/tree with the minimal cost is selected for $\mathcal{Z}_{[A \; B \; C]}^g$. Note that an optimal solution cannot be obtained by overlapping subproblems that have a high cost, i.e., we need to ensure that every subproblem is optimally solved. Note is that overall

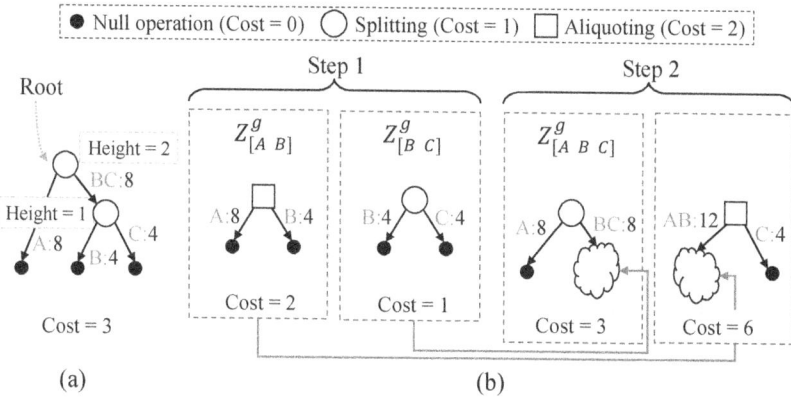

FIGURE 7.9
Construction of a tree $\mathcal{Z}^g_{[A\ B\ C]}$ that models the generation of 3 aliquots: (a) the minimal-cost tree; (b) solution methodology.

optimality of the problem cannot be guaranteed unless f_{sort} is optimal; this condition may not be achieved in our solution. Therefore, the above approach is said to provide a *pseudo-optimal* solution to the physical-level synthesis problem.

According to the above example, we obtain the following result (the proof can be found in Appendix E).

Lemma 7.1 *An aliquot-generation tree* $\mathcal{Z}^g_{[o_1...o_N]}$ *has the "overlapping-subproblems" and "optimal-substructure" properties.*

Based on the above properties, the construction of $\mathcal{Z}^g_{[o_1...o_N]}$ can be mapped to a dynamic programming problem [58], and the the mapping is defined as follows (proven in Appendix F).

Theorem 7.1 *An optimal aliquot-generation tree* $\mathcal{Z}^g_{[o_1...o_N]}$ *with root* $\mathcal{Z}^g_{([p_1...p_n],RT)}$ *can be constructed using dynamic programming, where the recurrence relation is defined as follows:*

$$
CT^g_{[o_1...o_N]} =
\begin{cases}
\min_{1 \le j < N}\{CT^g_{[o_1...o_j]} + CT^g_{[o_{j+1}...o_N]}\} \\
\quad + (ct^g_{([o_1...o_N],RT)} \cdot ht^g_{([o_1...o_N],RT)}) & \text{if } N \ge 1 \\
\\
0 & \text{if } N < 1
\end{cases}
$$

A description of our dynamic programming implementation is shown in Algorithm 7.2. We use a memoized method that performs recursion to compute $\mathcal{Z}^g_{[o_1...o_N]}$ (Lines 5–16), but it looks into a lookup table before computing an intermediate solution (Line 6). The merging of two subtrees (Line 14) is performed by selecting a suitable MEDA-enabled operation. If an aliquoting operation needs to be implemented but the aliquoting constraints cannot be

---

Algorithm 7.2 Construction of $\mathcal{Z}^g_{[o_1...o_N]}$

1: **procedure** MAIN( )
2:     $LT \leftarrow NULL$;   ▷ Initialize lookup table
3:     $\{\mathcal{Z}^g_{[o_1...o_N]}, CT^g_{[o_1...o_N]}\} \leftarrow$ CREATETREE($[o_1...o_N], LT$);
4:     **return** $Sol \leftarrow \{\mathcal{Z}^g_{[o_1...o_n]}, CT^g_{[o_1...o_N]}\}$;
5: **procedure** CREATETREE($[o_l...o_k], LT$)
6:     **if** HASRECORD($LT, [o_l...o_k]$) **then**
7:         **return** $\{\mathcal{Z}^g_{[o_l...o_k]}, CT^g_{[o_l...o_k]}\} \leftarrow$ USELOOKUPTABLE($LT, [o_l...o_k]$);
8:     **if** $[o_l...o_k]$.Length $== 1$ **then**   ▷ **p** contains one element
9:         SAVETOLOOKUP($LT, [o_l...o_k]$); **return** $\{[o_l...o_k], 0\}$;
10:     **for each** $o_j \in [o_l...o_k]$ **do**
11:         $\{\mathcal{Z}^g_{[p_l...p_j]}, CT^g_{[p_l...p_j]})\} \leftarrow$ CREATETREE($[p_l...p_j], LT$);
12:         $\{\mathcal{Z}^g_{[p_{j+1}...p_k]}, CT^g_{[p_{j+1}...p_k]})\} \leftarrow$ CREATETREE($[p_{j+1}...p_k], LT$);
13:         **if**   ISMINCOST($CT^g_{[p_l...p_j]}, CT^g_{[p_{j+1}...p_k]}, ct^g_{([o_l...o_k], RT)}), ht^g_{([o_l...o_k], RT)})$
    then
14:             $\{\mathcal{Z}^g_{[o_l...o_k]}, CT^g_{[o_l...o_k]}, \mathcal{Z}^g_{([o_l...o_k], RT)}\} \leftarrow$ MERGE($\mathcal{Z}^g_{[o_l...o_j]}, \mathcal{Z}^g_{[o_{j+1}...o_k]}$);
15:             $Sol \leftarrow \{\mathcal{Z}^g_{[o_l...o_k]}, CT^g_{[o_l...o_k]}\}$;
16:     SAVETOLOOKUP(LT, Sol); **return** Sol;   ▷ Memoization

---

satisfied, then a single aliquoting operation is not adequate to generate the target volume.

To explain the steps needed to overcome the above challenge, suppose that an aliquot with radius of curvature $W_2 = 140 \ \mu$m is needed, and the source droplet has radius of curvature $W_{src} = 2.004$ mm. Based on the relation ($W_{src} \approx \sqrt{(W_1^2 + W_2^2)}$), the radius of curvature of the bigger droplet is $W_1 = 2$ mm; see Figure 7.10(a). However, based on Figure 7.7, the aliquoting constraints for $W_2 = 140 \ \mu$m are $W_1 > 140 \ \mu$m and $W_1 < 724 \ \mu$m; thus aliquoting based on the above values of W_1 and W_2 is not possible[3].

To overcome this challenge, we choose a different value for W_2 in the interval $(140, W_{2.max})$ so that aliquoting can be performed. This implies that the target aliquot may be obtained after executing two aliquoting operations. Figure 7.10(b-c) explains the steps needed to address this challenge. For convenience, we consider the steps in Figure 7.10(b) as a single operation $\mathcal{Z}^g_{([o_1...o_N], j)}$, named *fragment*. The cost value $ct^g_{([o_1...o_N], j)}$ of this operation is determined based on the constituent sequence of aliquoting operations.

Post tree construction, we use a one-pass algorithm to compute the start time $st^g_{([o_1...o_N], j)}$ and finish time $ft^g_{([o_1...o_N], j)}$ of each node $\mathcal{Z}^g_{([o_1...o_N], j)}$ [280].
Construction of $\mathcal{Z}^a_i$: An aliquot-mixing tree $\mathcal{Z}^a_i$ represents a hierarchy of mixing operations that merge aliquots to form a target mixture MX^t_i; see Figure 7.8. The primitive aliquots are modeled as leaf nodes, whereas the last mixing operation before the target mixture is modeled as a root node. An intermediate tree node corresponds to a MEDA-enabled mixing operation

---

[3]Note that W_{src} is fixed.

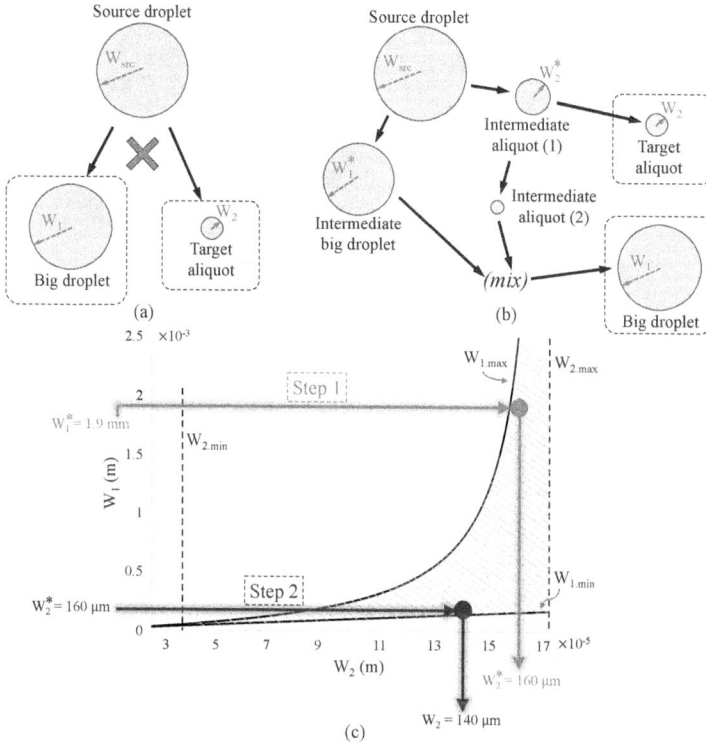

FIGURE 7.10
MEDA-enabled aliquoting: (a) aliquoting is not possible when the constraints (Figure 7.7) are not satisfied; (b) an alternative sequence of MEDA operations for aliquoting (fragment operation); (c) illustration of fragment steps using the aliquoting-constraints curve.

that merges two intermediate aliquots. We assume that all mixing operations are equivalent with respect to the processing time, i.e., cost.

Recall that the primitive aliquots are generated by the aliquot-generation trees, and therefore their finish times, which indicate their availability, may not be equal. Such a variation in the availability of the input primitive aliquots causes the generation of the aliquot-mixing trees $\mathcal{Z}_i^a$ to be computationally challenging. In other words, using a naive method that determines the mixing of aliquots randomly, without considering their availability, may unnecessarily increase the completion time, whereas adopting a priority scheme can enhance the synthesis performance. Figure 7.11 shows an example that compares the completion time based on the two approaches.

Formally, a tree $\mathcal{Z}_i^a$ consists of a set of nodes $\{\mathcal{Z}_{(i,1)}^a, \mathcal{Z}_{(i,2)}^a, ...\}$, where $\mathcal{Z}_{(i,j)}^a = \{st_{(i,j)}^a, ft_{(i,j)}^a\}$; the parameters $st_{(i,j)}^a$ and $ft_{(i,j)}^a$ represent the start time and finish time of the corresponding operation, respectively. Note that if $\mathcal{Z}_{(i,j)}^a$ is a leaf, then $st_{(i,j)}^a = ft_{(i,j)}^a$. Moreover, if a leaf node $\mathcal{Z}_{(i,j)}^a$ in $\mathcal{Z}_i^a$ is

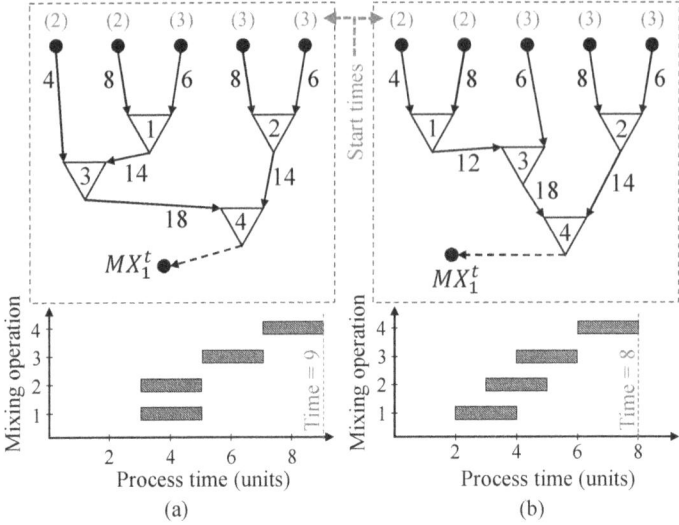

FIGURE 7.11
Computing an aliquot-mixing tree and the corresponding completion time using (a) a naive method; (b) a priority-based method. Each mixing operation takes two time steps.

directly connected to a leaf node $\mathcal{Z}^g_{([o_1...o_N],k)}$ in $\mathcal{Z}^g_{[o_1...o_N]}$ (Figure 7.8), then the start time of $\mathcal{Z}^a_{(i,j)}$ is defined as $st^a_{(i,j)} = ft^g_{([o_1...o_N],k)}$. Our goal in this stage is to minimize $\max_j(ft^a_{(i,j)}) \ \forall i$ in order to minimize the completion time.

To construct $\mathcal{Z}^a_i$, we develop a greedy method based on a priority queue QU that stores MEDA-enabled operations and sort them based on their finish time (Algorithm 7.3); more specifically, the queue uses a *nearest-finish-time-first* approach. Initially, the primitive aliquots and their finish times are inserted into the queue (Line 2). A mixing operation can be inserted into the queue if two primitive aliquots are ready for processing or if two intermediate mixing operations are completed (Lines 9–10). Next, the completed operations are then removed from the queue and are used by a tree-constructor function to properly insert new nodes into a tree (Lines 11–12). The algorithm, which runs as a time-wheel, stops when the queue becomes empty, meaning that the target mixture is successfully formed (Lines 4–7).

7.7 Simulation Results

We implemented BioScan using C++. We solved the ILP model using *lpsolve* [22], which was integrated in our C++ environment. All evaluations were carried out using a 3.4 GHz Intel i5 CPU with 4 GB RAM.

Algorithm 7.3 Construction of $\mathcal{Z}_i^a$

1: $Te \leftarrow \emptyset$;   ▷ Te: Tree $\mathcal{Z}_i^a$
2: $QU \leftarrow$ INITIALIZEPRIORITYQUEUE();   ▷ QU: priority queue
3: **repeat**
4:    **if** QU.Length $== 1$ **then**
5:        $\{\mathcal{Z}_{(i,j)}^a\} \leftarrow$ REMOVEFROMQUEUE(QU);   ▷ $\mathcal{Z}_{(i,j)}^a$: Operation
6:        $Te \leftarrow$ INSERTNODEINTOTREE($\mathcal{Z}_{(i,j)}^a$);
7:        **return** $\{Te,$Completion Time$\}$;
8:    **else**
9:        $\{ft_l, ft_k\} \leftarrow$ GETTWONEARESTFINISHTIMES(QU);   ▷ $ft_l \geq ft_k$
10:       INSERTMIX($QU, \mathcal{Z}_{(i,j)}^a, ft_l, ft_k+$ Mix Time);   ▷ ft_l+ Mix Time: Finish
      time
11:       $\{\mathcal{Z}_{(i,l)}^a, \mathcal{Z}_{(i,k)}^a\} \leftarrow$ REMOVEFROMQUEUE(QU);
12:       $Te \leftarrow$ INSERTNODEINTOTREE($\mathcal{Z}_{(i,l)}^a, \mathcal{Z}_{(i,k)}^a$);
13: **until** QU.Length $= 0$

To simulate the behavior of the iterative decision-making process in BioScan (Step 6), we used a regression-analysis model, known as Dirichlet regression, which performs regression analysis based on simplex spaces [164]. Since Dirichlet regression is a supervised-learning technique, gene-expression labels of the CF profiles were generated using a symmetric Dirichlet distribution function. This approach simulates the collection of gene-expression data using on-chip fluorescence detectors (Step 5).

We assess three aspects of the proposed flow: (1) space filling of the CF profiles, evaluated based on the E_m metric; (2) performance of the synthesis method, evaluated in terms of the synthesis imprecision Π and the average reagent volume consumed by a target mixture ($\Xi = \frac{\sum_{i,j} \Xi_{(i,j)}}{tm}$); (3) prediction accuracy of regression analysis, evaluated using the root-mean-square error (RMSE) metric.

While mixture preparation has been studied earlier [215], the scope of our work is different since it introduces a new sampling and optimization problem that has recently arisen in synthetic biology. Therefore, the trade-off analysis presented in this section cannot be applied to these prior methods and a meaningful comparison is not possible.

Analysis of CF Sampling

Recall that the sampling of the CF space is accomplished in two steps: regular sampling within the interval $[0, 1]$ followed by mapping of samples to the simplex CF space. Therefore, we evaluate the space filling of the CF profiles based on these two steps; we compare four sampling approaches: (1) OA-LHS (stratified sampling) followed by Dirichlet-based mapping where $\widetilde{\alpha} = 3$; (2) OA-LHS followed by Dirichlet-based mapping where $\widetilde{\alpha} = 20$; (3) OA-LHS followed by scaling-based mapping; (4) uniform sampling followed by

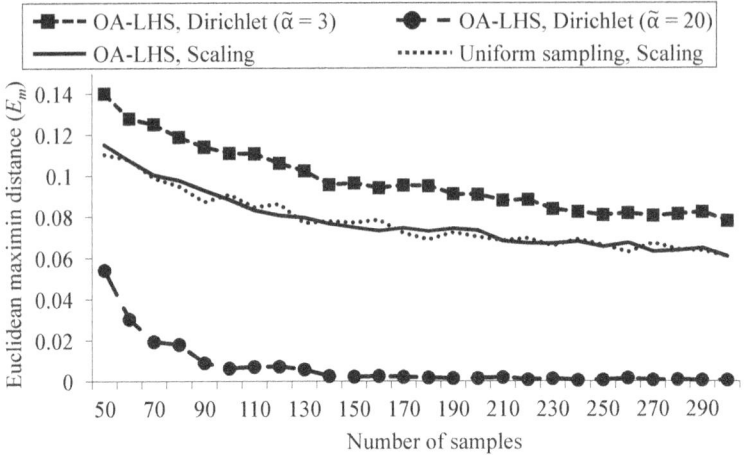

FIGURE 7.12
Impact of sampling and mapping techniques on the space filling.

scaling-based mapping. The number of CFs, i.e., reagents rg, is set to 8, and the number of sampling trials in each case is 1000. Results based on other values of rg also lead to the same conclusion, showing that our methodology is scalable with rg.

Figure 7.12 compares the above sampling approaches using E_m as a metric while varying the number of targeted samples tm. We observe that scaling-based mapping degrades the space-filling property, i.e., reduces E_m, regardless of which sampling method is utilized. This is expected since scaling-based mapping tampers with the stratification property. We also observe that space filling can be severely degraded if the Dirichlet-based mapping is not properly tuned. As shown in Figure 7.12, Dirichlet-based mapping with $\tilde{\alpha} = 20$ leads to the lowest E_m. Large values of $\tilde{\alpha}$ cause the generated samples to be highly concentrated instead of being uniformly distributed. In the following evaluations, we use OA-LHS with Dirichlet-based mapping where $\tilde{\alpha} = 3$.

Impact of β and H^t

We next evaluate the impact of the parameters β and H^t on the performance of the high-level synthesis using the example in Figure 7.5. For this purpose, we compare four sets of synthesis solutions associated with $H^t = \{8, 16, 24, 32\}$; comparison is performed in terms of the synthesis imprecision Π. In each set, we compute 11 synthesis solutions by varying the value of β from 0 to 1 using 0.1 as the increment between two subsequent values.

Based on Figure 7.13, we observe that increasing the value of H^t from 8 to 32 gradually lowers the synthesis imprecision. This result corroborates our argument that using a small value of H^t may limit the search space of all

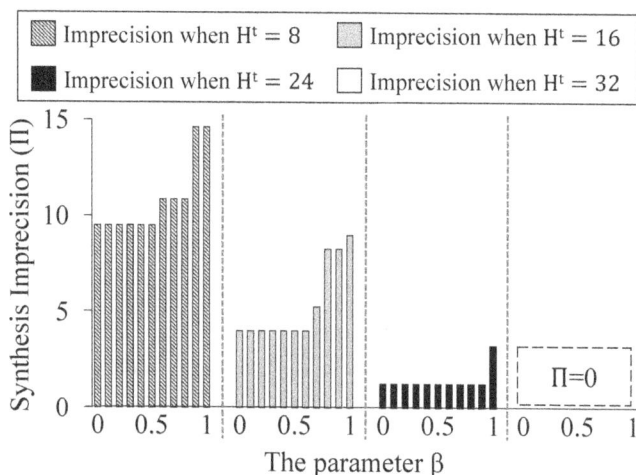

FIGURE 7.13
Impact of β and H^t on the synthesis performance.

feasible aliquots, thus increasing synthesis imprecision (Section 7.4). We also note that the impact of β becomes less significant when H^t is increased.

Synthesis Imprecision vs. Reagent Usage

We investigate the trade-off between the synthesis imprecision and the reagent usage. For this study, we consider a more complex setting where the number of reagents rg is 6 and the degree of accuracy δ is 128. The number of source mixtures sm is 10 and the number of target mixtures tm is 20. To generate the CF profiles of the source and target mixtures, BioScan was executed for two iterations. The first iteration generates the source mixtures (with volume equal to 128), which were then used in the second iteration to generate the target mixtures. The sampling in both iterations is performed using OA-LHS. We compute the optimal values of Π and Ξ while varying the target volume $H^t = \{8, 9, 10, ..., 1000\}$ and $\beta = \{0, 0.1, ..., 1\}$ in the second iteration. We then sort these results based on Ξ and plot the values of Π against the associated values of Ξ to perform the trade-off analysis.

Figure 7.14 shows the trade-off curve and the associated feasible region obtained using the above steps. A first observation is that the lowest values of Π, denoted by Π^*, is 28.3; this indicates that the problem setting described above is complex enough that the optimal synthesis solution cannot reach zero synthesis imprecision. Reducing the problem complexity (by minimizing tm, sm, and rg) can result in $\Pi = 0$. We also observe that increasing the amount of reagents beyond $\Xi(\Pi^*)$ causes increase in the synthesis imprecision Π. This counter-intuitive result is observed when H^t becomes significantly large, and

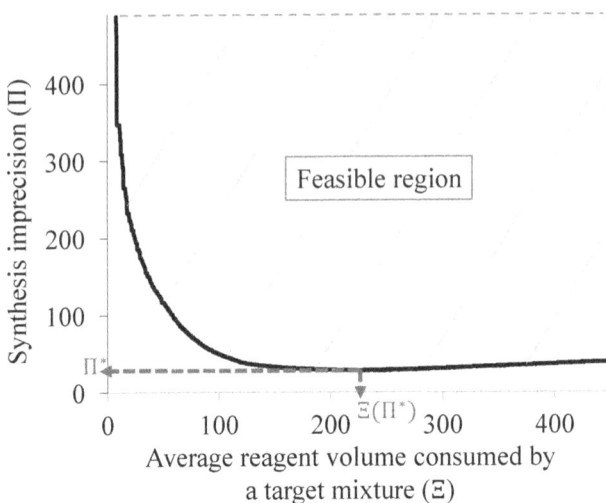

FIGURE 7.14
Illustration of the trade-off between Π and Ξ.

even though $\hat{CX}^t_{(i,j)} \approx CX^t_{(i,j)}; \forall i, j$. The reason for such a finding is that when H^t is increased, a large volume difference $\sum_i(\hat{H}^t_i) - tm \cdot H^t$ outweighs the impact of $\hat{CX}^t_{(i,j)} \approx CX^t_{(i,j)}$. This observation is confirmed by computing $\frac{\Pi}{\sum_i(\hat{H}^t_i) - nH^t}$ instead of Π. We use Π to analyze the performance because this allows us to determine the global minimum Π^* and the associated $\Xi(\Pi^*)$.

Regression Analysis

We finally investigate the prediction accuracy of the regression analysis and its role in decision making. For this purpose, BioScan is configured to execute 10 iterations, and the prediction accuracy (RMSE) is reported at the end of each iteration. Flow execution is performed considering one of two modes: (1) cost-effectiveness mode, which allows the target mixtures generated during the ith iteration to be used as source mixtures during the $(i + 1)^{th}$ iteration; (2) low-imprecision mode, which does not allow target mixtures to be re-used in a following iteration. We consider $rg = 8$, $\delta = 64$, and $\beta = 0.5$. The number of target mixtures tm generated at the first iteration is 10 mixtures, and the value of tm is incremented by 10 for every new iteration.

Based on Figure 7.15, we observe that adopting the cost-effectiveness mode significantly reduces reagent usage compared to the low-imprecision mode. We also observe that the RMSE gradually decreases when more iterations are executed. Note that the rate at which the RMSE decreases depends on: (1) the increment in the number of target mixtures, and (2) the shape of the actual parameter-space function. By specifying a maximum threshold Υ on

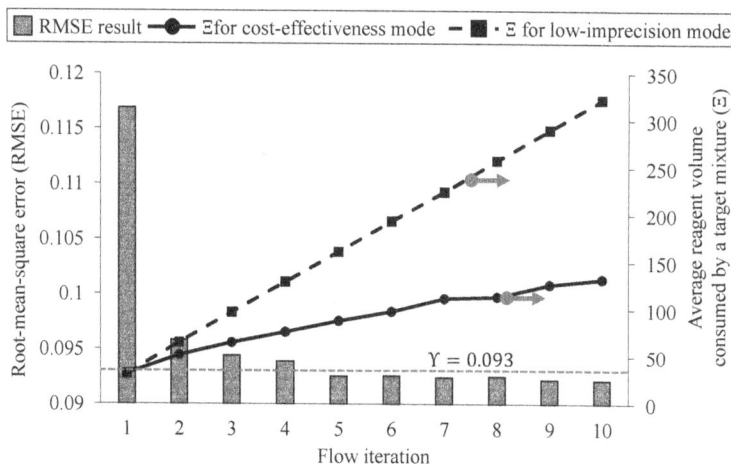

FIGURE 7.15
Prediction accuracy of the regression analysis and reagent usage based on 10 iterations of flow execution.

the prediction error, the procedure terminates when the prediction error goes below Υ. For example, if $\Upsilon = 0.093$ in Figure 7.15, BioScan stops execution after the fifth iteration.

7.8 Chapter Summary

We have introduced an optimization framework for PSE in synthetic biology. We formulated the PSE problem in terms of a BRS assay that is implemented on a MEDA biochip. The proposed framework uses statistical sampling to select reagent mixtures, ILP-based synthesis to generate the reagent mixtures needed for a BRS assay, physical-level synthesis that implements the BRS assay on MEDA, and regression analysis to assess the prediction of the parameter space. The proposed flow has been evaluated based on the synthesis imprecision, reagent usage, and the prediction error. Simulation results have shown the effectiveness of the proposed framework in implementing BRS while minimizing reagent usage and synthesis imprecision.

8

Fault-Tolerant Realization of Biomolecular Assays

Digital microfluidics has achieved remarkable success in miniaturizing PoC quantitative analysis testing [109]. However, a major stumbling block in the monitoring and controlling of diseases via such PoC systems is the lack of reliable diagnostic tests that can recover from unexpected errors. In addition, diagnostic tests, similar to all other quantitative-analysis procedures, are inherently stochastic systems that exhibit complex interactions among their constituent biochemical components [159]. Such characteristics signify the need for real-time error-recovery methods that verify the correctness of on-chip fluidic interactions during bioassay execution.

In this chapter, we present a hardware-based set-up for error detection and recovery [103]. The proposed set-up is designed based on the C^5 architecture described in Chapter 1; thus providing an integrated, portable demonstration of cyber-physical digital microfluidics, where errors in droplet transportation on the biochip array are detected using capacitive sensors. The interpretation of the test outcome and the decision-making process are fully accomplished using a hardware system. Experimental results are reported based on a fabricated silicon device.

The remainder of this chapter is organized as follows. Section 8.1 presents an overview of fault models and prior cyber-physical error-recovery methods. The adaptation of the multi-layer C^5 architecture is explained in Section 8.2. Next, Section 8.3 describes the experiment design and the back-end software support required by the hardware controller. Details of the hardware dictionary-based error recovery system are presented in Section 8.4. The process of biochip fabrication and the experimental results are illustrated in Section 8.5. Finally, Section 8.6 concludes the chapter.

8.1 Physical Defects and Prior Error-Recovery Solutions

A DMFB device is said to have a failure if its operation does not match its specified behavior. In order to detect defects using electrical methods, fault models that efficiently represent the effect of physical defects at some level of abstraction have been described in [102]. These models can be used to capture the effect of physical defects that produce incorrect behavior. Faults can be caused by manufacturing imperfections, or by degradation during use as electrodes are actuated (i.e., *operational fault*). Some possible causes of physical defects are listed below.

- *Dielectric breakdown:* High voltage levels during actuation cause dielectric breakdown, which creates a short between a droplet and the underlying electrode. In this case, droplet transportation cannot be controlled since the droplet undergoes electrolysis.

- *Degradation of the insulator:* The actuation of electrodes for long durations causes charge to be irreversibly concentrated near the electrodes (i.e., trapping of charge [270]). This causes an operational error since it impedes droplet transportation because of the undesired variation of interfacial surface tension along the droplet flow path.

- *Short-circuited electrodes:* A short between two adjacent electrodes effectively forms one longer electrode. When a droplet resides on this electrode, it is no longer large enough to overlap the gap between adjacent electrodes. As a result, the actuation of the droplet can no longer be achieved.

- *Open in the metal connection between the electrode and the control source:* This open can be on the path between the electrodes and the chip pads, or through the external wires between the biochip and the adjacent controller. This defect results in a failure in activating the electrode for droplet transport.

Besides the physical defects mentioned above, protein fouling may lead to the malfunction of multiple electrodes in the biochip since it renders the surface permanently hydrophilic [240]. The above issues and many others are key reasons for blocking droplet transportation during the execution of a bioassay, thus degrading system reliability if no action is taken. Hence, PoC quantitative-analysis systems need to be empowered with the capacity of recovering from errors.

Over the past decade, several methods have been proposed for enabling error recovery [296, 102, 159, 158, 14]. The work described in Zhao *et al.* [296] and Luo et al. [159] represent the state-of-the-art for dynamically adapting to error occurrences during assay execution. However, among all the previous studies, only the work in [102] has presented an experimental demonstration of hardware/software interface for error detection and recovery in cyber-phyiscal DMFBs . This approach, which uses a software-based control system, is considered a significant step towards fault-tolerance in DMFBs.

Although the software-based approach in [102] introduces flexible reconfiguration that supports error recovery, it requires a desktop machine to be involved in the control system to generate the recovery actuation sequences. The desktop transmits the actuation sequences to a low-end microcontroller board which is in charge of communicating these recovery sequences to the biochip actuation circuit. The drawbacks of this approach are twofold: (1) the use of a desktop machine for performing dynamic reconfiguration increases the complexity of the cyber-physical system and it does not lead to realizing a portable solution for PoC quantitative-analysis testing; (2) it requires a software-resynthesis step which leads to increased bioassay response time when errors occur.

Instead of using a software-based solution, a dictionary-based hardware-assisted error-recovery method can be adopted. In this method, for a given error that can occur in a bioassay, simulations are used to generate an error dictionary before the experiments are conducted. During the simulations, erroneous fluidic operations are considered and the corresponding error recovery plans are determined and then stored as entries in the dictionary. A computer performs both the simulation and dictionary generation in the offline data-preparation stage. Once an error is detected during the experiment, the control system will trigger the corresponding error-recovery plan stored in the error dictionary for error recovery.

The concept of dictionary-based error recovery was introduced earlier in [160], but it has not been demonstrated using realistic experimental settings. In this chapter, we use this concept to present the first integrated, fault-tolerant microfluidic platform, which is particularly useful for PoC quantitative analysis.

8.2 Adaptation of the $\mathcal{C}^5$ Architecture to Error Recovery

Recall that the $\mathcal{C}^5$ architecture, which is designed to streamline cyber-physical adaptation for quantitative analysis, consists of 5 layer: (1) Connection level (Level I), which includes biochip components and sensors; (2) Conversion level (Level II), which executes signal conditioning; (3) Cyber level (Level III), which collects and analyzes data; (4) Cognition level (Level IV), which performs synthesis based on the received signals; (5) Configuration level (Level V), which offers protocol-level decisions. See Chapter 1 for more details. This multi-layer architecture can be adapted for the purpose of error recovery, which can be designed using a software-based and a hardware-based solution. Figure 8.1 shows the architecture of $\mathcal{C}^5$-compliant designs of cyber-physical DMFBs for error recovery.

The key-idea of the hardware-based error recovery, which represents our target in this chapter, is to pre-compute and store recovery actuation sequences for all errors of interest that can occur during the execution of a bioassay. The detection signal can be used to trigger a "transition" in the system finite state machine (FSM). In other words, when an error is detected, the cyber-physical system simply looks up the recovery solution in the memory that matches the new recovery state, rather than performing online re-synthesis on a desktop computer.

This hardware-based solution relies on a capacitive sensor (at Level I) to examine the appearance of a droplet at a detection cell that is employed as a checkpoint. If a missing droplet is reported via a sensing signal, then a failure must have been encountered during droplet operation. The analog signal is captured via a signal-conditioning circuit (at Level II), which modulates the sensing signal as a digital signal. Once the digital signal is generated, it is transmitted to an FPGA-based processing unit that performs the function of Level III, i.e., it meters the digital-signal frequency, compares the resulting data with a pre-specified error threshold, and issues a suitable error-recovery procedure.

The Configuration level (Level V) as well as the Cognition level (Level IV) are excluded in the hardware-based solution to eliminate the computer-in-the-loop, clearly sacrificing a large degree of system reconfigurability; see Figure 8.1(b). It is possible that a DMFB design problem, such as error recovery, can be tackled using a $\mathcal{C}^5$-compliant cyber-physical system in different ways; each with its own advantages (e.g., highly reconfigurable vs. highly responsive and portable). The system proposed in this chapter favors responsiveness and portability to enable PoC quantitative-analysis testing. Details of the implemented hardware-based, also referred to as dictionary-based, error recovery are explained in Section 8.4.

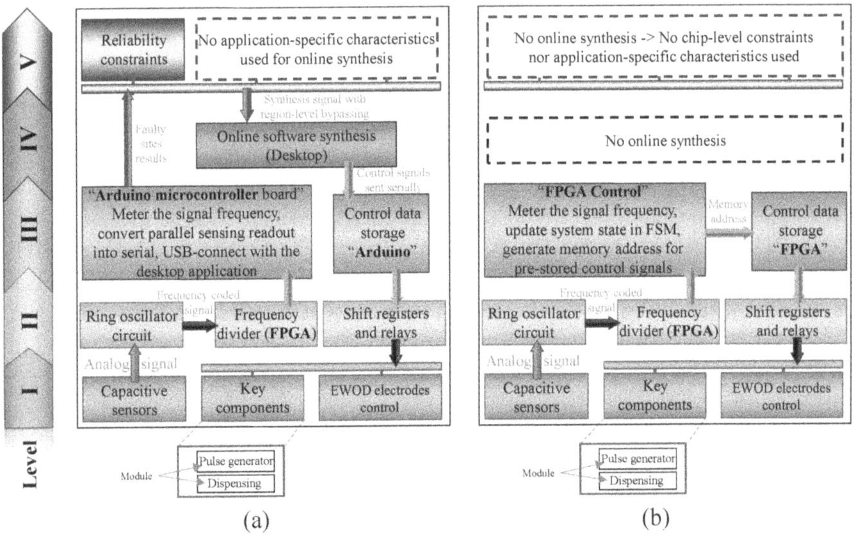

(a)   (b)

FIGURE 8.1
$\mathcal{C}^5$-compliant architecture design of cyber-physical DMFBs: (a) for software-based error recovery, (b) for integrated hardware-based error recovery.

8.3 System Design

In this section, we describe the experiment design to demonstrate the proposed dictionary-based cyber-physical error-recovery system. The required algorithmic implementation to support error recovery is also shown.

Protocol Design and Error Recovery

Our system has been designed to autonomously execute typical assay protocols empowered with the capacity to perform error recovery. In this work, a typical dilution assay is used to exemplify the utilization of our control system. This assay is widely used in many settings including sample preparation for gene-expression analysis (Chapter 2).

The executed dilution protocol aims at using a set of reagents to dilute an input sample into a targeted mixture/concentration. The description of the executed protocol is given by the sequencing graph in Figure 8.2(a). The vertices represent the bioassay operations, whereas the edges represent the interdependencies among these operations. As shown in the figure, the protocol requires dispensing three droplets (i.e., *"disp"* operation) for execution. After each *"split"* operation, one droplet is considered as a backup droplet and is routed into one of the chip boundaries to be stored, whereas the other droplet is driven into the detector cell. This cell is linked to a capacitance measurement circuit used to test for droplet presence/absence status. A typical biochip can be equipped with several sensor circuits to form multiple checkpoints; however increasing the number of checkpoints increases the cost of the system hardware. Therefore, we monitor the dilution protocol droplets in our chip using only one checkpoint.

When the sensor circuit reports a "missing droplet," we infer that a fault exists on the path used for driving the droplet. This failure may be caused by

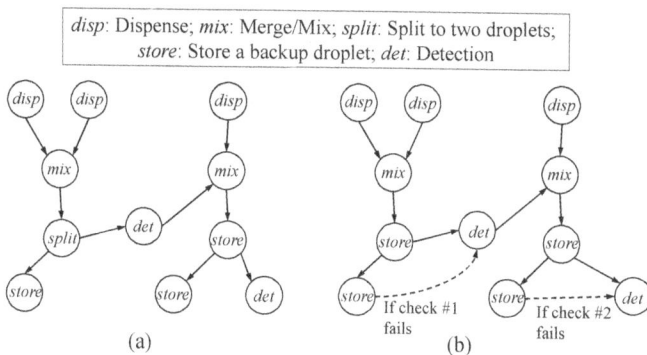

FIGURE 8.2
Sequencing graph of dilution protocol: (a) Fault-free execution. (b) Utilization of on-chip backup droplets for error recovery.

a physical defect, which can be the key reason for blocking droplet transportation during bioassay execution. In many cases, the sensing circuit reports a "missing droplet," thereby an action needs to be taken. Figure 8.2(b) shows how the dilution protocol sequence is updated so that recovery actions are considered in missing-droplet cases. After each split, if the to-be-detected droplet is reported to be missing, the control system loads recovery actuation sequences from the hardware memory. In this action, the stored backup droplet is used as replacement for recovering from this error and continuing the operations. If no backup droplet exists, a new droplet has to be dispensed from the reservoir and the sequence must be repeated.

Biochip Layout and Checkpoint Localization

Our experimental chip (fabrication details are discussed in Section 8.5), whose layout is shown in Figure 8.3, consists of 32 electrodes (including reservoir electrodes) that are independently controlled by external pins. The detector cell shows the location of the checkpoint. Based on the pre-computed actuation sequences, we know the clock cycles when the droplets are expected to reach the detector cell. If a droplet is reported to be missing, a recovery action is invoked in the manner that was explained in the previous subsection.

Back-end Algorithmic Support for the Control System

The error-recovery system depends solely on the hardware for generating recovery actuation sequences whenever an error is detected. The dictionary-based approach, however, requires a back-end software support that is capable of extracting protocol specification from the user (including covered errors), modeling the dictionary components to generate the recovery sequences, and feeding the hardware controller with the required scripts and files for bioassay execution and error recovery. Figure 8.4 shows the inputs/output of the

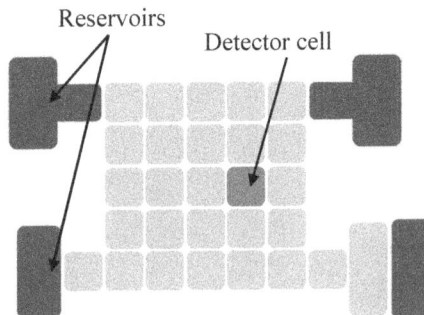

FIGURE 8.3
Schematic of the biochip layout.

back-end software. The inputs to the software tool are: (1) the protocol description which can be provided by the user or extracted from a droplet-routing computer-aided design (CAD) tool. (2) The pin-mapping file which is used to map the electrodes to the actuating FPGA pins. The output is a memory file whose values are stored in the FPGA memory.

In order to incorporate error recovery in our control system, the software tool must be aware of all possible recovery scenarios that need to be included. This information can either be driven directly by the user or as a result to an offline synthesis by a high-level synthesis tool [287]. Hence, the back-end software models the underlying hardware finite-state machine by flattening it into a decision tree. To illustrate this concept, suppose that it is required to provide the control information needed to execute the dilution protocol, shown in Figure 8.2(b), while considering recovery from a potential error detected during check #1. In this case, the control information for both error-free as well as erroneous execution are communicated to the back-end software tool to build a decision tree (we call it execution tree). Note that this tree is translated later to a finite-state machine in the hardware. Figure 8.5(a) displays the top part of the execution tree that results from the above scenario.

Having identified the nodes of the execution tree, a file is generated following the structure shown in Figure 8.5(b). Using this file, the back-end software tool is capable of generating the memory file and logging the start addresses of every node represented in the tree; thus forming a dictionary of error-recovery plans. These addresses are used by the hardware control system during its state transitions in order to adapt the system to the sensor readouts. Illustration of state transitions based on the error-recovery dictionary is shown in Figure 8.6.

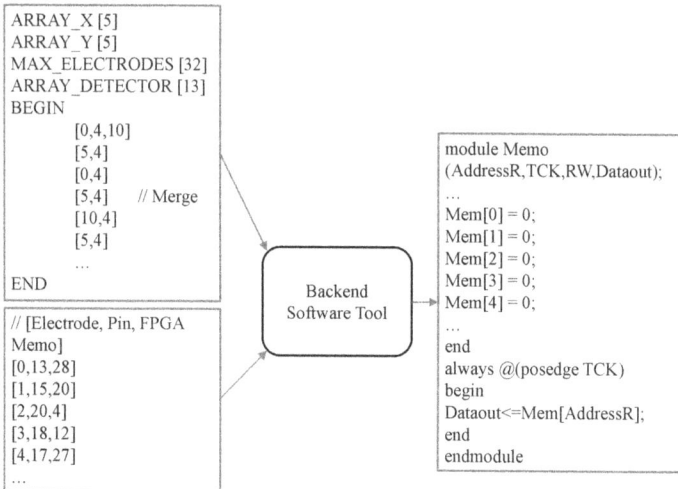

FIGURE 8.4
Inputs and output of the back-end software tool.

8.4 Dictionary-Based Error Recovery

The integrated system consists of three parts: a relay board, a connection board, and an FPGA with integrated memory. Since no software execution is needed, the computer and its related interfaces are no longer necessary. Figure 8.7 shows a schematic of the integrated system setup. The implementation and functionality of each component are presented below.

Relay Board

The relay board, whose image is shown in Figure 8.8(a), is used to synchronously actuate the 32 electrodes in the biochip. The input to the relay board is the low-voltage control signal provided by the FPGA (0-5 V), and the output is the high-voltage actuation voltage for the biochips (40 V, 1 kHZ AC voltage). Four 8-bit serial-in-parallel-out shift registers (Part No.: 74HC595) and 32 photoMOS relays (Part No.: AQW610S) are integrated on the relay board. The shift registers are used to realize serial-to-parallel conversion of the control signal, and the photoMOS relays are used to convert the signal from the low-voltage control signal to the high-voltage actuation signal for the biochip.

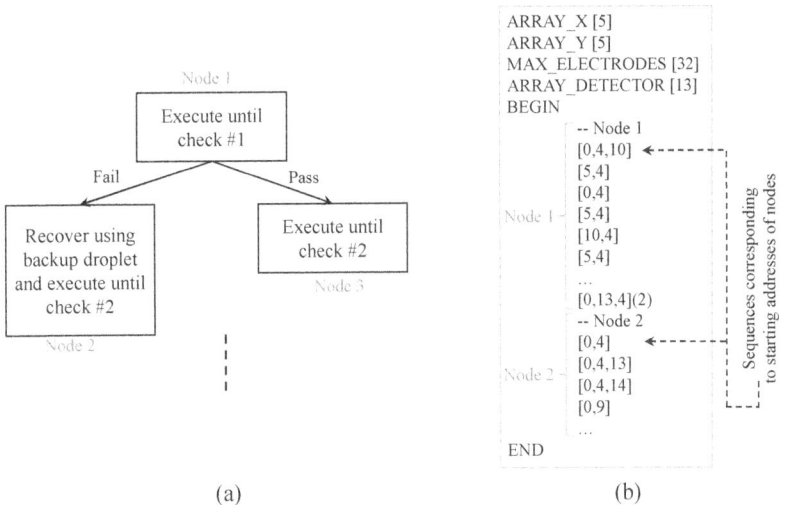

(a)   (b)

FIGURE 8.5

Steps needed to generate input files of the software tool: (a) Generation of the execution tree for the dilution protocol; (b) Mapping the execution tree to a suitable input format that simulates error-recovery dictionary.

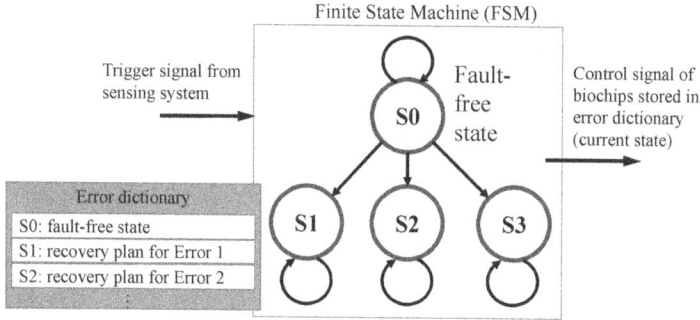

FIGURE 8.6
The implementation of dictionary-based error recovery using finite-state machine.

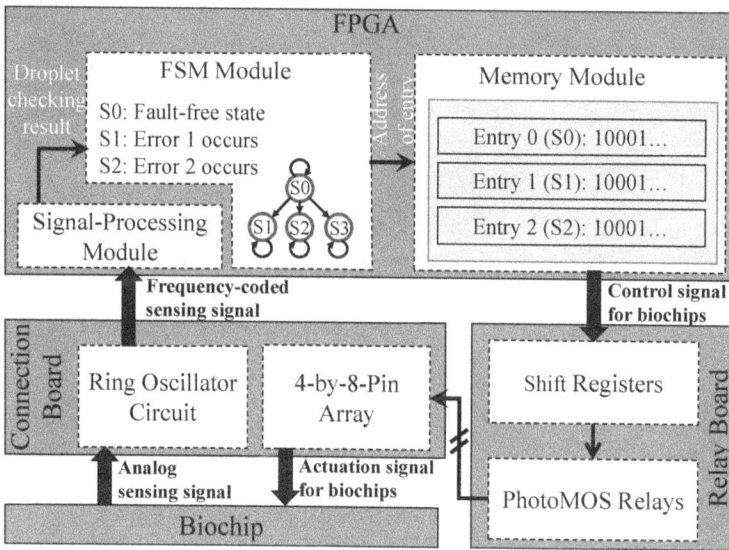

FIGURE 8.7
The implementation of dictionary-based error recovery using finite-state machine.

Connection Card and Capacitive Sensing Circuit

The connection board, whose image is shown in Figure 8.8(b), contains two modules: a 4-by-8-pin array and a ring oscillator circuit used for capacitive sensing. The pin array on the back of the connection board can contact the metal pads on the biochip so that the high-voltage actuation signal is transferred from the relay board to the chip. The circuit diagram for the capacitive sensing can be found in [102]. The ring oscillator circuit is used

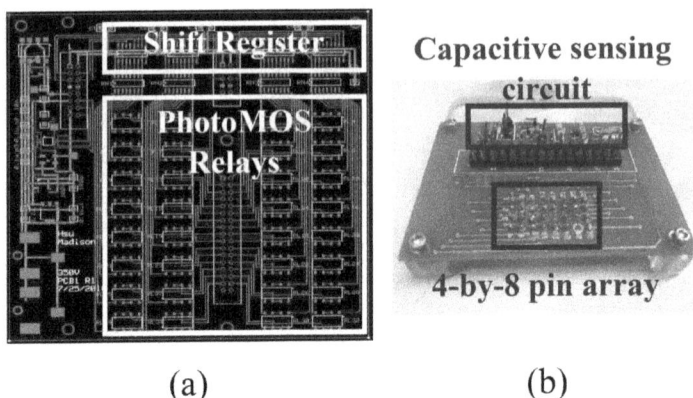

(a)   (b)

FIGURE 8.8
Relay board components: (a) The layout of the relay board. SR1 - SR4 in the upper white rectangle are four 8-bit serial-in-parallel-out shift registers. RL1 - RL32 in the lower white rectangle are 32 photoMOS relays; (b) Image of the connection board, where the circuit in the upper rectangle is the capacitive sensing circuit.

to detect the presence or absence of droplets by monitoring the capacitance change between the control and ground electrodes. When a droplet of ionic liquid is present between the electrodes, the majority of the applied electric field forms over the dielectric layer of the biochip. This decreases the charge separation distance in the device and thus increases the capacitance for the overall structure, which lowers the detected output frequency of the ring oscillator circuit. Conversely, if a droplet is absent, the electric field is established over both the dielectric layer and the filler medium (e.g., silicon oil or air). This increases the distance over which charge is separated in the device, thereby decreasing the capacitance measured for the overall structure.

FPGA

The circuitry for cyber-physical error recovery consists of three main modules: (1) the signal-processing module that analyzes the frequency feedback signal from the capacitive sensing circuit and determines if an error has occurred, (2) the memory module that stores the error dictionary, and (3) the FSM module that dynamically adjusts the actuation sequences when an error is reported. Figure 8.7 illustrates the interconnection between these modules.

1. Signal-processing module: The signal-processing module is essentially a frequency counter. It is used to: (1) process the frequency signal collected from the ring oscillator circuit on the connection board, and (2) determine whether an error occurs in the bioassay and report the sensing results to the FSM module. However, an important design challenge arises due to the fact that a

1 kHz AC power source is directly connected to the biocell for the actuation. This AC power source is in close proximity to the capacitance sensor, which leads to significant interference with the capacitance sensing output, which is a frequency-encoded square wave. To reduce this interference, an exact 1 s window is set in the frequency counter to accumulate frequency signals in 1000 periods of the AC source before updating the output.

2. Memory module: The memory module is used to store the error dictionary of the bioassay. During the fault simulation of the bioassay, entries of the error dictionary are generated. Note that each entry in the dictionary contains only the resynthesis result from the time that the error is detected to the end of the bioassay. The error dictionary is then written into a memory initialization file (MIF). The MIF specifies the content for each memory cell in the read-only memory (ROM) of the FPGA. The starting and ending addresses of each dictionary entry are simultaneously generated and loaded into the FSM module.

3. FSM module: The FSM is used to implement automatic error recovery. The sensing results from the signal-processing module are sent as the input to the FSM, and the outputs from the FSM are the starting and ending storage addresses of the dictionary entry that should be applied to the biochip. When the signal-processing module sends a signal indicating that an error has occurred, the FSM transitions to the corresponding faulty state for error recovery and the starting and ending storage addresses of the dictionary entry are updated accordingly. When the FSM sends the addresses of the dictionary entry, the data that are stored in the corresponding memory cells are sent to the output of the ROM. In the experiment, we limit the maximum number of errors in the bioassay to one in order to prevent both the size of the error dictionary and the CPU time required to generate the error-dictionary from becoming prohibitively high.

8.5   Experiment Results and Demonstration

In this section, we present biochip fabrication details, experimental results, and a demonstration of error recovery.

Biochip Fabrication and Chip Specifications[1]

As illustrated Figure 8.9(a), the DMFB device used in the experiment consists of three major components. The gasket is sandwiched between the top plate and the bottom plate to create the operation area for the

---

[1]The fabrication and assembly of biochip components have been executed by Prof. Richard Fair's research group.

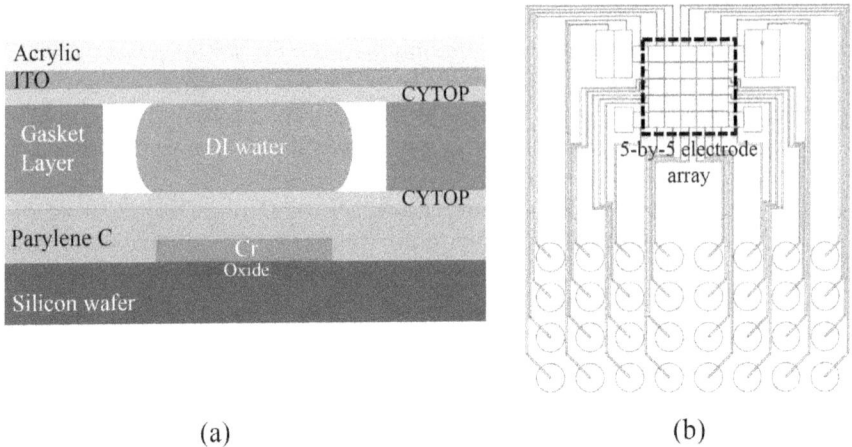

(a)   (b)

FIGURE 8.9
The fabricated biochip which is used in the experiment: (a) Cross section; (b)
layout.

electrowetting-on-dielectric (EWOD) function. The acrylic top plate is laser
cut into the appropriate shape and the pipette holes are also laser cut into
the top plate. A layer of ITO is then applied on top of the acrylic to provide the ground plane when the EWOD function is active. The ITO layer is
sputtered onto the acrylic plate and its thickness is controlled to be 140 nm.
The surface of the top plate needs to be hydrophobic in order for the EWOD
function to work. Therefore, CYTOP is spun onto ITO layer of thickness 80
nm. The bottom plate is fabricated based on a silicon substrate of 500 μm
thick. On top of the silicon substrate, there is a thin layer of thermal oxide of
thickness 1 μm. The electrode is patterned out of the Cr layer. The Cr layer
is deposited using EBPVD technique and the thickness of that layer is 100
nm. The electrode is patterned using lithography, resulting in a total of 32
electrodes and connection wires to the contacting pads. Above the Cr layer,
the dielectric layer is deposited using Parylene C with thickness of 1.1 μm.
The surface is also kept hydrophobic by being spun on a layer of CYTOP of
thickness 80 nm.

The device is completed by bonding the top plate and the bottom plate
using the gasket layer. The gasket layer is made out of a 120 μm thick adhesive
SecureSeal. The SecureSeal is laser cut to define the operation area. The three
components is aligned and bonded together after the gasket is ready. Over
night vacuum desiccation is used to evaporate the solvent used during the
fabrication.

The chip layout is shown in Figure 8.9(b). The size of electrode used in
the experiment is 1mm×1mm. Actuation signals of electrodes are transferred
through the contact between the contact pads on the biochip and the pin
array on the connection board. The channel height is determined by the gasket

FIGURE 8.10
The experiment setup in the laboratory.

thickness, which is 120 μm in our experiment. The volume of the droplet can be hence calculated to be 120 nL. The liquid in the droplet is DI water. During the experiment, the device is filled with 2 cS silicone oil first to prevent the water droplet form evaporation.

Results and Videos

The FPGA device used in the experiment was an Altera Cyclone IV with 6.3 Mb embedded memory. The synthesis results show that the signal-processing modules, memory modules, and FSM modules on the FPGA consume a total of 597 logic elements (2% of the available elements), 180 Kb (3% of the available memory) and 10 user I/Os (3% of the available pins). The demo setup is shown in Figure 8.10. Several experimental runs were carried out using the fabricated chip and the control system. A CCD camera was used to capture the video, and images were extracted from the recorded video.

Videos of the error-recovery experiments are available online [7]. Droplets were routed as discussed in Section 8.3. The fault-free video shows the initial droplet route. If an error occurs, the droplets cannot reach the checkpoints as planned in the experiment. At that point, the corresponding entry in the error dictionary is activated and the error-recovery solution is triggered. Successful re-routing of droplets was observed in all cases, and snapshots of the experiment are shown in Figure 8.11.

8.6 Chapter Summary

In this chapter, we have demonstrated error detection and recovery in a cyber-physical digital-microfluidic biochip. Compared to previous work, fault tolerance was achieved in the proposed system via a fully integrated

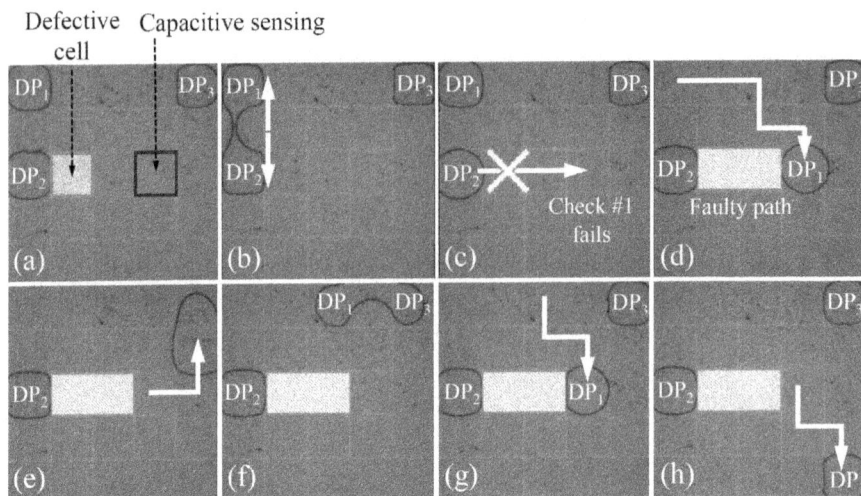

FIGURE 8.11
Snapshots extracted during the execution of the dilution-protocol experiment described in Figure 8.2(b). These snapshots show automated error recovery when check #1 fails: (a) Start of the experiment, where three droplets DP_1, DP_2, and DP_3 are dispensed onto the array; (b) a dilution operation is performed using DP_1 and DP_2; (c) check #1 fails indicating that DP_2 is stuck due to a certain defect, therefore the control system invokes error recovery through the hardware dictionary; (d) error recovery is realized by moving the backup droplet DP_1 to the detector; (e) a further step of dilution is performed by mixing DP_1 and DP_3 (while avoiding the faulty path); (f) splitting operation is performed; (g) DP_1 is moved to the detector to perform check #2; (h) after a successful check, the final product DP_1 is moved for collection.

hardware-based control system. Hence, the approach described in this chapter is a significant step towards reliable PoC quantitative-analysis protocols. Detection of errors was performed using a capacitive sensor, the test outcome was interpreted by a hardware signal processing module, and error recovery was accomplished through a hardware dictionary. We have presented the details of the fabricated biochip. We have also reported the experimental results based on this biochip and a link to videos has been given.

Part IV

Security Vulnerabilities and Countermeasures

9

Security Vulnerabilities of Quantitative-Analysis Frameworks

Despite the widespread use of biomolecular quantitative assays in clinical diagnostics, environmental monitoring, and DNA forensics; there has been no study on the potential security implications of biomolecular platforms particularly cyber-physical microfluidic systems. Recent cyberattacks have revealed the vulnerabilities of automated systems [96, 147].

If an attacker gets control of the quantitative-analysis framework, he can tamper with the collected samples or maliciously modify the assay operation to either manipulate the assay outcome or to disrupt the assay operation. The attacker could be a person who wants to jeopardize another person's health by manipulating his/her clinical diagnostic results, a malicious organization that wants to disrupt the products from a specific vendor, or a criminal who aims to mislead forensics-based judicial decisions [13].

This chapter provides an assessment of the following security vulnerabilities [13, 11, 10]: (1) bioassay-result manipulation on DMFBs, which can be triggered by bioassay-level attacks; (2) insecure cyber-physical integration in DMFBs, which can form a backdoor for system-level sensor-spoofing or information-leakage attacks; (3) forgery of DNA samples, which can be carried out as a biochemical-level attack during DNA preparation. In Chapter 10, we introduce security countermeasures against bioassay-level attacks and DNA-forgery attacks. *Research that investigates additional countermeasures for microfluidic systems, especially for securing cyber-physical integration and sample differentiation, can be found elsewhere* [263, 255, 257, 251, 258, 254, 260, 256].

The key contributions of this chapter are as follows:

- We define a comprehensive attack framework and assess the security of a DMFB from all possible malicious aspects that relate to a potential attacker. We identify strengths and weaknesses of an attacker based on his role in the design, manufacturing, and use of the DMFB.

- We identify assay-outcome manipulation attacks and describe the basic security vulnerabilities associated with these attacks. A case study based on a colorimetric assay for glucose testing is used.

- We identify the challenges that arise from attacks in the presence of error-correction and error-recovery mechanisms in cyber-physical DMFBs. We show how to overcome these challenges and develop attacks on cyber-physical DMFBs with built-in error recovery.

- We discuss our benchtop experimental study that investigates threats associated with DNA preparation in gene-expression analysis. We present real-life demonstration of DNA-forgery attacks along with the resulting outcomes.

The rest of the chapter is organized as follows. Section 9.1 assesses threats associated with DMFBs. Section 9.2 describes attacks of assay-outcome manipulation on enzymatic glucose assay. Section 9.3 discusses attacks on DMFBs in the presence of built-in cyber-physical error-recovery mechanisms. Section 9.4 demonstrates security threats associated with DNA preparation in gene-expression analysis. Section 9.5 concludes the chapter.

9.1   Threats Assessment of DMFBs

Advances in digital-microfluidics technology offer tremendous benefits for enzymatic analysis, DNA analysis, immunoassays, toxicity monitoring, and clinical diagnostics. It has been also considered as a means to counter bio-terrorism [69]. On the other hand, since standard CMOS is an attractive technology option for DMFBs [187], they may be a target of attacks demonstrated on CMOS ASICs [130] and FPGAs [267]. Similar to their counterparts in CMOS chip (ASIC or FPGA) design, attacks on DMFBs can target stealing of hardware intellectual property (IP) [97], chip and IP reverse engineering [43, 207], and chip counterfeiting [89]. However, in this section, we assess the security of DMFBs against attacks that are unique to biochip-based bioassays such as bioassay result-manipulation attacks.

Who is the Attacker?

The attacker could be a user of the DMFB or anyone associated with the design and manufacturing flow of a DMFB. Considerable interest has been generated in recent years in remote-access laboratories that can implement microfluidics-based or robotics-based automated procedures for running bio-chemistry protocols [266, 142]. For such remotely accessible lab-automation systems, the attacker can be anybody with internet access who can compromise the service provider's cybersecurity system. We consider the following two DMFB design flows.

General-Purpose DMFB Design Flow: This is an FPGA-like design flow, where a general-purpose DMFB is procured, i.e., it can run any bioassay. The sequencing graph of a bioassay is synthesized onto the DMFB, i.e., the corresponding actuation sequences is generated. Figure 9.1(a) shows the different participants in this design flow. This design flow is also applicable when designing cyber-physical DMFB systems where the synthesis step is repeated and a new set of actuation sequences is generated based on feedback from the sensors on the DMFB (Chapter 8). In this design flow, it is reasonable to assume that the biocoder—who converts an assay protocol into a sequencing graph and provides it to the designer—the DMFB designer, the tester, and the user are the same individual. Hence, one needs to consider two adversaries namely, the biocoder/designer and the CAD tool vendor.

Custom DMFB Design Flow: This is an ASIC-like design flow (Figure 9.1(b)), where the biocoder who controls the DMFB platform sends the biochemical protocol to the design house as a sequencing graph. He gets the actuation sequences and the fabricated application-specific DMFB from the design house and programs the DMFB with it. The DMFB platform runs the assay on the DMFB according to the actuation sequences. In this design flow, the CAD tool vendor, the biocoder, the designer, the foundry, or the tester could be a potential adversary.

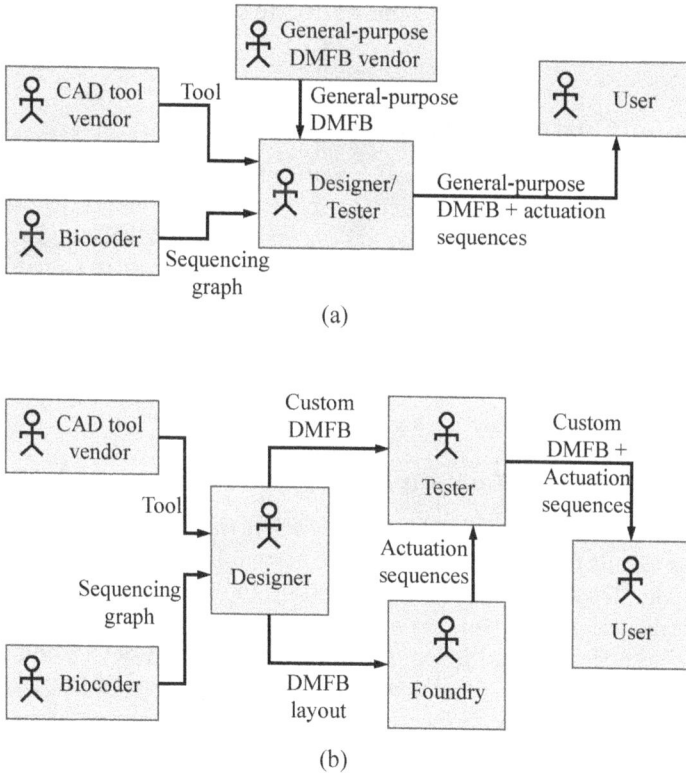

FIGURE 9.1
Participants in (a) a general-purpose DMFB design flow, and (b) custom DMFB design flow.

Attacks on General-Purpose DMFBs

We discuss the security implications of a general-purpose DMFB when the attacker is one of the following two individuals.

Malicious Biocoder/Designer: A malicious biocoder/designer can launch the strongest attack as he can tamper with the assay or other system software. The attack steps that a malicious designer can follow are: (1) tamper with the assay operation by modifying the assay or by altering the CAD steps; (2) generate the actuation sequences for the golden assay and the malicious assay; (3) deploy the golden actuation sequences and opportunistically replace it with the malicious actuation sequences. Since the designer is also the user of the DMFB, he can manually substitute the golden actuation sequences with the malicious actuation sequences. He can also do the same by using a DMFB control software controlled trigger. Therefore, he can launch manipulation and denial-of-service attacks.

Malicious CAD Tool Vendor: A compromised CAD tool can add malicious operations into the assay. For example, it can add operations that corrupt the assay outcome. This attack is similar to the always-active hardware Trojan attack [130]. The assay outcome will always be wrong and can be easily detected by a user.

Attacks on Custom DMFBs

In a custom DMFB design flow, there are numerous parties that could be malicious. Only a malicious biocoder can incorporate different types of triggers with the actuation sequences as he has access to the DMFB platform. Other parties, even if malicious, cannot incorporate different types of triggers as they do not have access to the DMFB platform. In the absence of triggers, malicious operations will be always active during the assay execution and can be easily detected by a user. Therefore, only the malicious biocoder can launch stealthy attacks.

Relation to Prior Work on Hardware Trojans

A hardware Trojan is a malicious modification that can be inserted into an integrated circuit to disable or destroy the system based on specific inputs or a specific time. Trojan threats are facilitated by the fact that the CMOS design flow is horizontal and distributed among different actors. A Trojan taxonomy classifies these threats and helps develop frameworks to detect and mitigate them [130]. Since DMFB design flows are similar to those of CMOSs; therefore, most of the contemplated hardware Trojans for CMOS chips are applicable to DMFBs as well. A DMFB Trojan taxonomy is shown in Figure 9.2. Trojans can be broadly categorized based on Trojan insertion phase, abstraction level, trigger mechanism, effect, and location [10]. We use this taxonomy to describe the attacks on DMFBs, assuming a horizontally distributed design flow.

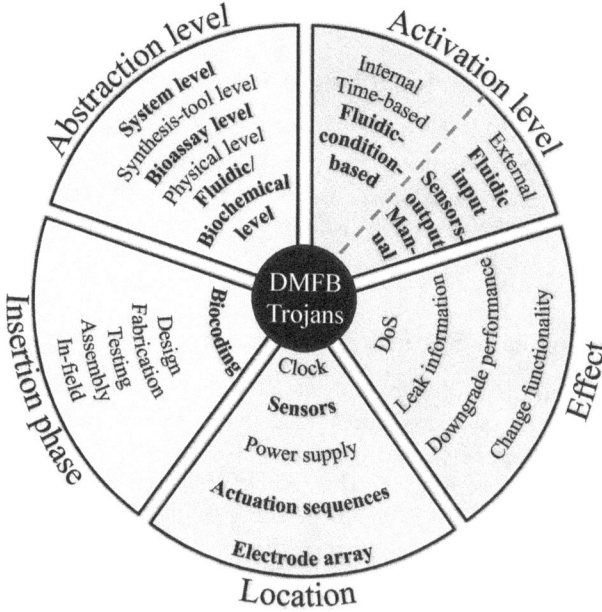

FIGURE 9.2
DMFB Trojan taxonomy. Trojan categories in bold text are unique to DMFBs. Note that the abstraction-level classification is aligned with the security assessment performed in Section 1.5.

The result-manipulation attacks launched during the general-purpose DMFB design flow can be classified as hardware Trojans inserted during the design phase and at the bioassay-level. Furthermore, these Trojans are externally triggered by a malicious biocoder/designer and the trigger mechanism is manual. Finally, these Trojans change functionality and is inserted into the DMFB processor. The only difference for a custom DMFB design flow is the insertion phase. Unlike the general-purpose DMFB design flow, the Trojan is inserted in the specification phase when the malicious biocoder provides the assay sequencing graph to the design house.

The investigated attacks against DMFBs explore Trojans inserted during the specification phase and that are manually triggered. As the vertically-integrated DMFB design flows evolve to a horizontally-distributed design flow, additional Trojan attacks demonstrated on ASICs and FPGAs [130] become relevant.

9.2   Manipulation Attacks on Glucose-Test Results

In this section, we highlight result-manipulation attacks on DMFBs wherein a malicious biocoder/designer manipulates an assay outcome by maliciously altering different parameters of the assay, such as sample concentration, incubation time, and mixing time. We also present a case study based on a glucose-testing assay.

Compromising the Sample Preparation

In DMFB-enabled quantitative analysis, one of the major tasks is to prepare samples and reagents of desired concentrations, which are then mixed together to perform the assay operation (Section 1.6). A typical DMFB uses LED-photodiode sensors. In an assay, the rate of reaction is proportional to the concentration of a specific element in the sample. The rate of reaction is equivalent to the rate of change of the optical absorbance [239]. Hence, if the concentration of the sample is altered, the optical absorbance measured by the photodiodes will be different.

A malicious biocoder/designer can tamper with the sequencing graph of the assay in order to manipulate the assay outcome. We show two such attacks that alter the sample concentration by tampering with the sequencing graph. Consider the case, where the required concentration of the sample is $\frac{1}{2^m}$, where m is an integer. The required concentration is achieved during the sample preparation phase of the assay by performing a series of m mix-and-split (dilution) operations using the original sample and the buffer[1] [214]. In each mix-and-split operation, the sample concentration is reduced by half. Let us assume that the test result will be considered to be positive only when the concentration of the sample is greater than or equal to CF_{high}. In another case, the test result will be negative when the sample concentration is less than CF_{low}.

Suppose the malicious biocoder/designer wants to manipulate the result to change it to positive. In this case, he saves one of the discarded sample droplets, e.g., the waste droplet of the i-th mix-and-split operation, such that the concentration $\frac{1}{2^i}$ of the discarded waste droplet is greater than CF_{high}. At the end of the sample preparation, the saved intermediate waste droplet is replaced by the target droplet. The malicious biocoder/designer knows about the assay operation corresponding to the sample preparation, and hence, he can simulate the sequencing graph of the assay and interpret the value of i. On the other hand, to produce a negative test outcome, the malicious biocoder/designer can perform additional mix-and-split operations on the target sample. Therefore, instead of the m-th mix-and-split operation, the target

---

[1]A buffer solution, such as 1 M NaOH, is mixed with a sample droplet in order to dilute the constituent sample.

droplet is generated at the n-th mix-and-split operation ($n > m$), such that the final concentration $\frac{1}{2^n}$ is less than CF_{low}.

Figure 9.3(a) shows the sequencing graph for sample preparation. The concentration of the sample droplet is diluted to $\frac{1}{2^4}$ by performing four mix-and-split operations. In each mix-and-split operation, the sample droplet is mixed with a buffer droplet and then split into two droplets of half the concentration. One of the two droplets (W) is discarded and the other one (I) is used for the next mix-and-split operation.

In this example, $CF_{high} = \frac{1}{2^4}$ and $CF_{flow} = \frac{1}{2^5}$. The dotted line in Figure 9.3(a) shows how the target droplet can be replaced by the waste droplet of the third mix-and-split operation, enforcing positive test result ($\frac{1}{2^3} > CF_{high}$). On the other hand, Figure 9.3(b) introduces an additional mix-and-split operation (highlighted by the dotted rectangle) to further reduce the concentration to $\frac{1}{2^5}$. Therefore, one can manipulate the assay outcome by adding only a few edges or nodes in the original sequencing graph.

The sample concentration can also be tampered with during DMFB synthesis. A malicious designer can alter the concentration either during the architectural-level or during the physical-level synthesis [118]. During the architectural-level synthesis, the malicious designer can modify the timing of scheduled operations. In the physical-level synthesis, malicious modifications

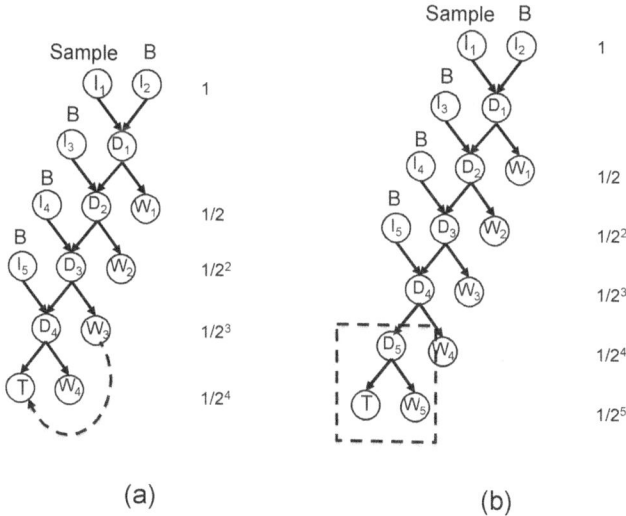

(a)   (b)

FIGURE 9.3
Malicious modification of the sequencing graph for sample preparation: (a) Replacing the final droplet with the waste droplet. (b) Inserting additional mix-and-split cycle. B, W, and T are the buffer, the waste, and the target droplets, respectively. The nodes I and D represent the dispensing and the dilution operations, respectively.

can be done in two different ways: (1) by changing the placement of mixing/dilution operations, such that the operations are assigned to mixers with unsuitable characteristics (Chapter 2); and (2) by tampering with the droplet routing to extend a droplet route, increasing the rate of liquid evaporation in DMFBs (Chapter 3). Such tampering can manipulate the results when a general-purpose DMFB is used. Malicious modification of results is not possible in custom DMFB design flow, due to lack of trigger mechanism.

Compromising the Incubation Time and Mixing Time

The incubation time and the mixing time of an assay have a major impact on the sample preparation. Incubation is widely used for cell culture, lysis, immunoassays, etc. For example, immunoassays are based on reactions that perform antigen-antibody bindings. One can target a specific antibody by designing a sample with the corresponding antigen attached to magnetic beads, as described in Section 1.1. Therefore, the sample droplet is mixed with one droplet containing magnetic beads with primary capture antibodies and another droplet containing reporter antibodies to report the binding progress. This mix is then incubated for a specific duration depending on the target percentage of antigen capture [234].

If the incubation time is not sufficient, the required percentage of binding may not be reached. Mixing time also has a similar impact on sample preparation, i.e., an insufficient mixing time might lead to incorrect results [194]. A malicious biocoder can tamper with the incubation/mixing time while providing the assay specifications to the designer. He can develop a result-manipulation attack by getting the golden and the malicious actuation sequences from the designer, and then by opportunistically replacing the golden actuation sequences with the malicious one. This attack is applicable to general-purpose and custom DMFB design flows.

Instead of directly changing the mixing time, a malicious biocoder can change the mixer geometry during resource binding. Mixing via a 4×2 two-dimensional array takes less time than that via a 4×1 linear array. The malicious biocoder can also manipulate the incubation/mixing time by altering the timing of scheduled operations (architectural-level synthesis) or by tampering with droplet routing (physical-level synthesis).

Case Study: Attacks on Glucose-Test Results

We use in-vitro measurement of glucose, which is a widely used quantitative-analysis clinical-diagnosis method for diabetes mellitus (hyperglycemia), as a case study. According to data from the Centers for Disease Control and Prevention (CDC), in 2015 alone, 26.3 million people in the U.S. were diagnosed with diabetes [75]. A diabetic patient has to undergo regular glucose test for proper monitoring. Based on the blood glucose level, the amount of insulin to be injected into the patient is determined.

We demonstrate two attacks on the in-vitro measurement of glucose in serum and show that a malicious biocoder/designer can manipulate the assay outcome. We consider a general-purpose DMFB design flow. The malicious biocoder/designer can generate either positive or negative test results irrespective of the original glucose concentration in the serum. An erroneous positive test result (high blood glucose) could trigger a high dose of insulin injection into the patient's body, which may lead to a life threatening hypoglycemia. Similarly, an erroneous negative test result (low blood glucose) may further worsen the hyperglycemia (the patient remains untreated). In both cases, the patient's life is endangered. Moreover, the clinical laboratory under the attack may be subjected to litigation and lawsuits due to inaccurate test results [127].

In-Vitro Glucose Test: In-vitro glucose test is used to determine the concentration of glucose in human physiological fluids, such as serum. An obvious application of this economical test is the determination of the blood sugar level[2]. The bench-top sequence for this test, known as glucose assay, is realized on a DMFB as a colorimetric assay, in which the color change is detected using an absorbance measurement system consisting of a light emitting diode and a photodiode [239].

This assay measures the glucose concentration level in a blood sample by constructing the glucose calibration curve (Figure 1.12) via serial dilutions of the standard glucose solution, as described in Section 1.6. The X-axis represents the different concentrations formed by these dilutions (in mg/dL) and the Y-axis represents the rate of reaction quantified by the change in absorbance degree reported as AU/sec (absorbance unit per second). This curve helps interpolate the concentration of the glucose sample under test. As shown in Figure 1.12, the reaction rate of the sample is a point on the Y-axis and the corresponding point on the X-axis is the sample concentration.

DMFB Attack Trigger Mechanism: Suppose the malicious biocoder/ designer wants to manipulate test samples of a specific patient. In this case, the malicious biocoder/designer can trigger it using the source id. He can mark the targeted source ids in the database. The trigger program is then activated when the targeted sample is injected to the chip (a barcode scan system can be used to register/identify the sample before operation starts [80]) and the registered identity matches the target source id. Once activated, the malicious actuation sequences are executed instead of the original one.

Attack Model: The malicious biocoder/designer can manipulate the assay by tampering with the calibration curve shown in Figure 1.12. The calibration curve can be tampered with when the DMFB is calibrated using standard glucose solutions. The malicious biocoder/designer alters the concentrations for one or more of the standard glucose dilutions and produces a wrong calibration curve. He can also manipulate the assay by changing the concentration of the sample itself. Note that the LED-photodiode sensor in this assay is not designed to identify malicious modifications. Its sole purpose is to report the rate change in absorbance in the enzyme-kinetic reaction [239]. As described

---

[2]Normal/safe levels of blood glucose (measured in mg/dL) depend on several factors, such as age, food activity, interference with other diseases, etc.

earlier, known concentrations of glucose solution are used to plot the calibration curve. On this curve, the rate of reaction measured by a fluorescence detector is plotted against the glucose concentration. After drawing the curve, the rate of reaction for a glucose sample is used to determine the sample concentration. To demonstrate the attacks, we consider the following three scenarios:

1. *Golden execution:* No attack is carried out.

2. *Attack 1:* The concentration of the glucose sample is modified via a malicious dilution operation.

3. *Attack 2:* The calibration curve is manipulated by tampering with the concentrations of the glucose solution during calibration.

Golden Execution: The sequencing graph shown in Figure 9.4 describes the golden execution for the glucose assay. The sequencing graph consists of four independent reaction chains 1, 2, 3, and 4, measuring the rate of reaction for a blank/buffer droplet (chain 1), glucose solution concentrations 800, 400, 200, 100, 50, 25 mg/dL (chain 2), glucose solution concentrations 300, 150, 75 mg/dL (chain 4), and the glucose sample (chain 3). The calibration curve is generated using the reaction chains 1, 2, and 4, while the reaction chain 3 is used to determine the glucose concentration of the sample.

Attack 1: The malicious biocoder/designer tampers with the assay result by changing the concentration of the glucose sample as shown in Figure 9.5.

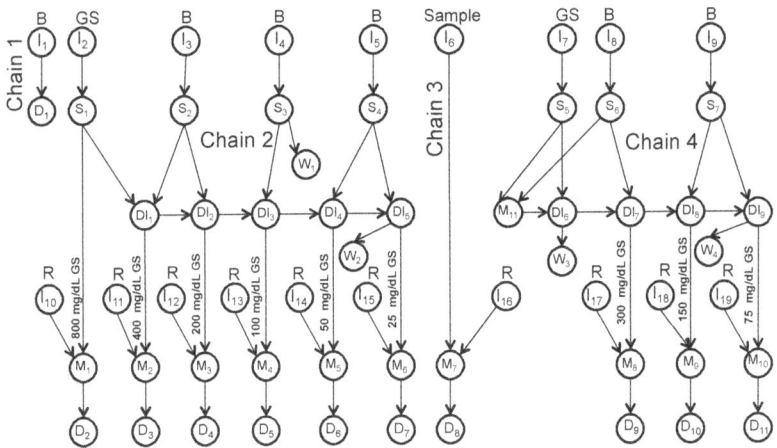

FIGURE 9.4

Golden execution: B is the 1.4 μL buffer droplet, Sample is the 0.7 μL glucose sample droplet, R is the 0.7 μL reagent droplet, GS is the 1.4 μL 800 mg/dL glucose solution droplet, and W_i is the waste droplet. D_i, Dl_i, S_i, M_i, and I_i are the detection, dilution, splitting, mixing, and dispensing operations, respectively.

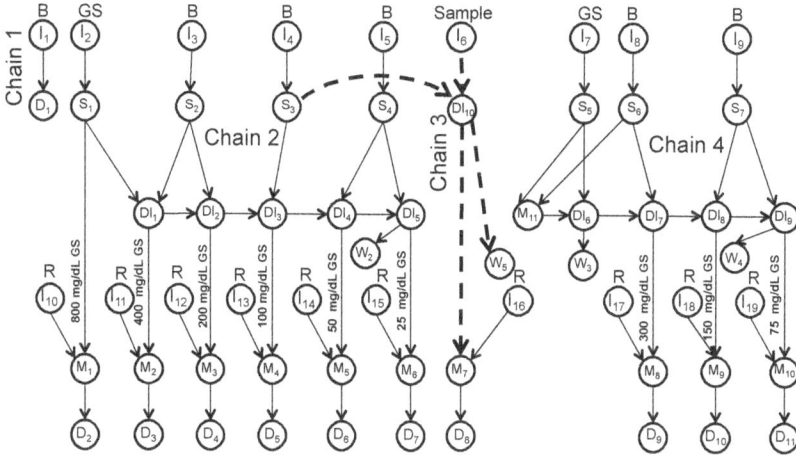

FIGURE 9.5

In Attack 1, the waste buffer droplet generated by the splitting operation S_3 is used to dilute the sample droplet in Dl_{10}. The thick dotted lines show the changes with respect to the golden sequencing graph.

The thick dotted lines show the changes in the sequencing graph compared to the golden sequencing graph. The waste buffer droplet W_1 generated from S_3 is mixed with the glucose sample droplet of I_6 and then diluted in Dl_{10}. Since the concentration of the glucose sample is halved, the result of the assay execution will be wrong. The user is unaware that a waste buffer droplet is used for tampering with the sample concentration. Using the golden calibration curve shown in Figure 9.6, the user will interpret the result as follows. In the golden calibration curve, the dots are the standard sample points corresponding to glucose solution concentrations (75, 150, ..., 800 mg/dL). The user will interpret the sample concentration as 110 mg/dL instead of the original concentration of 220 mg/dL. Hence, the patient may not be treated with the medication for high blood sugar, which could be life threatening.

If the patient or the medical practitioner wants to verify the result then there are two possible options: either to repeat the same test on the same DMFB, or to test it on a different DMFB by a different lab. It is highly likely that the test, when repeated by a different lab, will yield the correct result.

Attack 2: The malicious biocoder/designer tampers with the golden calibration curve to have the resulting reported concentration of the glucose sample different from (either higher or lower than) the golden value. We will show how the reported concentration of the glucose sample can be made higher than the golden value. The attack is performed by tampering with the sequencing graphs for reaction chains 2 and 4 to generate a malicious calibration curve. The two waste buffer droplets generated from D_1 and S_3 in the golden

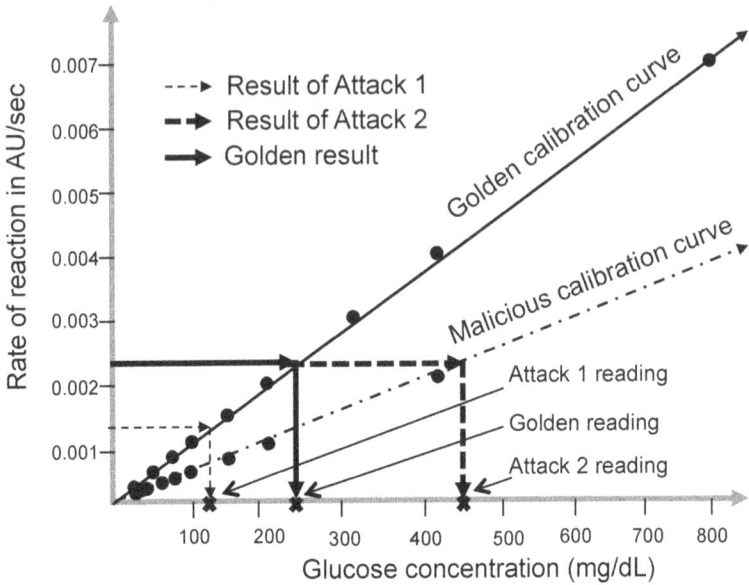

FIGURE 9.6
Malicious glucose concentrations: The thin dotted lines represent the modified glucose concentration measured on the golden curve. The thick dotted lines represent the glucose concentration measured on the malicious curve.

sequencing graph are used for this purpose. The malicious sequencing graph for such an attack is shown in Figure 9.7.

The thick dotted lines show the changes with respect to the golden sequencing graph. The waste buffer droplet (after D_1) in the reaction chain 1 is merged with the glucose solution (the droplet generated from I_2) in the reaction chain 2, thus diluting the entire reaction chain 2. The glucose solution concentrations in the reaction chain 2 are reduced to (400, 200, 100, 50, 25, 12:5 mg/dL) half of their golden values. Similar effect can also be seen in the reaction chain 4, where the waste buffer droplet generated from S_3 is mixed with the glucose solution droplet generated from I_7. The dotted curve in Figure 9.6 shows the malicious calibration curve generated by Attack 2.

The DMFB user is unaware that the calibration curve is malicious. The user will interpret the result using the malicious calibration curve (the dotted curve in Figure 9.6). The result will show a higher concentration of glucose compared to the golden result. As the figure shows, the original concentration is 220 mg/dL when the golden calibration curve is used. Following Attack 2, the measured concentration is 440 mg/dL since the malicious calibration curve is used. Hence, the patient will be falsely alarmed and may receive a high dose of insulin, if this is the only test that he relies on.

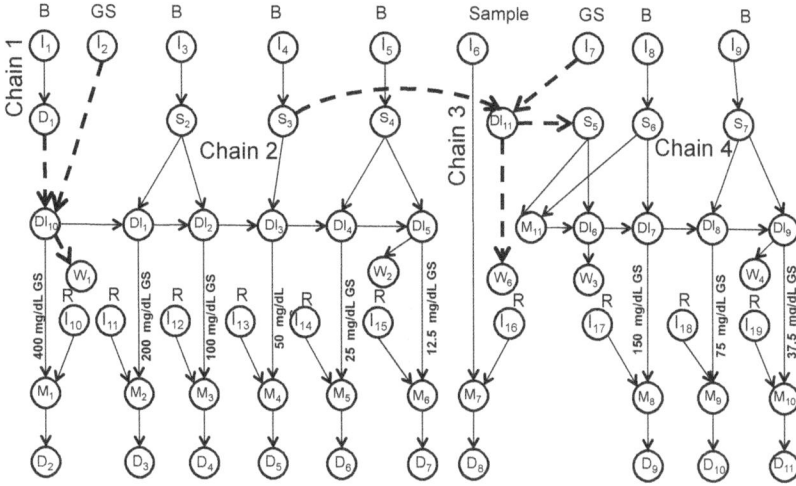

FIGURE 9.7
In Attack 2, the discarded buffer droplets of D_1 and S_3 are mixed with the droplets of I_2 and I_7, respectively, diluting the reaction chains 2 and 4, respectively.

A practical step-by-step scenario for the attack described above is as follows.

- *Step 1:* The patient visits the pathology department and his blood sample is collected, labeled with the patient's barcode, and forwarded to the diagnosis lab.

- *Step 2:* In the lab, the pathologist scans the sample using the barcode reader (connected to the target system) and then selects the in-vitro diagnostic test (actuation sequences) in the system.

- *Step 3:* The trigger program is activated upon the scan of the barcode and alerted that the sample belongs to the target patient. The trigger program then selects the Attack 2 actuation sequences instead of the original actuation sequences.

- *Step 4:* When the malicious actuation sequences are executed, a high glucose concentration in the sample will be detected by the assay operation.

- *Step 5:* The patient will be falsely treated with high dose of insulin.

Experimental Results

The golden, Attack 1, and Attack 2 sequencing graphs are executed using an open-source DMFB tool [86] on a 17×31 electrode-array DMFB with 7 input reservoirs[3]. A 100 Hz clock was considered for actuating the electrodes. The DMFB design times for the golden, Attack 1, and Attack 2 assays are 35, 39, and 62 milliseconds, respectively, while the assay execution times are 8.5, 9.26, and 10.46 seconds, respectively. Attack 1 is difficult to detect, since the difference in the DMFB synthesis time (35 ms versus 39 ms) and assay execution time (8.5 s versus 9.26 s) are negligible. This is because the attack alters the glucose sample using one additional dilution operation. Attack 2 impacts a large portion of the glucose assay, since it alters the concentrations of the glucose solution. The difference between the golden assay execution time and Attack 2 assay execution time is 1.9 seconds (8.5 s versus 10.46 s). The difference in the DMFB synthesis time is only 23 ms (39 ms versus 62 ms). It is unlikely that the user can notice such a small change in the DMFB assay execution times.

9.3 Attacks in the Presence of Cyber-Physical Integration

As explained in Chapter 8, error recovery can be introduced to DMFBs via cyber-physical integration. Runtime operational errors related to droplet mobility, size, or sample concentration can be detected using an integrated sensor or a connected charge-coupled device (CCD) camera. A feedback signal from this monitoring device can be used to invoke a procedure for error correction if needed.

From a biocoder's perspective, error-recovery techniques might expose the malicious modifications in the assay. In order to make the attack stealthy, the biocoder not only needs to manipulate the sample as shown in Attack 1 and Attack 2 in Section 9.2, but also needs to bypass the error-recovery techniques. However, from a designer's perspective, the current error-recovery techniques are relatively easy to bypass and, in fact, can be used as a backdoor for sensor-spoofing or information-leakage attacks [231] (Figure 9.2).

Attacks on a General-Purpose Cyber-Physical DMFB

The malicious designer needs to bypass the checkpoints and he can do so by removing them from the malicious sequencing graph. The checkpoints (similar to checks #1 and #2 in Figure 8.2) are meant for the control software and not for the users, and they can be incorporated with the help of back-end software tools, as demonstrated in Chapter 8. When a fault-free DMFB is used, the removal of checkpoints will not impact the assay outcome. However, if a faulty DMFB is used and the user knows the expected assay outcome *a priori*, only

---

[3]We used list scheduler, left-edge placer, and maze router for the DMFB design.

then he will infer that the DMFB is faulty and the deployed error-recovery technique will fail to detect the fault. It should be noted that the attack does not permanently bypass the error-recovery mechanism. Rather, it is bypassed only for the specific samples from the targeted individual as mentioned in the five steps of the last paragraph of Section 9.2. We highlight that this is a novel attack within the taxonomy shown in Figure 9.2 and it has not been reported or explored in the context of ICs. Therefore, the user will not have prior knowledge of the attack.

Alternately, the malicious designer can tamper with the threshold against which intermediate results, e.g., frequency-coded signals in Figure 8.7, are compared so that the error-recovery system does not report an error.

Attacks on a Custom Cyber-Physical DMFB

In this design flow, the malicious biocoder provides a malicious sequencing graph to the design house. The design house incorporates checkpoints into the malicious sequencing graph by following their standard design process for the cyber-physical DMFB with error recovery [159, 296, 108]. Therefore, the malicious biocoder will receive a set of actuation sequences with support for error recovery for the malicious assay. Now, although the malicious biocoder can replace the golden actuation sequences with the malicious actuation sequences, the checkpoints remain in the malicious assay and the error-recovery component remains in the control software. Therefore, while executing the malicious assay, the error-recovery software will compare the intermediate results generated at the checkpoints with the golden values. Since the intermediate results for the malicious assay will be different from those for the golden assay, an error is generated, triggering re-synthesis of the golden assay. This in turn overwrites the malicious actuation sequence with the golden actuation sequence. Thus the attack will fail.

Attacks on the Control Software

Error-recovery techniques in a cyber-physical DMFB pose new challenges to an attacker (malicious biocoder/ designer), specially in a custom DMFB design flow. For a cyber-physical DMFB that uses a control software for error recovery, the only option for the malicious biocoder is to tamper with the error-recovery software. In this case the attack will be the same for both design flows. A malicious biocoder can tamper with the error-recovery portion of the control software to bypass the error-recovery mechanism. This requires reverse-engineering the control software binary [208].

9.4 DNA-Forgery Attacks on DNA Preparation

In this section, we demonstrate DNA-Forgery attacks using a benchtop experimental study. We highlight the impact of such attacks on the resulting DNA amplification.

DNA Preparation, Amplification, and Analysis Using Traditional PCR

Recall that the objective of PCR-based gene-expression analysis is to study the transcriptional profile of a certain reporter gene, e.g., green fluorescent protein (GFP), in a given sample. In Chapter 2, we described our benchtop study for quantitative PCR (qPCR), also known as real-time PCR, where the amount of DNA can be determined in real-time using DNA-amplification curves. Herein, we describe a similar study for determining the amount of DNA, but using traditional PCR instead. Unlike qPCR, traditional PCR measures the amount of accumulated DNA segments, referred to as amplicons, at the end of the PCR cycles by using agarose gel electrophoresis [146]. Traditional PCR is less expensive than qPCR and it is suitable for DNA fingerprinting that distinguishes between samples based on the genetic material. Hence, it is widely used as a reference model in many settings including DNA forensics and clinical diagnostics.

Similar to our qPCR study in Chapter 2, traditional PCR typically requires three sample pathways to be processed in parallel. One pathway represents PCR processing of the *sample under investigation* (SUI). A second pathway represents a *positive control* (PTC), which is expected to produce visible amplicons at the end of the reaction. A third pathway represents a *negative control* (NTC), which should not show any amplicons in the reaction result. PTC is used to examine the efficiency of the reagents and the PCR procedure, whereas NTC is used to detect carryover contamination. If amplicons are observed from SUI, then it is concluded that the given sample belongs to a cluster of sample types that exhibit gene-expression activity at the locus of this particular amplicon. Likewise, if no amplicons are observed, i.e., similar to the NTC pathway, then the sample belongs to a different cluster. Obviously, both PTC and NTC play a key role in the determination of the sample identity; therefore, they must be processed carefully.

Agarose gel electrophoresis is a routinely used method for separating protein, DNA or RNA molecules [146]. Hence, PCR products (amplicons) are size-separated by the aid of an electric field where negatively charged DNA molecules migrate toward an anode (positive) pole. The shorter the amplicon[4], the further the sample will reach on the gel. Figure 9.8 shows the outcome of gel electrophoresis for four samples used in our study. Samples S_2 and S_3 are

---

[4]It indicates a lower molecular weight.

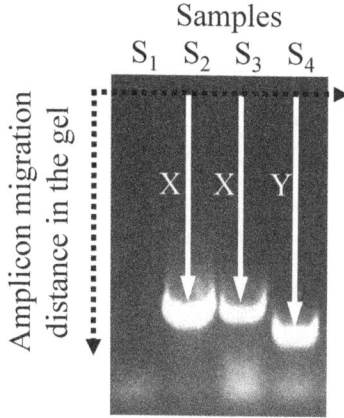

FIGURE 9.8
Measurement of DNA amplification using gel electrophoresis.

samples that exhibit amplicon generation through DNA amplification and the target amplion contains a large number of base pairs. Hence, the molecular-weight size markers (shown as white bands called "DNA bands") associated with these samples indicate that a short distance (X) has been migrated. Sample S_4 also exhibits DNA amplification, but the resulting amplicons contain a smaller number of base pairs[5]. Therefore, the DNA band associated with S_4 indicates that S_4 has traveled a longer distance; see Y. Sample S_1 represents an NTC pathways and it does not show a white band since no DNA amplification has occurred.

The process of DNA amplification is highly precise and it is sensitive to any modifications applied to the DNA-preparation process. Therefore, the results we obtain from agarose gel electrophoresis can be used to capture potential attacks on DNA-processing flows.

Attack Models for Gene-Expression Analysis

Despite stringent precision requirements of PCR implementation, DNA-preparation steps for PCR can be regarded as potential attack surfaces that can be exploited by a malicious adversary, who may be interested, for example, in tampering with the final results of a DNA forensic analysis. The tampering can result in the complete destruction of evidence (the true DNA-amplification profile), or modification of evidence such that it produces a misleading result. To demonstrate this capability, we carried out a benchtop experiment that executed DNA-amplification analysis on 6 different samples. In this experiment, 4 runs of PCR amplification were performed in parallel. The first run

---

[5]A gene mutant with a smaller number of base pairs can be created from another by enabling enzymatic gene deletion.

represents the golden, i.e., mainstream, implementation of the protocol, where all the protocol settings are adjusted normally. The remaining three runs represent malicious implementation of PCR, where DNA preparation was deliberately altered. More specifically, we "simulated" three potential attacks on the PCR bioassay[6]; these attacks were designed to demonstrate the following malicious behavior:

- *Attack 1:* It is an attack that causes all collected samples to maliciously report a positive signal of DNA amplification. We refer to this attack as *positive denial-of-service (PDoS) attack.*

- *Attack 2:* It is an attack that causes all samples to maliciously report a negative value of DNA amplification (i.e., no DNA amplification occurs). We refer to this attack as *negative denial-of-service (NDoS) attack.*

- *Attack 3:* It is an unsophisticated attack that replaces samples. We refer to this attack as *sample switching.*

Attack 1 and Attack 2 cause complete destruction of the evidence, whereas Attack 3 results in modification of the evidence. Details of the implemented benchtop experiment are presented below.

Benchtop Experiment of Malicious DNA Preparation

As mentioned earlier, we performed an experiment that involved 4 parallel runs of PCR amplification. Recall that the first run was specified to include the only trusted implementation of PCR assay (golden assay). Herein, we first describe the experiment setup for the 4 parallel PCR runs. Next, we explain the steps we used to implement the golden as well as the malicious PCR reactions.

Experiment Setup and DNA Preparation: A typical implementation of a PCR run consists of four main steps: (1) sample preparation; (2) reagent preparation; (3) conducting PCR reaction and amplification; (4) analysis using gel electrophoresis (Figure 1.11). Hence, to start the experiment, we first prepared the crude DNA samples, which were isolated from a culture of *S. pombe* cells (via cell lysis and purification process) after the cell culture had been grown overnight. Figure 9.9 shows the starting DNA samples we used in the experiment. Second, for each PCR run, we prepared a PCR mixture tube, also known as a master mix, that consists of 36 μL of buffer, 18 μL of deoxynucleotides (dNTPs), 36 μL of primers, 7.2 μL of $MgCL_2$, 82.8 μL of H_2O, and 3 μL of the Taq polymerase, each quantity was pipetted carefully into the tube starting with the largest component, i.e., H_2O. Since our experiment included 4 parallel PCR runs, we prepared 4 PCR master mixes, as shown in Figure 9.10. To ensure complete diffusion of PCR reagents, PCR tubes were finally spun using a micro-centrifuge (Figure 9.10).

---

[6]Similar evaluation can be applied to microfluidic PCR.

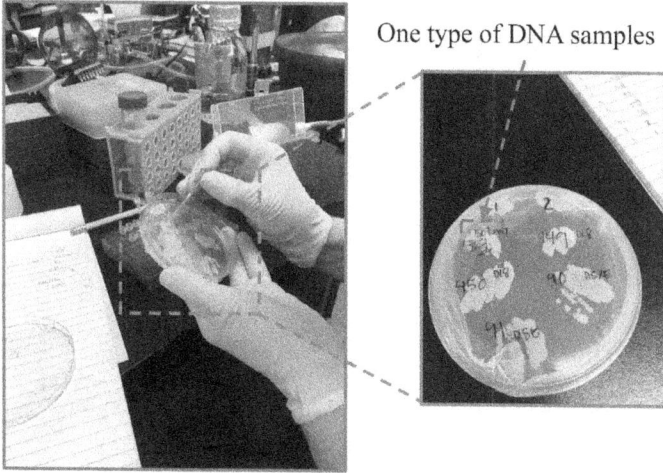

FIGURE 9.9
Crude DNA samples used in the experiment.

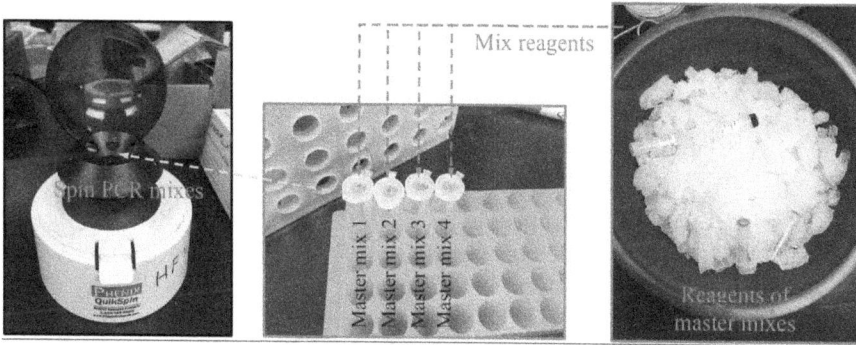

FIGURE 9.10
Preparation of 4 PCR master mixes.

Golden vs. Malicious Reactions: The steps that have been described thus far do not include any adversarial reactions. That is, the 4 master mixes were clean and they were prepared using the same quantities of reagents. To demonstrate the attacks described earlier, we opted to alter the contents of master mixes 2 and 3 to reflect Attack 1 (PDoS) and Attack 2 (NDoS), respectively, while master mixes 1 and 2 were kept clean to implement the golden reactions and to implement Attack 3 (sample switch), respectively[7]. To implement PDoS, we added a small quantity of PTC (ura4+) to master

---

[7]Sample-switch attack does not require alteration to the master mix.

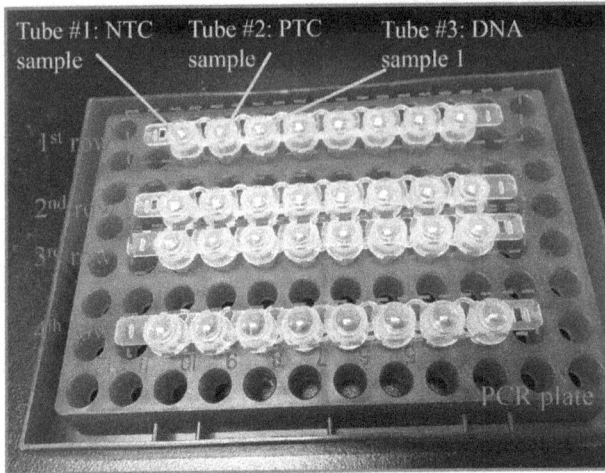

FIGURE 9.11
Preparation of the PCR plate.

mix 2. We also implemented NDoS by adding ethylenediaminetetraacetic acid (EDTA) to master mix 3.

Having prepared the samples and PCR master mixes, we prepared a PCR plate that included all the chemicals needed for the 4 parallel runs; see Figure 9.11. We therefore divided a set of empty tubes into 4 rows, where each row contained 8 tubes. The reason for including 8 tubes (not only 6 tubes associated with the 6 DNA samples) was that two additional tubes per row (i.e., per PCR run) were needed for PTC and NTC, as described earlier in this section.

The tubes that were located at the first row on the plate were used for chemical reactions associated with the golden PCR assay. In each tube, we pipetted 20 μL of PCR reagent, which was transferred from the first (golden) master-mix tube. Next, we added 1 μL of crude extract of each DNA sample to a separate tube, starting from tube #3. We also pipetted 1 μL of NTC (distilled water) into tube # 1 and 1 μL of PTC (ura4+) into tube #2; thus using tubes #1 and #2 for referencing (Figure 9.11). We repeated the same steps to fill in the tubes located at the second and the third rows, where master mix 1 was replaced with master mixes 2 and 3, respectively. Finally, the tubes located at the four row were filled in using DNA samples that were randomly selected; thus triggering Attack 3 (sample switching).

Having filled in all the tubes on the PCR plate, the tubes were closed, then placed in a balanced micro-centrifuge, and spun for a few seconds. Immediately after the spin, we placed the reaction tubes in the thermal cycler, which had been programmed to perform DNA amplification by continuously raising and lowering the temperature of the PCR-plate content in discrete, pre-programmed steps. Figure 9.12 shows the transfer of PCR samples into the thermal cycler. After the PCR program had finished, we transferred the

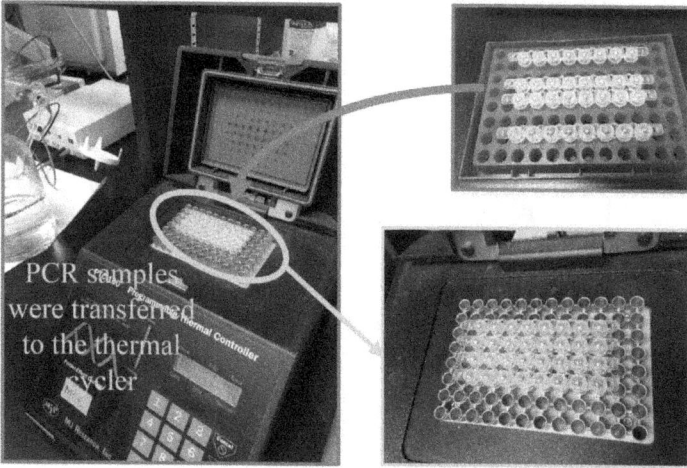

FIGURE 9.12
Transfer of PCR tubes into the thermal cycler.

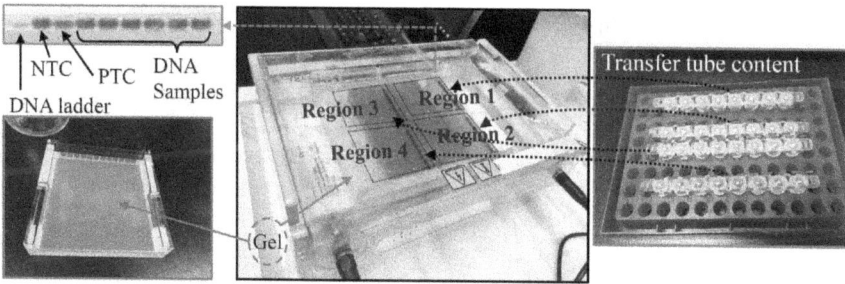

FIGURE 9.13
Transfer of PCR tubes into the gel electrophoresis device after thermal cycling.

PCR-plate contents into the gel electrophoresis apparatus, allowing the chemical solutions to migrate over the gel with the aid of an electric field; Figure 9.13 shows the transfer of 8 × 4 samples into the gel. Note that the gel was virtually divided into four sections, and each section was loaded with samples from a certain row of tubes on the PCR plates. This approach allowed us to easily compare the outcomes of all the PCR runs. Also, in each section, we pipetted a specific chemical reagent, known as DNA ladder, which has fragments of DNA of different sizes, allowing us to benchmark the results of the PCR-related chemical solutions. Gel electrophoresis was run for an hour, then the gel block was transferred to a chamber that contains a source of UV light in order to visualize the DNA bands. Figure 9.14 shows a view of the gel after the electrophoresis process, viewing the migration of PCR samples. Result interpretation is discussed next.

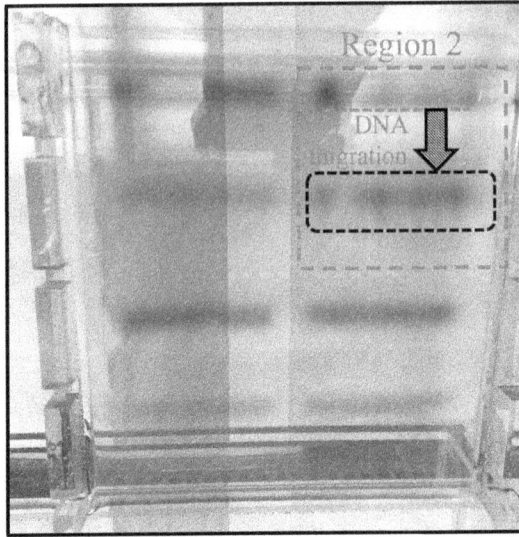

FIGURE 9.14
Migration of PCR samples through the gel.

Interpretation of Gel-Electrophoresis Results

Figure 9.15 shows the results obtained by gel electrophoresis. As expected, the DNA-ladder plot is divided into four regions: (1) the top-left region (Region 1) contains the DNA bands associated with the golden PCR reaction; (2) the top-right region (Region 2) represents the DNA bands based on Attack 1 (PDoS); (3) the bottom-left region (Region 3) shows the DNA bands for Attack 2 (NDoS); (4) the bottom-right region (Region 4) contains the DNA bands based on Attack 3 (sample switching). In all of these sections, the result of the benchmark (the DNA ladder) is located at the leftmost column (column 1).

By analyzing Region 1 (trusted result), we observe that column 2 does not show any DNA bands; this result is expected since this column is associated with the NTC tube where DNA amplification does not occur. In contrast, column 3 shows a DNA band, capturing the effect of DNA amplification within the PTC tube. The remaining columns show various results of DNA amplification, depending on whether the target gene expresses at the target amplicon locus and also the length of the amplicon (number of base pairs). The DNA samples represented by columns 4 and 8 exhibit DNA amplification and they have amplicon mutants that are longer (contain more base pairs) compared to those in the DNA samples represented by columns 5 and 9.

In Region 2, we observe that all the columns exhibit the same high DNA-amplification profile, indicating that all the samples were deliberately manipulated to express the target gene. This behavior shows the impact of maliciously

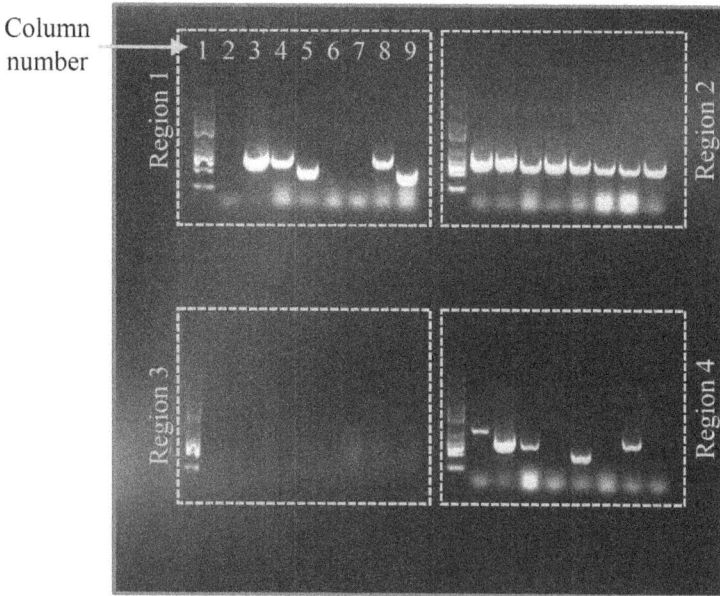

FIGURE 9.15
Results of DNA amplification for the 4 PCR reactions.

adding PTC reagent to all samples, causing PDoS attack. This result is invalid and cannot be used by clinical lab for assessing a patient condition or by a forensic lab to distinguish DNA samples.

Similarly, in Region 3, no DNA amplification is reported at all samples since no DNA bands exist. This result indicates that all the samples, including the reference PTC sample at column 3, were tampered with to suppress the expression of the target gene, causing NDoS attack. This result is also invalid and cannot be used to distinguish DNA samples. Note that both PDoS and NDoS attacks can be easily detected.

Finally, the different DNA bands observed in Region 4 indicate that no PDoS or NDoS attacks were launched. However, the profile of DNA bands shown in Region 4 is different from that of Region 1, indicating that DNA samples were likely switched or replaced with other samples, i.e., subjected to sample-switching attack. Note that this attack is hard to detect unless the switching action impacted either the PTC or the NTC samples, causing abnormal behavior at either column. For instance, in column 1 of Region 4, we observe a DNA band, meaning that DNA amplification has occurred. Since column 1 is associated with the NTC sample, which is supposed to disallow DNA amplification, this observation is sufficient to prove that either the NTC sample was tampered with/contaminated or the samples (including NTC) have been switched.

In many cases, Attack 3 can be stealthy and hard to observe, especially if the attacker is aware of the locations of the NTC and PTC samples. Therefore, in Chapter 10, we present a benchtop study that provides an efficient countermeasure technique against Attack 3 based on DNA barcoding.

9.5   Chapter Summary

We have reported an assessment of the security risks associated with today's quantitative-analysis frameworks that can jeopardize the healthcare or defense industry. We have presented three forms of threats, two of which are related to DMFBs: (1) assay-result manipulation threats, which can have a catastrophic effect on the integrity of the DMFB assay outcomes; (2) threats related to cyber-physical system components, which may result in improper sensor calibration or adversarial control behavior; (3) DNA-forgery threats, which can deliberately destroy or switch DNA samples. We have described a benchtop study that evaluates the security implications of malicious DNA analysis.

An important next step in our research is to develop fluidic defenses against attacks on basic biomolecular analysis.

10

Security Countermeasures of Quantitative-Analysis Frameworks

In Chapter 9, we examined security vulnerabilities of quantitative-analysis frameworks. This chapter continues our security analysis by presenting two effective countermeasures that secure the execution of biomolecular assays. The proposed countermeasures target: (1) securing the execution of a DMFB-enabled assay against intellectual property (IP) piracy [12], and (2) protecting genomic samples against forgery threats. The former defense mechanism introduces the concept of microfluidic encryption based on microfluidic logic gates, and the latter is developed based on a benchtop experimental study. The key contributions of this chapter are as follows:

- We introduce the application of microfluidic logic gates to secure DMFB designs. Next, we develop microfluidic multiplexer logic to enable fluidic input-based control of droplets.

- We propose a microfluidic encryption technique to protect assays against bioassay-level Trojans, DMFB IP piracy, overproduction, and counterfeiting. Analysis is performed to highlight the correlation between DMFB design parameters and a security metric.

- We introduce the concept of DNA barcoding to secure collected samples against biochemical-level forgery attacks. Basic security analysis is performed based on the characteristics of a DNA barcode, e.g., sequencing information.

- We describe a benchtop experimental study that demonstrates the barcoding defense mechanism.

The rest of this chapter is organized as follows. Section 10.1 presents our DMFB-enabled encryption technique. Section 10.2 explains the role of biochip's aging characteristics in reinforcing security and the associated security metric. Security analysis and simulation results corresponding to microfluidic encryption are presented in Section 10.3. Section 10.4 describes the principles of the DNA-barcoding defense strategy. The details of the barcoding benchtop study are provided in Section 10.5. Finally, Section 10.6 concludes the chapter.

10.1   Microfluidic Encryption

Robust countermeasures are needed to address the threats described in Section 9.2. Our premise is that protection at the microfluidic bioassay level will be especially effective to ensure the integrity of bioassay outcomes. The concept of hardware metering-based protection of IP with a secret key has been explored for CMOS chips [15, 42]. We propose to adapt this solution for DMFBs and incorporate microfluidic encryption into the synthesis of biochemistry protocols. This approach can be viewed as *encryption at the microfluidic level*, whereby the sequencing graph G_s for a bioassay protocol is transformed to a different sequencing graph $\widetilde{G}_s (G_s \subset \widetilde{G}_s)$ through a sequence of control data (the "secret key") that is known only to an authorized user. If the correct key is provided during assay execution, the desired sequence of microfluidic operations will be permitted and the bioassay will be enabled. On the other hand, if the key provided by a user does not match the secret key, the flow of droplets through the DMFB will be blocked and no detection results will be provided by the system.

Microfluidic Encryption Methodology

For microfluidic encryption, we propose the use of 2-to-1 fluidic multiplexers with two fluidic data inputs, one fluidic control input, and one fluidic data output. The control input can be viewed as one bit of the "secret key." A number of such multiplexers can be inserted into G_s to form $\widetilde{G}_s$. The sequencing graph $\widetilde{G}_s$ executes the correct assay operations only if these multiplexers are controlled by the correct secret key input. Since only one droplet is forwarded to downstream assay operations, the additional droplets at some point must be routed to an on-chip waste reservoir. To prevent an attacker from making an inference about droplet transportation to the waste reservoir, the additional droplets can be held in randomly chosen temporary locations on the DMFB, and discarded with the waste resulting from the biochemical procedures at the end of the assay.

Figure 10.1(a) shows a segment of the sequencing graph of a generic bioassay, where the two dispensed droplets in O_1 and O_2 are mixed in O_3. The droplet resulting from O_3 is again mixed in O_5 with the droplet generated from O_4. A detection operation O_6 is performed on the droplet that results from O_5. The encrypted sequencing graph is shown in Figure 10.1(b). The two input droplets to mixer O_3 are controlled by the control inputs of the multiplexers. The presence of a control droplet (logic 1, as explained below) is known only to the authorized user. An inverse implementation can also be considered, whereby the absence of a control droplet (logic 0) forwards the required droplet. We envision strategic insertion of several such fluidic multiplexers in the sequencing graph; the associated presence (logic 1) or absence

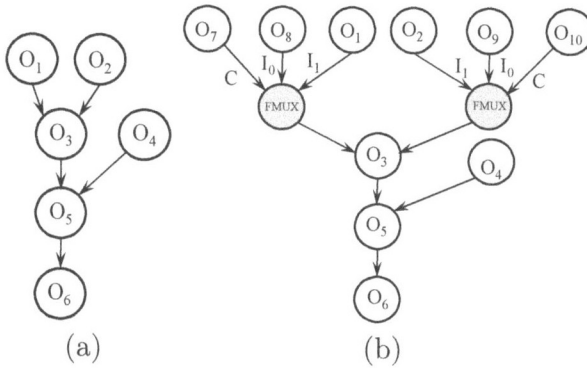

(a)   (b)

FIGURE 10.1
Microfluidic encryption: (a) An assay that mixes three droplets and then the detection operation is performed on the resultant droplet. (b) Encrypted sequencing graph with two fluidic multiplexers (FMUX) to control the droplet generated from O_1 and O_2. The correct control input, i.e., the key, ensure that the droplet O_1 and O_2 are selected.

(logic 0) of control droplets constitute the secret key. Only an authorized user would know the bit pattern that "opens" all the fluidic multiplexers to forward droplets as required.

Realization of a Fluidic Multiplexer

Digital microfluidic logic gates were introduced in [294] to enable built-in self-test (BIST) for DMFBs. A number of key logical operations (e.g., AND, OR, and inverter) were experimentally demonstrated using a fabricated DMFB. The main idea behind the design for microfluidic logic is that in a DMFB platform if a droplet volume is not large enough to have sufficient overlap with an adjacent electrode, it cannot move under a nominal actuation voltage. A logic '1' defines the presence of a droplet of unit or larger volume at an input or output port, while a logic '0' defines the absence of a droplet at an input or output port. Using the principle of electrowetting-on-dielectric, the basic gates AND, OR and NOT are implemented through basic droplet-handling operations such as transportation, merging, and splitting.

We adopt the concept of microfluidic logic gates to develop a fluidic multiplexer. Figure 10.2 shows an example of the fluidic multiplexer and its fluidic operations, where I_{FMUX} and II_{FMUX} are the inputs for the data droplets and C_{FMUX} is the control input. The output droplet is produced at Z_{FMUX}. The multiplexer takes 33 cycles to produce the output. Actuation sequences are crafted in such a way that it will first merge and split the droplets at electrodes 1 and 3. Then the droplet at 3 is transported to 1 to mix it with the

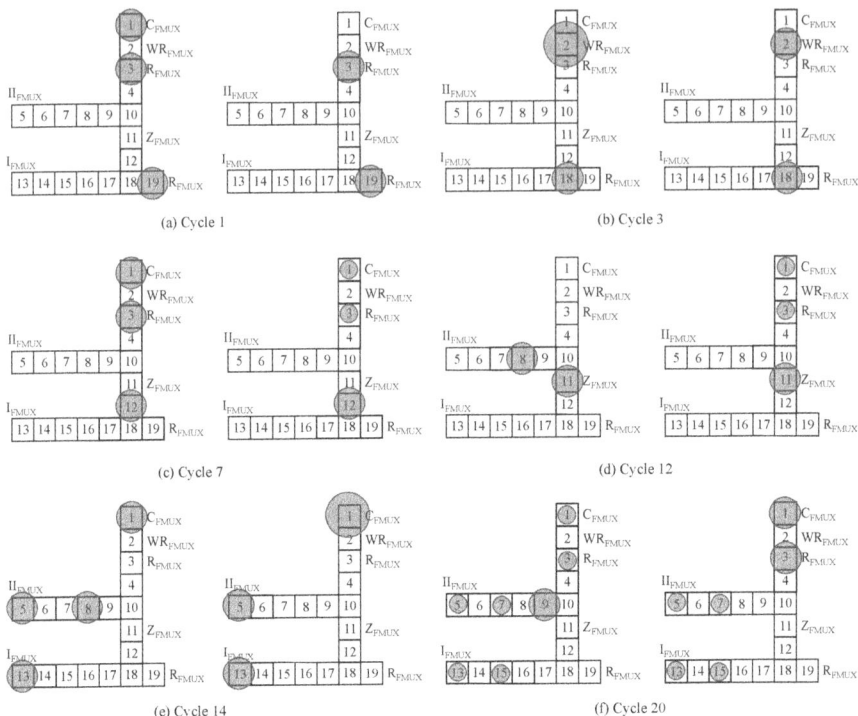

FIGURE 10.2
Fluidic multiplexer operation sequence. In each pair of figures, the figure on the left illustrates the presence of a control droplet while the figure on the right illustrates the absence of the control droplet. (a) Shows the initial state, (b) droplets at 1 and 3 merge, (c) droplet at 2 splits, (d) droplets at 3 and 12 move towards 9 and 1, respectively, (e) droplet at 1 splits, (f) droplet at 3 moves towards 13.

input droplet generated through I_{FMUX} and carry it to the output Z_{FMUX} while the other droplet of the split operation is discarded.

Next, the reference droplet at 19 is transported to 1, so that it can merge with any untransferable droplets (half-volume droplets). The reference droplet is split at 2 and one of the droplets is transferred to 13 to mix with the input droplet generated through II_{FMUX} and carry it to the output.

Figure 10.2(a) shows the initial state of the multiplexer for two different scenarios: the figure on the left shows the presence of a control droplet and the figure on the right shows the absence of the control droplet. Therefore, the presence of control droplet will have valid mix and split operations (shown on the left of Figure 10.2(b)). On the other hand, in the absence of the control droplet, only the split operation will take place and the resultant droplets will be of half-unit volume, which cannot be actuated (shown on the right of

Figure 10.2(b)). Similarly, the second reference droplet is used to transport the leftover droplets of the split operation. Again, if there are leftover droplets at 1 and 3, only then will the second reference droplet mix with them and split into unit-volume droplets (shown on the right of Figure 10.2(e) and (f)). Thus, the presence of a control droplet will lead to the transportation of the input droplet II_{FMUX} to the output and while I_{FMUX} will remain static in the form of two half-unit volume droplets. In the absence of the control droplet, the effect will be vice versa. Note that the input droplets are split into half-unit volume so that they cannot be directly actuated/transported.

Fluidic Multiplexer Synthesis

The fluidic multiplexer implements conditional fluidic operations, which is not compatible with the existing DMFB design synthesis framework. For example, as shown in the operations in Figure 10.2(b), if C_{FMUX} is present, then the corresponding mix-and-split of C_{FMUX} and R_{FMUX} will produce two unit-volume droplets. In that case, one droplet moves to electrode 9 to mix with droplets in 5 and 7, and the other droplet is transported to wash reservoir (WR_{FMUX}) as shown in Figure 10.2(c). On the other hand, in the absence of the droplet C_{FMUX}, the corresponding mix-and-split operation will produce two half unit-volume droplets, which cannot be transported and later mixed with the second reference droplet. The droplet volume, which could be half or full unit-volume, dictates the resulting mix; half-unit volume implies a mix with the second reference droplet, while full unit-volume implies a mix with input droplets at II_{FMUX}. This scenario is not supported by existing CAD tools [118].

Motivated by our synthesis methods in Chapters 2 and 3, which support "if-then-else" conditions, we propose a new synthesis approach to support conditional fluidic operations by incorporating certain routing information; e.g., a route has to include certain predetermined electrodes in the sequencing graph. Figure 10.3 shows the sequencing graph of the fluidic multiplexer. A dotted edge $\tilde{e}_s$ between nodes (v_{s1}, v_{s2}) defines that during routing, droplet v_{s1} has to be transported to the electrode storing v_{s2}, where $v_{s1}, v_{s2} \subset G_s$ and G_s is the sequencing graph. For example, in Figure 10.3(a), the droplet $R_{FMUX,2}$ will be transported to electrodes storing two droplets generated by the $S_{FMUX,1}$ split operation. This implies that $R_{FMUX,2}$ will mix in $W_{FMUX,1}$ and $M_{FMUX,2}$ if and only if $S_{FMUX,1}$ produced two half unit-volume droplets. In the other case, the droplets will follow the sequencing graph without the dotted edges. Figure 10.3(b) shows the fluidic operations when the droplet C_{FMUX} is present; the execution of the leftmost subgraph will transport the input droplet I_{FMUX} to Z_{FMUX} while the other two subgraphs will result in half unit-volume droplets. Figure 10.3(c) shows the fluidic operations in the absence of C_{FMUX}, where the execution of the subgraph on the right will lead to the transportation of the input I_{FMUX}.

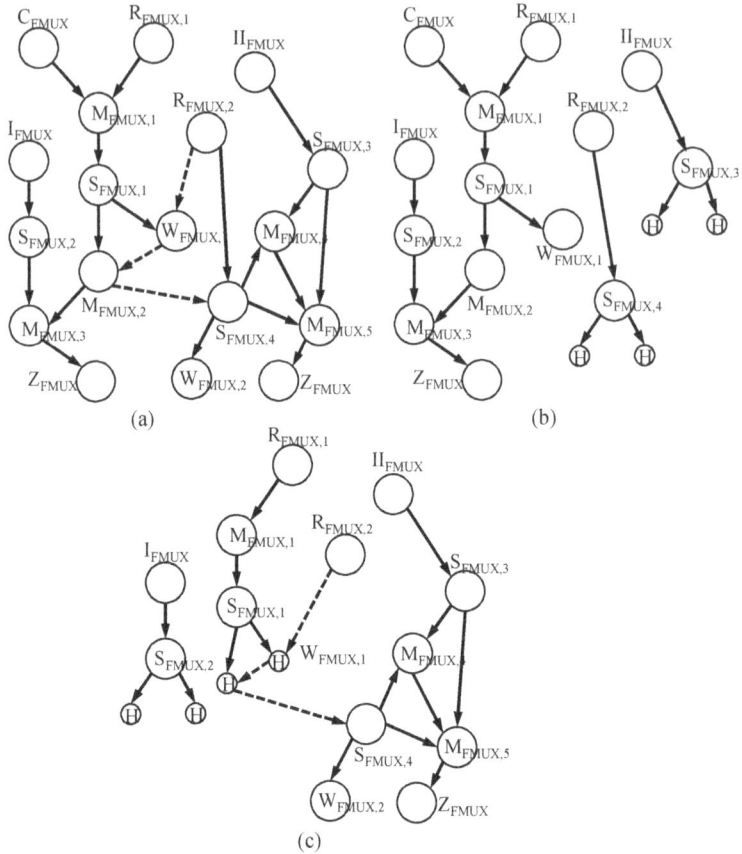

FIGURE 10.3
Sequencing graph of the fluidic multiplexer: (a) Sequencing graph corresponding to the multiplexer. C_{FMUX}, R_{FMUX}, and $\{I_{FMUX}, II_{FMUX}\}$ are control, reference, and input droplets, respectively. The dotted lines represent additional condition, i.e., during routing, droplet R_{FMUX} will be transported to the electrodes storing droplets generated from $S_{FMUX,1}$. (b) Multiplexing in the presence of the control droplet. I_{FMUX} is transported to Z_{FMUX} by one of the droplets generated from $S_{FMUX,1}$. Droplet $R_{FMUX,2}$ will not be able to mix with the droplets generated from $S_{FMUX,1}$, hence, splits into half volume droplets (shown as H). (c) Multiplexing in the absence the control droplet. In this case, $R_{FMUX,2}$ will mix with two half-volume droplets generated from $S_{FMUX,1}$, hence, II_{FMUX} will be transported to Z_{FMUX}.

Selection of Multiplexer Locations

From a biochemistry perspective, the execution time of an assay is a crucial parameter. Each droplet in an assay has a specific lifetime (dictated by physical constraints such as degradation time and evaporation rate) and the droplet has to be used within its lifetime (Chapter 3). Hence, the insertion of the multiplexer should have minimum effect on the critical execution time of the assay. We define the critical timing delay t_{crit} of an assay as the longest path in the assay from a dispense node to the output node. For example, in Figure 10.1, the critical timing delay is 3 ($O_1 \rightarrow O_3 \rightarrow O_5 \rightarrow O_6$). Therefore, the objective of multiplexer-insertion optimization is to determine the locations of a given number of fluidic multiplexers such that the critical timing delay of the encrypted sequencing graph is minimum. The critical timing delay of a fluidic multiplexer itself in Figure 10.1 is 5 ($C_{FMUX} \rightarrow M_{FMUX,1} \rightarrow S_{FMUX,1} \rightarrow M_{FMUX,2} \rightarrow M_{FMUX,3} \rightarrow Z_{FMUX}$). Algorithm 10.1 describes a greedy approach for multiplexer insertion, where $nMUX$ is the maximum number of multiplexers to be inserted into the sequencing graph.

Secure Biochemistry Synthesis

The microfluidic encryption procedure is invoked between the biocoding and the DMFB design phases (see Figure 9.1). The input to the encryption algorithm is the high-level design specification of the assay in the form of a sequencing graph. The output of this algorithm is the encrypted sequencing graph. The designer receives an encrypted sequencing graph and synthesizes it to generate the DMFB layout and the actuation sequences. The actuation sequences represent the locked form of the assay, which can be unlocked by applying the correct control input pattern (the secret key) to obtain a valid (fluidic) data flow.

Algorithm 10.1 Greedy Microfluidic Encryption

Input: Sequencing graph $G_s = (V_s, E_s)$
Output: Encrypted sequencing graph $\widetilde{G}_s$
1:    $\widetilde{G} \leftarrow G_s$;   ▷ Initialization
2:    $V_{sel} \leftarrow \emptyset$;   ▷ Nodes selected for multiplexer insertion
3:    **for** $i = 0$ to $nMUX$ **do**
4:       $\mathcal{G}_{FMUX} \leftarrow \emptyset$;   ▷ A list of candidate encrypted graphs
5:       $\mathcal{K}_{FMUX} \leftarrow \emptyset$;   ▷ Timing delays associated with $\mathcal{G}_{FMUX}$
6:       **for each** $v_{si} \in V_s - V_{sel}$ **do**
7:          $G_{sv_i} \leftarrow$ INSERTMULTIPLEXER($i, v_{si}, \widetilde{G}_s$);
8:          $t_{crit,v_{si}} \leftarrow$ COMPUTEDELAY(G_{sv_i});
9:          ADDMODIFIEDGRAPHLIST($\mathcal{G}_{FMUX}, G_{sv_i}$);
10:         ADDTIMINGDELAYTOLIST($\mathcal{K}_{FMUX}, t_{crit,v_{si}}$);
11:      $t_{crit(min)} \leftarrow$ GETMINIMUMDELAY($\mathcal{K}_{FMUX}$);
12:      $\widetilde{G}_s \leftarrow$ SELECTENCRYPTEDGRAPH($t_{crit(min)}, \mathcal{G}_{FMUX}$);
13:      ADDSELECTEDNODETOSET(V_{sel}, v_{si});
14: **return** $\widetilde{G}_s$;

10.2   Aging Reinforces DMFB Security

Aging has a greater impact on DMFBs as compared to their CMOS counterpart. As described earlier in Section 2.3, it is known that DMFBs degrade quickly and must be discarded within a few hours [62]; the short lifetime can be attributed to the rapid degradation of electrodes during DMFB operation. Several experimental and analytical methods have studied the causes of electrode degradation in order to identify DMFB fabrication methods that can enhance reliability [187]. In our framework, we take advantage of DMFB electrode degradation to enhance system security against potential attacks. We exploit the fact that electrodes can withstand only a limited number of actuations before dielectric breakdown occurs [179]. Therefore, an attacker can make only a limited number of attempts to break the security scheme (i.e., guess the secret key through trial and error) before the DMFB fails.

Next we examine two DMFB-related parameters in more detail to characterize and evaluate the security countermeasure: (1) the number of electrode actuations per electrode; (2) the thickness of the dielectric layer. Electrode degradation (or lifetime) can be analyzed on the basis of the threshold voltage needed to transport a droplet between adjacent electrodes based on the electrowetting phenomenon.

Number of Electrode Actuations

The degradation model for the electrodes in Section 2.3 describes the impact of excessively using an electrode on its lifetime. According to this analysis, an electrode's lifetime can be divided into three regions: reliable operation, safety margin, and breakdown (Figure 2.7). In the reliable operation region, the threshold voltage needed for actuation is constant. Then, it increases linearly in the safety margin region. Finally, in the breakdown region, a significant increase in the threshold voltage is required to transport a droplet; this increase, in turn, quickly leads to dielectric breakdown and electrode failure [270].

Thickness of the Dielectric Layer

Thickness of the dielectric layer plays a crucial role in determining the lifetime of a DMFB. It is desirable to make the dielectric thinner in order to reduce the voltage required for actuation—the thinner the dielectric, the lower is the actuation voltage. However, the breakdown voltage imposes a lower limit on the dielectric thickness.

We can design and fabricate a DMFB such that it permits reliable actuation only for a certain duration, thus limiting the usability of the DMFB if an attacker attempts to obtain the secret key through brute-force trial and

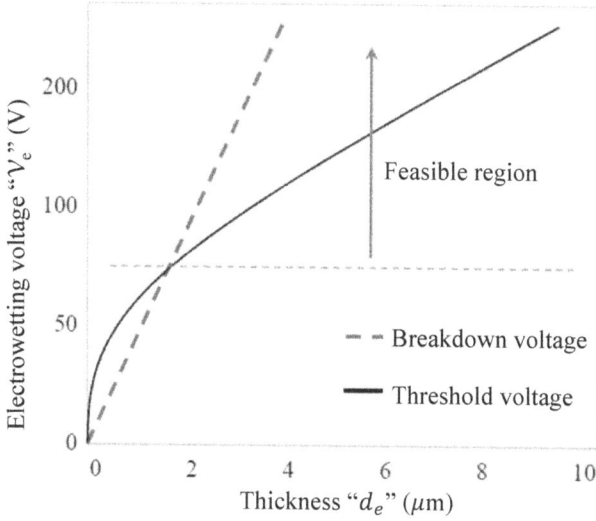

FIGURE 10.4
Electrowetting threshold and dielectric breakdown voltage versus dielectric thickness.

error. For example, given the dielectric thickness $d_e = 2.3$ μm for the dielectric material considered in Figure 10.4, the designer can derive the breakdown voltage (120 V) and the threshold voltage (90 V) from Figure 2.7. Using these two values, we can use Figure 2.7 to determine the maximum number N_{act} of allowable electrode actuations ($N_{act} = 260$, in this case) for reliable execution of the DMFB. Since each attempt to run the target bioassay with a random key leads to a known number of electrode actuations, N_{act} can be used to derive an upper limit n_{act} ($n_{act} << N_{act}$) on the number of attempts that an attacker can make before the chip breaks down.

10.3 Encryption Security Analysis and Simulation Results

In this section, a detailed security analysis is provided to evaluate the effectiveness of microfluidic encryption. Microfluidic encryption is applied to three benchmark assays—in-vitro, PCR, and Protein [286]—and area and performance overheads are obtained. We have used a custom C++ program to encrypt a given assay by optimally inserting multiplexers into the sequencing graph. Two multiplexer-insertion algorithms have been implemented: (i) a baseline method in which multiplexers are randomly inserted at various positions in the sequencing graph; (ii) the proposed greedy algorithm. The open-source DMFB synthesis tool [86] is used to synthesize assays. For the synthesis flow, we used list scheduler, left-edge placer, and the modified maze router [86].

Security Analysis

In this subsection, we examine the security benefits associated with the use of fluidic multiplexers.

Number of Electrode Actuations: Figure 10.5 shows the maximum number of electrode actuations corresponding to the number of multiplexers being used. Without encryption, in-vitro, PCR, and Protein assays require 8, 2, and 22 actuations, respectively. These numbers increase in a linear fashion with the number of multiplexers, necessitating an increase in the dielectric thickness in order to retain the same lifetime of the DMFB. On the other hand, a fixed dielectric thickness will degrade the life-time of the DMFB with an increase in the number of multiplexers. For example, in-vitro assay requires a maximum of 8 actuations per electrodes. If the dielectric thickness is chosen as 2.3 μm, the maximum number of actuations that is possible before breakdown can be calculated (from Figure 2.7 and Figure 10.4) to be 250. This implies that with 2.3 μm dielectric thickness, a DMFB can reliably execute the in-vitro assay $\frac{250}{8} \approx 31$ times. However, with eight multiplexers, the number of times the DMFB can be used is reduced to $\frac{250}{60} \approx 4$. Hence an attacker will

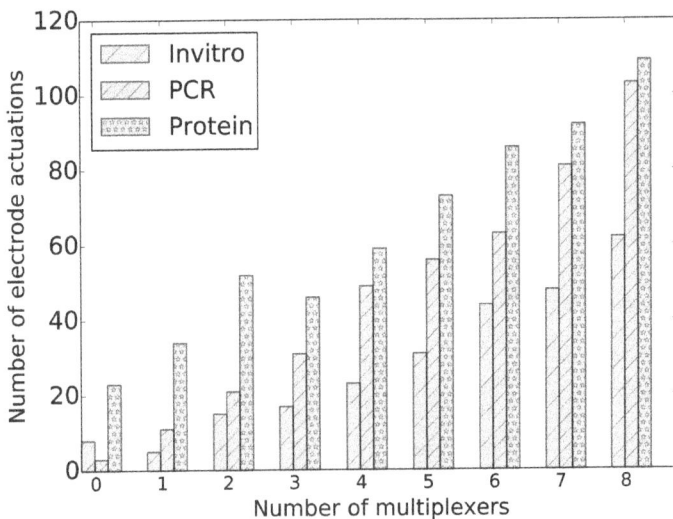

FIGURE 10.5

Change in the number of electrode actuations with an increase in the number of multiplexers.

get very few attempts to guess the secret key by trial and error, and security of the DMFB is significantly enhanced. By carefully choosing the dielectric thickness and the number of multiplexers, the designer cannot only protect the DMFB against a brute-force attack, but also quantify the strength of this countermeasure. Since DMFBs are disposable and intended for one-time use, a reduction in the number of times that it can be used does not affect its applicability in practice.

Protection Against Brute-Force Attacks: The number of multiplexers defines a security metric for microfluidic encryption. As discussed in Section 10.2, the designer can carefully choose the number of multiplexers and the DMFB dielectric thickness to thwart attacks. For example, with a 2.3 μm dielectric thickness and an eight-bit key, an attacker can be limited to only five brute-force attempts. Therefore, the attacker cannot exhaustively try all 256 possible keys.

It may be noted that the microfluidic encryption is based on one common secret key to activate all the DMFB chips, as all these chips are generated based on the same encrypted sequencing graph. As shown in [103, 97], there exists significant chip-to-chip variability in DMFB fabrication, characterization, measurements. Such inherent variability can be incorporated into the proposed fluidic encryption framework, via *side-channel fingerprinting* schemes [10], to ensure unique key for DMFBs. Efficient designs of side-channel fingerprinting is left for future work.

Protection Against Hardware Trojan Attacks: The hardware Trojan attacks described in Chapter 9 manipulate the assay outcome by altering the sequencing graph. In order to launch such a manipulation-based attack, the attacker must have a prior knowledge of the assay. The proposed microfluidic encryption obfuscates the assay; therefore, the attacker can no longer alter the assay to get a meaningful outcome that can pass scrutiny.

DMFB Supply-Chain Security: In the proposed framework, any party in the DMFB supply chain other than the biocoder can be malicious. To ensure security, the biocoder will provide the designer only an encrypted sequencing graph for the assay, but does not hand over the secret key. Without the secret key, a malicious designer is thwarted from extracting the assay protocol, and hence, cannot steal the IP. A malicious foundry can overproduce DMFBs, but without the secret key, overproduced DMFBs will be useless. In the same way, it is evident that the proposed microfluidic encryption provides protection against counterfeiting.

Area Overhead

The area overhead is calculated as the number of electrodes in the electrode array. Figure 10.6 shows the area overhead corresponding to the number of multiplexers, where "0 multiplexer" represents no encryption. We have considered up to eight multiplexers. The number of electrodes increases linearly with the number of multiplexers. Eight multiplexers lead to 286%, 139%, and

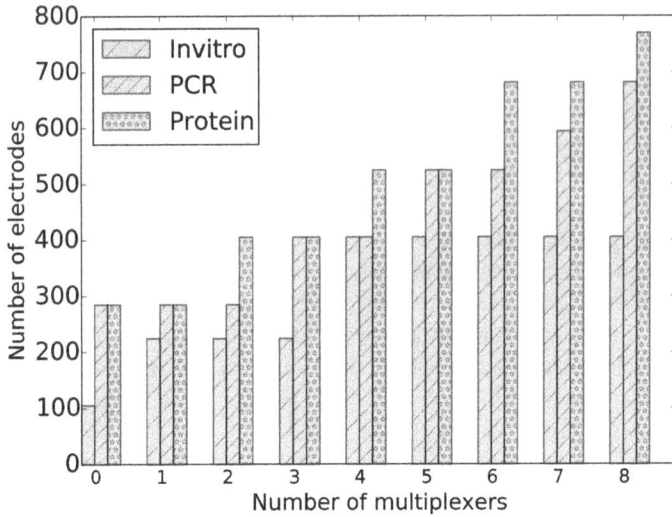

FIGURE 10.6
Change in number of electrodes with an increase in the number of multiplexers.

170% increase in the number of electrodes for in-vitro, PCR, and Protein assays, respectively. The increase in the DMFB footprint will be much less than the increase in the number of electrodes, because in real DMFBs, the input/output pads are much larger than the actual microfluidic array. As shown in Figure 8.9(b), for a fabricated biochip (described in Chapter 8), 32 input/output pads consume more area than the 5×5 electrode array. We can keep the number of I/Os the same for the larger number of electrodes by sharing control pins. As a result, the impact on DMFB area can be minimized. Note that these results are specific to the synthesis tool that we have used. While the results are likely to be slightly different for other synthesis tools, we expect the trends to hold.

Bioassay Execution-Time Overhead

Figure 10.7 and 10.8 show the execution-time overhead corresponding to the insertion of multiplexers using the baseline (random) method and the proposed optimization method, respectively. As shown in Figure 10.7, the random insertion of multiplexers result in a sharp increase in the execution time for all the three assays. For example, with 8 randomly inserted multiplexers, the execution time of in-vitro and PCR assays increase by 10X, whereas it is only 2X for Protein assay. Each additional multiplexer adds as much as one second of execution-time overhead. For the proposed optimization method, the execution-time overhead is considerably less for the largest assay (20% increase in execution-time overhead for Protein assay), but it is comparable

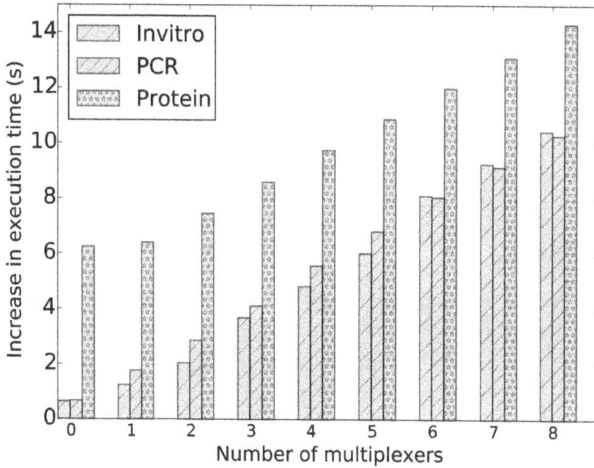

FIGURE 10.7
Execution-time overhead associated with the random insertion of multiplexers.

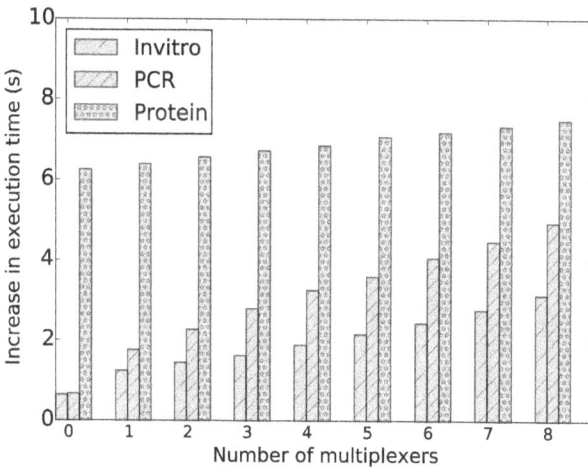

FIGURE 10.8
Execution-time overhead associated with the greedy insertion of multiplexers.

to the overhead for random insertion for the two smaller assays. This is because for smaller assays, there are only a few possible locations where multiplexers can be inserted and random insertion is as effective as optimized placement. For emerging cyber-physical on-chip bioassay protocols that target complex applications, e.g., gene-expression analysis (Chapter 2), random insertion will be significantly more intrusive than the proposed optimized placement.

10.4   DNA Barcoding as a Biochemical-Level Defense Mechanism

As described in Section 9.4, a major security threat arising in quantitative-analysis settings, e.g., forensic analysis, is sample forgery; that is, a real DNA sample collected from a field can be replaced/switched with another false sample with the intent of deceiving investigators or misleading identification results. This type of threats is especially serious because it is stealthy and it may not be detected easily. In this section, we describe a defense strategy against DNA-forgery attacks. The proposed strategy is inspired by a technique from microbiology called DNA barcoding [92].

DNA-Barcoding Methodology

The use of DNA barcoding to identify a sample is similar to the optical identification of an object using a machine-readable barcode. Each sample can be assigned one or a set of DNA barcodes that can be used to confirm the identity of the sample during down-stream analysis. This technique can be implemented in the field (by collectors) and be applied once the samples are collected. Barcoded samples are then shipped to a clinic or an investigation lab to perform regular quantitative analysis (by analyzers) on the sample. However, to ensure that no sample-forgery attacks have occurred[1], the embedded DNA barcode has to be sequenced first and be compared with a database. If the sequenced barcode matches the database, then the sample is considered genuine and it can be analyzed. On the other hand, if no barcode or an incorrect barcode is detected, then the associated sample is likely false and it has to be discarded. Figure 10.9 explains the above strategy using a schematic representation.

Note that the proposed countermeasure can be effective only if the following conditions are satisfied:

1. Only authenticated collectors are allowed to barcode the samples. Similarly, only authenticated analyzers can sequence/detect the barcodes before conducting quantitative analysis.

2. All the information related to the barcodes is stored in a secure database.

Our biochemical-level defense technique fulfills (1), whereas (2) can be achieved at the software or kernel level [289]. Security characteristics of DNA barcoding based on (1) and (2) are described next.

---

[1]DNA forgery can be performed by a man-in-the-middle (MITM) attacker who secretly relays and possibly alters transferred samples.

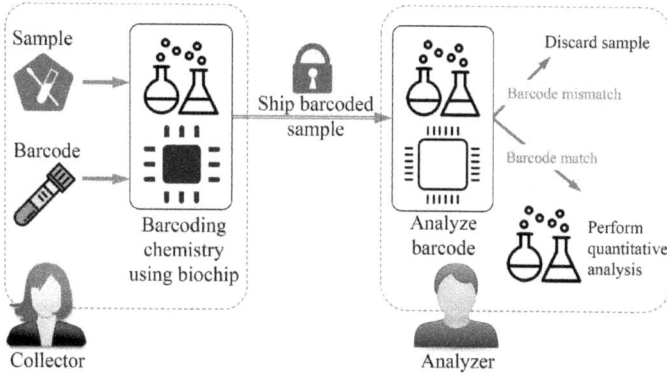

FIGURE 10.9
Illustration of the DNA-barcoding methodology.

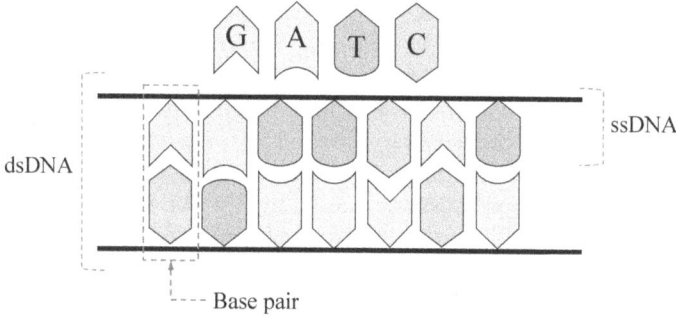

FIGURE 10.10
Structure of ssDNA and dsDNA.

Security Analysis of DNA Barcoding

DNA barcoding is a form of side-channel fingerprinting, which ensures that only genuine (trusted) samples can be used. The key principle that enables this security scheme is that each DNA barcode is initially designed with a unique sequence of nucleotides whose order forms the identity of a barcode. Each nucleotide can be one of the following four options: adenine (**A**), cytosine (**C**), guanine (**G**), and thymine (**T**). This sequence forms a single-stranded DNA (ssDNA) that is chemically coupled with another complementary ssDNA to form a double-stranded DNA (dsDNA). Figure 10.10 shows the basic structure of a dsDNA, which is comprised of base pairs of nucleotides.

Hence, to design a barcode of length L_b base pairs, the number of possible dsDNA structures n_{ds} is 4^{L_b}. For example[2], if $L_b = 280$ base pairs,

---

[2]Note that 4^{L_b} is a rough estimate since some nucleotide sequences may not be possible, e.g., AAAAAA or GGGGG. Nevertheless, we can still get a very large number of combinations.

which is a typical length of a small barcode, then $n_{ds} \approx 3.773962e + 168$. Since each barcode has a specific sequence of nucleotides, the barcode can be sequenced or detected (using PCR) at the down stream only using two specific primers (forward and reverse primers) that are tightly coupled with the nucleotide sequence. Hence, the number of primers is $2 \times 4^{L_{pmr}}$, where L_{pmr} is the primer length ($L_{pmr} = 21$ in our experiment). This large number ensures that launching a brute-force attack with the intent of identifying the sequence of a used barcode is impractical.

The above genetic characteristics can be exploited to design an effective security scheme; see Figure 10.11. In the first step, a set of barcodes with varying lengths are randomly designed by a trusted party, which also develops barcode-specific primers. These barcodes and the associated primers are registered in a secure database. Users who have access to this database belong to one of two classes: authenticated collectors and authenticated analyzers. Each one of these classes has only a specific view of the database. Authenticated collectors can obtain the barcodes and their associated secret identification numbers, whereas authenticated analyzers can receive primers and their secret identification numbers. Note that the identification of a barcode-specific primer must match the identification of the barcode itself. Also, authenticated users are not required to have detailed knowledge of the barcoding sequences. For example, an authenticated analyzer can have access to the barcode-specific primers and their identities, but not the barcode information itself. The security interactions described thus far, denoted by A and B in Figure 10.11, ensure that both types of users, i.e., collectors and analyzers, are individually trusted, therefore the process of DNA barcoding and detection can be trustworthy.

While collectors and analyzers can work independently based on the secret information and material obtained from the trusted party, they still need to interact because barcoded samples are prepared by collectors and are delivered to analyzers. A significant advantage of the proposed countermeasure is that both types of users can interact securely and semi-anonymously. A collector communicates with an analyzer by sending two types of material: (1) a barcoded DNA sample (biological material), which encapsulates a secret barcode obtained from the trusted party; (2) a public key-encrypted message (information material), denoted by C in Figure 10.11, which includes secret information about the barcode and the sample identification numbers. Using public-key cryptography, also known as asymmetric cryptography [241], ensures that analyzers can decrypt the message, using a private key, select the right type of primers based on the received identification message, and then perform DNA-barcode analysis using DNA sequencing or DNA amplification (i.e., gel electrophoresis analysis). By analyzing the identities of the barcode and the sample along with the result of the DNA-barcode analysis, the sample analyzer can verify the genuineness of the DNA samples.

Note that this scheme does not prevent an adversarial attacker from intercepting an encrypted message and replacing it with an encrypted false

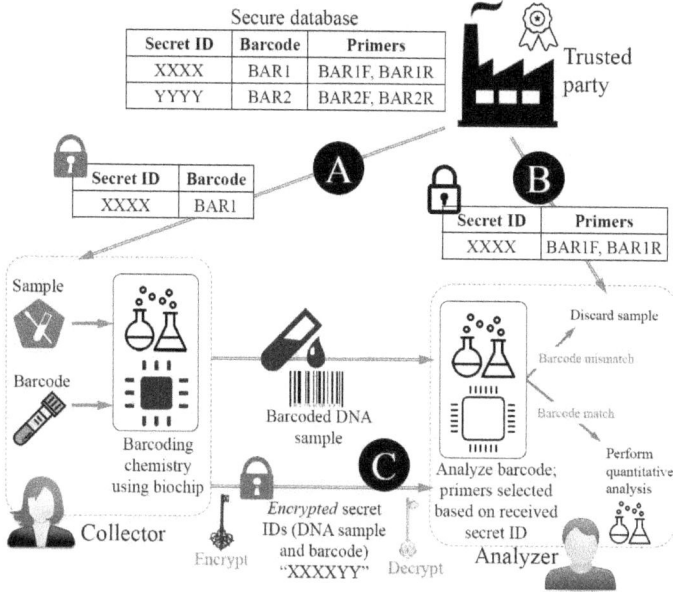

FIGURE 10.11
Details of the proposed security scheme.

message[3]. However, such an action is intrusive and is easy to detect since the end analyzer will not be able to identify the barcode. Therefore, the sample will eventually be discarded. Nevertheless, to avoid such a scenario, an end-to-end encryption protocol can also be adopted to prevent eavesdroppers from being able to access the encrypted message [66].

By comparing the above flow with the conditions described in the previous subsection, we observe that both conditions can be satisfied.

10.5   Benchtop Demonstration of DNA Barcoding

In this section, we present our benchtop study that implements DNA barcoding for two *S. pombe* samples. We describe the experimental procedure and discuss the obtained results.

---

[3]A public key is known to everyone, and it is tightly coupled with a private key that is only known to an analyzer.

Specifications of DNA Barcodes

Recall that each sample is associated with a specific barcode, and each DNA barcode must have unique characteristics such as nucleotide sequence or fragment size. For this purpose, we use two types of DNA barcodes (gBlocks® Gene Fragments from Integrated DNA Technologies). The first barcode, denoted by BAR1, consists of 280 base pairs, whereas the second barcode, denoted by BAR2, consists of 190 base pairs. Table 10.1 shows the nucleotide sequence information for BAR1 and BAR2. The difference in fragment size allows us to distinguish barcodes at the down stream after DNA amplification is performed (Figure 9.8).

Each barcode is associated with two specific primers that are designed to anneal to a specific region; one primer to is designed to anneal at the 5'-end, called forward primer, and the other primer, called reverse primer, is designed to anneal at the 3'-end of the barcode segment. The forward and reverse primers of BAR1 are denoted by BAR1F and BAR1R, respectively. Also, the forward and reverse primers of BAR2 are denoted by BAR2F and BAR1R, respectively. The nucleotide sequence information for BAR1F, BAR1R, BAR2F, and BAR2R are also shown in Table 10.1.

Identification of DNA Barcodes Using Traditional DNA-Analysis Flow (PCR)

In order to detect the presence of a certain DNA barcode in a sample, we use traditional PCR that allows us to measure the number of base pairs (or the molecular weight) using agarose gel electrophoresis, as described in Section 9.4. By using the right type of primers in the PCR reaction, the DNA barcode is amplified, allowing us to verify its presence in the associated reaction. Since BAR1 has a longer sequence of base pairs compared to BAR2, we expect that BAR2 amplicons migrates a further distance through the gel than BAR1 amplicons.

Experiment Procedure

The purpose of our experiment is twofold. First, we aim to verify that BAR1 and BAR2 exhibit different characteristics (fragment size) that can be identified using gel electrophoresis. This objective can be achieved without using *S. pombe* samples, i.e., only using DNA barcodes. Second, we aim to demonstrate that after an *S. pombe* sample is barcoded, the amplification of the DNA barcode can occur only if the right primers are used in the reaction. This important demonstration, despite its technical challenges, confirms that DNA barcoding functions as expected to secure samples against forgery attacks.

Experiment Design: To achieve the above goals, the experiment was designed considering 7 parallel PCR reactions. We describe these reactions and their associated objectives in Table 10.2 and Table 10.3, respectively.

TABLE 10.1

The Sequence information related to BAR1 and BAR2.

BAR1	5' - CAG TCA CTA TGG CGT GCT GCT AGC GCT ATA TGC GTT GAT GCA ATT TCT ATG CGC ACC CGT TCT CGG AGC ACT GTC CGA CCG CTT TGG CCG CCG CCC AGT CCT GCT CGC TTC GCT ACT TGG AGC CAC TAT CGA CTA CGC GAT CAT GGC GAC CAC ACC CGT CCT GTG GAT CCT CTA CGC CGG ACG CAT CGT GGC CGG CAT CAC CGG CGC CAC AGG TGC GGT TGC TGG CGC CTA TAT CGC CGA CAT CAC CGA TGG GGA AGA TCG GGC TCG CCA CTT CGG GCT C - 3'
BAR2	5' - GAC ATG AAG CTT TAA ATC AAT CTA AAG TAT ATA TGA GTA AAC TTG GTC TGA CAG TTA CCA ATG CTT AAT CAG TGA GGC ACC TAT CTC AGC GAT CTG TCT ATT TCG TTC ATC CAT AGT TGC CTG ACT CCC CGT CGT GTA GAT AAC TAC GAT ACG GGA GGG CTT ACC ATC TGG CCC CAG TGC TGC AAT GAT A - 3'
BAR1F	5' - CGC TAT ATG CGT TGA TGC AA - 3'
BAR1R	5' - AGA TGG TAA GCC CTC CCG TAT - 3'
BAR2F	5' - TGC TTA ATC AGT GAG GCA CCT - 3'
BAR2R	5' - AGA TGG TAA GCC CTC CCG TAT - 3'

Sample and Reagent Preparation: To start the experiment, we first prepared the DNA samples, DNA barcodes, and the reagents. We used two types of *S. pombe* samples, and they were both grown in yeast media so that they could be presented in the experiment in a liquid form (Figure 10.12(a)). Second, we prepared the DNA barcodes BAR1 and BAR2, shown in Figure 10.12(b), by performing serial dilution for each barcode[4]. The dilutions we obtained for each barcode were 1:10, 1:100, and 1:1000; see Figure 10.12(c). Third, we prepared barcoded samples by spiking the first DNA sample, labeled as 501, with BAR1 (1:100) and the second DNA sample, labeled as 90, with BAR2 (1:100), as shown in Figure 10.12(d). The diluted DNA barcodes (Figure 10.12(c)) were used to to implement the first and second PCR reactions, whereas the barcoded samples (Figure 10.12(d)) were used to implement the remaining reactions.

---

[4]During the execution of the experiment, we referred to BAR1 as (Code 1) and BAR2 as (Code 4).

TABLE 10.2

Description of PCR reactions designed in the experiment.

Reaction	Tube	Sample 1	Sample 2	BAR1	BAR2	BAR1F(R)	BAR2F(R)
				Barcodes		Primers	
1	1			✓(1:10)		✓	
	2			✓(1:100)		✓	
	3			✓(1:1000)		✓	
2	1				✓(1:10)		✓
	2				✓(1:100)		✓
	3				✓(1:1000)		✓
3	1			✓(1:100)	✓(1:100)	✓	✓
4	1						
	2	✓		✓(1:100)			
	3		✓		✓(1:100)		
5	1						
	2	✓		✓(1:100)		✓	
	3		✓		✓(1:100)		✓
6	1						
	2	✓		✓(1:100)	✓(1:100)	✓	
	3		✓	✓(1:100)	✓(1:100)		✓
7	1						
	2	✓		✓(1:100)			✓
	3		✓		✓(1:100)	✓	

TABLE 10.3

Objectives of PCR reactions.

Reaction	Objective
1	Verify PCR amplification of BAR1 and BAR2 at different
2	dilutions. Verify the difference in fragment size.
3	Verify the ability to use multiple "distinguishable" barcodes.
4	Verify that a barcode is not amplified if its specific primers are not used (the template is a barcoded sample).
5	Verify that a barcode is amplified if its specific primers are used (the template is a barcoded sample).
6	Verify that a barcode is amplified if its specific primers are used even if they are used along with other primers (the template is a barcoded sample).
7	Verify that a barcode is not amplified if the used primers are not the right ones (the template is a barcoded sample).

Next, for each PCR reaction, we prepared a PCR master mix, which consists of primers (of DNA sample and/or DNA barcodes), buffer, water, and Taq solution. The amounts of these components were adjusted based on the

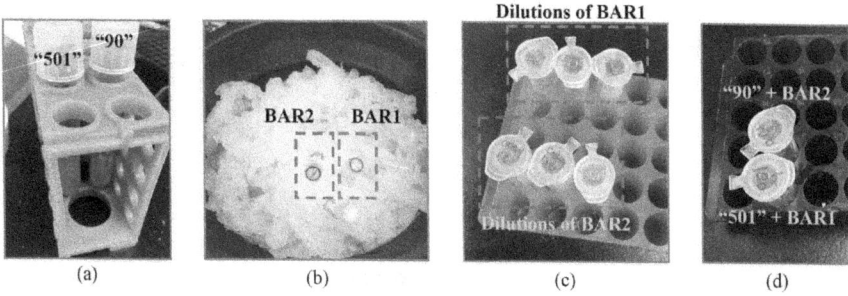

FIGURE 10.12
Sample and reagent preparation.

requirements of each PCR reaction. These components were pipetted carefully into 6 master-mix tubes, as shown in Figure 10.13. Note that Reaction 3 is similar to Reaction 1 and Reaction 2, therefore no specific master mix was designed for Reaction 3.

Having prepared the samples and PCR master mixes, we loaded all the chemicals into 6 strips of empty tubes on a PCR plate; see Figure 10.14(a). Each 8-tube strip was associated with a PCR reaction, except the second strip which included Reaction 2 and Reaction 3. Note that in each strip, we only used three out of eight tubes, as described in Table 10.2.

Running PCR Reaction and Gel Electrophoresis: Having filled in the assigned tubes on the PCR plate, the tubes were closed then spun for a few seconds using a micro-centrifuge. Next, we placed the strips in the thermal cycler and started the PCR reaction; see Figure 10.14(b). After the PCR program had finished, we transferred the PCR-plate contents into the gel electrophoresis apparatus, allowing the chemical solutions to migrate over the gel with the aid of an electric field. Gel electrophoresis was run for an hour, then the gel block was moved to a UV to a chamber that contains a source of UV light to visualize the DNA bands. Results interpretation is discussed next.

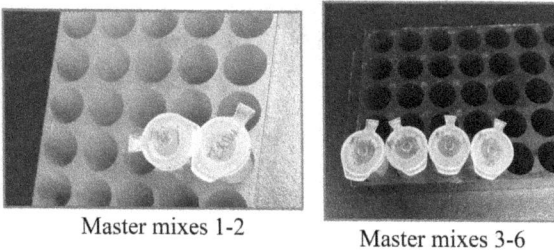

FIGURE 10.13
Six tubes containing master mixes.

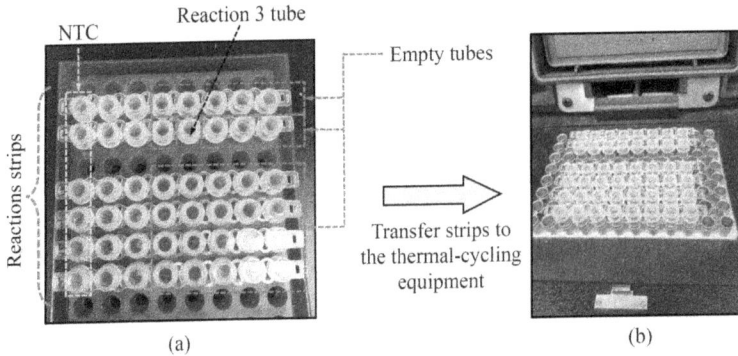

FIGURE 10.14
(a) Preparation of the PCR plate. (b) Moving PCR strips to the thermal-cycling equipment.

Interpretation of Gel-Electrohoresis Results

Figure 10.15 shows the results obtained by gel electrophoresis. The plot is divided into 7 regions. Region 1 and Region 2 show the results of Reaction 1 and Reaction 2, respectively. In other words, they are associated with the serial dilution for BAR1 and BAR2, respectively. Region 3 shows the result for a combined BAR1/BAR2 reaction, i.e., Reaction 3. The results of Region 1, Region 2, and Region 3 are located at the top row.

Furthermore, the results shown in Region 4, Region 5, Region 6, and Region 7 (bottom row) are used to assess the barcoding performance in the presence of DNA samples, and they are associated with PCR Reaction 4, Reaction 5, Reaction 6, and Reaction 7, respectively. Each one of these regions contains 3 columns, where the leftmost column is related to the reference negative control (Section 9.4), the middle and the rightmost columns are associated with *S. pombe* samples 501 and 90, respectively.

By comparing Region 1 and Region 2, we observe that the bands in Region 2 are located at a further distance compared with the bands in Region 1. This observation indicates that both BAR1 amplicons and BAR2 amplicons can be distinguished because of their different sizes, even if they are both included in a single tube (Region 3). The bands of BAR2 are located at a further distance since BAR2 (190 base pairs) has a shorter sequence of base pairs than BAR1 (280 base pairs).

Another observation is that the bands shown in Region 1, Region 2, and Region 3 are significantly bright, even after diluting the barcodes. This indicates that the copy number of DNA fragments that exist in BAR1 and BAR2, i.e., concentration, is significantly large. This result raises questions concerning the optimization of the copy numbers of a starting *S. pombe* sample and a DNA barcode when they co-exist in one tube. In other words, the *S. pombe* samples 501 and 90 produce amplicons of length 400 ∼ 700 base pairs, whereas

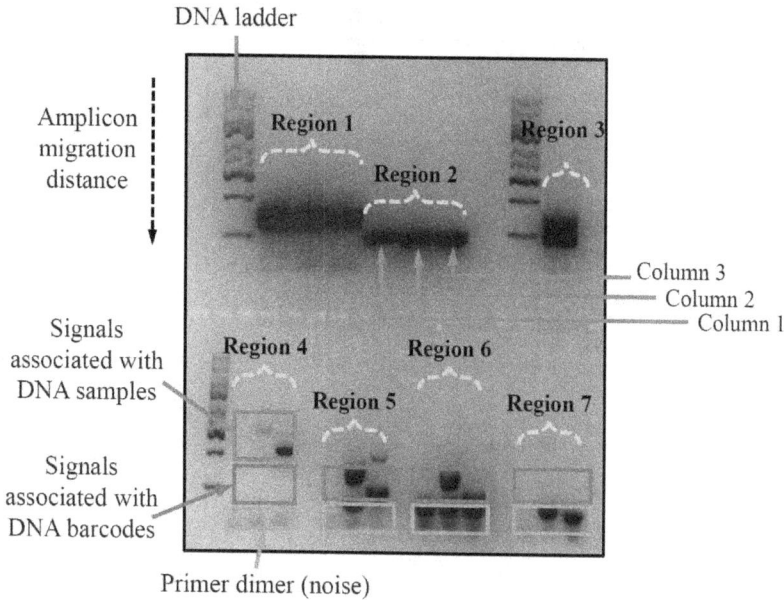

FIGURE 10.15
Results of DNA amplification for the 7 PCR reactions.

BAR1 and BAR2 produce amplicons of length 190 ~ 280 base pairs. With such a difference in molecular weights, a microliter volume of an *S. pombe* sample contains a lower copy number than a microliter volume of a DNA barcode. Hence, by introducing an excessive number of barcode amplicons, the yield of *S. pombe* amplicons is significantly reduced. This is why we observe that the bands related to the *S. pombe* amplicons are faint, as shown in Region 4, or they completely vanish, as shown in Regions 5-7. Discussion related to the optimization of copy numbers is presented later in this section.

In Region 4, no barcode bands are observed even though DNA barcodes were used in Reaction 4. This behavior is because no barcode-specific primers were used in the reaction. Similarly, in Region 7, no barcodes are observed even though barcodes and primers were used. Such behavior is because we deliberately added the wrong primers—we used BAR1-specific primers when we added BAR2 and vice versa. These results indicate that the detection of a certain barcode is possible only if a specific set of primers are used.

In contrast, in Region 5, the bands related to DNA barcodes are observed as the right type of primers were used. In Region 6, we also observe bands related to the barcodes since each tube contained all types of primers; a DNA barcode will bind with its specific primers during annealing.

Hence, despite the technical challenges related to the optimization of amplicon copy numbers, the above results show the effectiveness of the

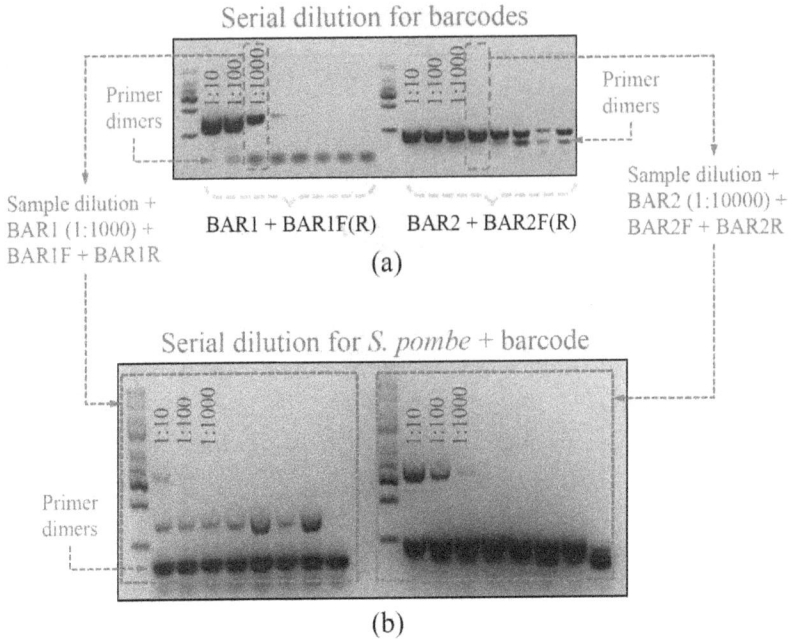

FIGURE 10.16

Results of serial-dilution experiments that investigate the co-optimization of DNA amplification for *S. pombe* samples and DNA barcodes.

DNA-barcoding scheme in identifying samples and securing them against forgery attacks. We carried out additional benchtop studies, based on serial dilution, to co-optimize DNA amplification for both *S. pombe* and DNA barcodes. These studies are explained next.

Optimization of Amplicon Copy Numbers Using Serial Dilution

To co-optimize DNA amplification for both *S. pombe* samples and DNA barcodes, we performed two PCR experiments for a 10-fold serial dilution of a genomic DNA sample and the barcodes BAR1 and BAR2. The goal of these experiments is to determine the optimal concentrations of the *S. pombe* sample and the barcodes such that the overall reaction yield is improved. In the first experiment, we prepared a serial dilution of BAR1 and BAR2 (similar to Reaction 1 and Reaction 2 in the previous experiment) followed by DNA amplification and gel-electrophoresis analysis. The outcome of this experiment is shown in Figure 10.16(a). We observe that the optimal dilution factors for BAR1 and BAR2 that result in clear bands with reduced primer dimers (noise) are 1:1000 and 1:10000, respectively.

Having determined the optimal dilution factors for the DNA barcodes, we conducted a second PCR experiment over a serial dilution of a multi-template setting, i.e., an *S. pombe* sample barcoded with (1:1000) BAR1 or (1:10000) BAR2. The result obtained from gel electrophoresis after running this experiment is shown in Figure 10.16(b). We observe that using dilution factors of (1:100) or lower for the *S. pombe* sample may lead to degradation in the sample yield, thus causing the bands associated with the sample amplicons to vanish. As a result, to co-optimize the DNA-amplification profile for the DNA sample and barcodes based on the above setting, the following conditions need to be fulfilled: (1) the volumes of the *S. pombe* sample and the barcode used in the same PCR reaction are equal; (2) by using (1:1000) BAR1 and (1:10000) BAR2, the DNA sample should not be diluted lower than 1:10 in order to get clear bands for both templates. This result is intuitive since the size of the *S. pombe* amplicons that we used in the experiments is significantly larger than the size of the barcodes amplicons.

A different approach for co-optimizing DNA amplification for the barcoded samples is to fix the dilution factors for both templates while examining the impact of using different volumes of DNA sample and the barcode. However, this approach however is more complex; therefore, it is not considered in this work.

Discussion

We highlight subtle aspects of this work in a Q&A format:

Q1. *Can an attacker separate the barcode amplicons from the S. pombe amplicons using microfluidic size-based separation techniques?*

Answer: This is possible if the difference between the size of the barcode amplicons and the size of the *S. pombe* amplicons is big. In our benchtop study, we used fragment size as a differentiator in for visualization purpose. However, this design approach is not recommended in real life.

Q2. *How can we prevent an attacker from separating barcodes from S. pombe?*

Answer: Technically, this attack can be prevented by designing barcodes whose amplicons have approximately the same number of base pairs as in *S. pombe* amplicons. By using this methodology, DNA fragments cannot be separated based on their sizes.

10.6 Chapter Summary

We have presented two defense mechanisms that secure quantitative-analysis frameworks against potential threats. First, we have described a microfluidic encryption strategy that protects biochemical protocols through the insertion of fluidic multiplexers in the DMFB design. These multiplexers are controlled by a secret key that is not revealed by the biocoder. Any attempt by an attacker to identify the secret key via repeated trials is thwarted by the short lifetime of DMFBs and the upper limit on the number of times an electrode can be actuated. The security provided by microfluidic encryption can be quantified in terms of the number of multiplexers inserted in the sequencing graph and the thickness of the dielectric layer. However, the security of the proposed microfluidic encryption is based on the assumption that each DMFB will have a unique key. In our future work, we plan to develop an efficient key management scheme to assign a unique key to each and every manufactured DMFBs.

Second, we have demonstrated, using a benchtop study, a biochemical-level countermeasure that protects DNA samples against DNA-forgery attacks. This technique employs unique DNA barcodes to secretly label DNA samples after being collected. By using a barcode management system, the DNA barcodes and their associated chemicals can be securely distributed and transferred along with their host samples. Complex genetic characteristics of DNA barcodes, e.g., sequence information, are key enablers for the success of this defense strategy. Although the DNA-barcoding countermeasure has been demonstrated using a benchtop study, it can be miniaturized and integrated with reconfigurable quantitative analysis using the cyber-physical framework described in Chapter 5.

11

Conclusion and Future Outlook

Cyber-physical microfluidics technology is transforming microbiology research by providing new opportunities for high-throughput sample preparation and point-of-care diagnostics. Over the past decade, several design-automation (synthesis) techniques have been developed for on-chip droplet manipulation. However, these methods oversimplify the dynamics of biomolecular protocols and they have yet to make a significant impact in biochemistry/microbiology research, leading to a large gap between advances in biochip design and the adoption of biochips for running biomolecular protocols. This book has covered an array of research that bridges this gap. The proposed research is the first comprehensive effort that is targeted at the realization of optimized and trustworthy microbiology-on-chip.

11.1   Book Summary

In this book, we have presented a comprehensive workflow for design optimization and security of biomolecular quantitative analysis using cyber-physical microfluidic systems. Contributions include automated synthesis of multi-sample biomolecular assays, optimization of scalable sample-differentiation techniques, optimization of protocol parameter-space exploration, development of an integrated error-recovery technique, and realization of security attacks and countermeasures of quantitative-analysis applications. In contrast to previous conceptual methods for miniaturized quantitative analysis, the proposed techniques address practical issues, e.g., scalability, reliability, and security, that arise due to contemporary microbiology research. The solutions investigated in this book have been supported with real-life benchtop and biochip experiments, showing that the proposed design paradigm can enable optimized, reliable, and trustworthy microbiology-on-chip.

Chapter 2 described an optimization framework for cyber-physical DMFBs to support quantitative gene-expression analysis. The proposed framework, which is inspired by a real-life benchtop study, allows multiple sample pathways to be dynamically processed based on an adaptive spatial-reconfiguration technique that efficiently manages resource sharing. A multi-layered software infrastructure has been developed to coordinate the interactions between the cyber-physical system components. More specifically, the software helps collecting data from sensors during runtime, performing analysis, and making decisions about further experimental steps. Support for inherent variations in biomolecular protocols was incorporated to the framework, and evaluations were performed using data obtained from the benchtop study.

Chapter 3 extended the previous optimization method to support quantitative epigenetic gene-regulatory analysis. The proposed design provides dynamic adaptation to protocol decisions under spatio-temporal constraints imposed by the experimental environment. An optimization solution based on real-time multi-processor scheduling has been proposed, thus introducing the concept of "fluidic tasks." Corresponding real-time interactions of fluidic and system tasks were leveraged to construct a hierarchical framework that illustrates the coordination between the synthesis of multiple pathways, electrode actuation, and firmware computation. Such interactions were demonstrated using a micro-controller board.

Chapter 4 incorporated the above cyber-physical optimization methods and others in a large-scale Internet-of-Things (IoT)-enabled framework that enables collaborative and coordinated molecular biology studies. The proposed framework is a research effort towards: (1) enabling integrative methods of microbiology tracking and analysis across thousands of network-connected microfluidic and benchtop "nodes," (2) standardized sharing of genomes and related omics data among researchers, and (3) dynamic reconfiguration of microbiology lab settings (e.g., to refocus the analysis scope) based on the dynamic restructuring of computational models. The proposed design, named BioCyBig, was the first to introduce Microfluidics-as-a-Service (MaaS) for genomic studies. By combining the capabilities of big-data computation, cyber-physical integration, and biochip design into a unified system, BioCyBig is well-suited for challenging experimental settings such as those related to cancer research.

Chapter 5 presented the first cyber-physical optimization technique for integrated cell differentiation and single-cell analysis using a hybrid microfluidic platform. Automated reconfiguration of diverse microfluidic components was developed using efficient graph-theoretic algorithms, thus allowing a large number of cells to be processed concurrently. Also, based on real-life experimental data, we have been able to support realistic settings where protocol characteristics can only be specified probabilistically. Therefore, the proposed design, named CoSyn, is a significant step forward in addressing challenges related to dynamic single-cell analysis.

Chapter 6 introduced a timing-driven design method for a pin-constrained reconfigurable flow-based biochip, thus enabling scalable single-cell screening. This method addresses pin-interfacing limitations of CoSyn. A fluidic multiplexed-control scheme has been proposed to actuate biochip valves independently (to support dynamic reconfiguration) while using shared control pins. By exploiting this scheme, an optimization framework has been developed to minimize the number of control pins, but concurrently satisfy throughput constraints related to single-cell screening. The proposed framework analyzes timing specifications of the single-cell application using a circuit model that was adopted based on computational fluid-dynamics simulations.

Chapter 7 described a comprehensive framework for biomolecular parameter-space exploration (PSE). The proposed optimization framework, named BioScan, was developed to explore gene-expression parameters of synthetic biocircuits using mixtures of suitable reagents, thus allowing efficient and fine-tuned modulation of circuit functions. The PSE problem was therefore formulated in terms of a biomolecular assay that systematically generates an array of reagent mixtures on a MEDA biochip. The proposed flow addressed optimization problems related to statistical sampling of reagent mixtures, automated realization of the above-mentioned biomolecular assay on a MEDA biochip, and eventually prediction of the biocircuit's parameter space. This framework is a significant milestone for enabling complex gene-regulatory circuits to be designed and tuned using microfluidic biochips.

Chapter 8 explained a hardware-based design methodology for reliable and fault-tolerant execution of biomolecular protocols. An experimental study based on a fabricated digital microfluidic biochip demonstrated error detection and recovery during bioassay execution. In contrast to previous error-recovery methods, the proposed approach fully integrates cyber-physical system components in the hardware, thus turning reliable point-of-care quantitative analysis into reality. We have described the details of the fabricated chip, the hardware control system, and the serial-dilution protocol that has been employed to test the integrated platform.

Chapter 9 presented an assessment of the security vulnerabilities of the current platforms used for biomolecular quantitative analysis. Discussion on adversarial threats and their implications and categories highlighted the need for a Trojan/attack taxonomy that can be applied to microfluidic or benchtop platforms. We described three forms of attacks on quantitative-analysis assays: (1) assay-result manipulation attacks on a digital microfluidic biochip; (2) attacks on cyber-physical system components, which have been integrated to enable error recovery or protocol decision making; (3) DNA-forgery attacks on DNA-analysis flow. Severity of DNA-forgery attacks has been investigated based on a benchtop experimental study. Our contribution in this chapter is a call-to-action for security researchers to engage in the transition towards trustworthy microbiology-on-chip.

Chapter 10 complemented our discussion on the security analysis of biomolecular protocols by presenting effective countermeasures. Two defense mechanisms have been proposed to avoid malicious attacks attributed to: (1) outcome manipulation of an assay implemented on a DMFB, and (2) DNA forgery through sample switching. In the first defense mechanism, we developed a microfluidic encryption strategy that protects biomolecular protocols through the insertion of fluidic multiplexers in the DMFB design flow. In the second defense mechanism, we designed a sample-barcoding strategy that allows researchers identify the DNA samples *in vivo* even before regular genomic analysis starts. This strategy can be the backbone of any side-channel fingerprinting scheme against various types of attacks including piracy actions or forgery attacks. Discussion on DNA barcoding-based defense was accompanied with a benchtop study that illustrates the capabilities of this technique.

11.2 Future Research Directions

The contents of this book open up a number of new research directions related to the optimization and security of biomolecular quantitative-analysis protocols based on microfluidic biochips. Contemporary microbiology research focuses on the creation of "wetware" (e.g., DNA elements introduced *in-vivo* or *in-vitro*) that can be controlled via electrical methods. This research has enabled new co-design methodologies in synthetic biology, thus presenting new optimization challenges that need to be addressed appropriately.

On the other hand, with the wide adoption of networked cyber-physical microfluidics in DNA forensics, foodborne pathogen detection, and clinical diagnostics, proactive defense mechanisms are necessary to counter cyber threats including sensor-spoofing attacks. The effectiveness of such mechanisms should also be independent of the design (bioassay) scale. Some potential new directions are summarized below.

Hardware/Wetware/Software Co-Design Methodology for Synthetic Biology

Synthetic biology research results in the creation of wetware that control living systems using a microfluidic biochip. Examples include transcriptional/translational Boolean algebraic circuits [60], recombinase based state-machines [23], and various communication mechanisms [47]. Biological DNA building blocks not only can sense environmental changes via gene-expression alterations, but they can also be actuated if combined with an electrical communication pathway (e.g., Mtr electron transfer pathway) that creates a *de novo* electronic interface between a host microorganisms and an electrode [125, 264]. The control of the electrical communication can be performed using a control software. Such a hardware/wetware/software co-design methodology opens up new venues where synthetic bio-electronic devices can be deployed and used in various fields, e.g., clinical diagnostics, bioenergy, and climate-smart agriculture. Figure 11.1 shows a block diagram of the co-design environment.

The control of the above co-design environment requires two feedback systems: (1) a hardware/software co-design system that tracks the progress of microfluidic operations (hardware) and reacts based on the results of the biochemical reactions (e.g., fluorescent reporters or presence/absence of a droplet); (2) a wetware/software co-design system that monitors biological activity and reacts by actuating DNA building blocks via genetically encoded electron conduits. The presence of two feedback loops imposes new constraints at the software level.

Therefore, a new synthesis methodology is required to provide a dual-control configuration for synthetic biocircuits. The $\mathcal{C}^5$ architecture, described in Section 1.4, needs to be revised to support coordinated sensing/actuation interactions between the two feedback loops.

Paramater-Space Exploration of Biocircuits in the Presence of DNA Memory

Current techniques of parameter-space exploration (PSE) in synthetic biology, e.g., the method presented in Chapter 7, are only applicable to transcriptional/translational (combinational) circuits [60]. Yet many strategies exist nowadays to engineer memory in biological systems including: (1) the use of recombinases to invert the orientation of specific pieces of DNA (DNA rearrangement); (2) constructing positive feedback systems using transcription factors. An additional mechanism for achieving biological memory is through epigenetic modifications to chromatin [131]. With synthetic memory, biological systems can be produced to have the ability to record events, and turn ON/OFF genetic programs without the requirement of a sustained input [191].

To solve the PSE problem for a biocircuit with DNA memory, an optimization strategy is required to : (1) encode the states and transitions of a biomolecular memory using a suitable computational model (e.g., graph model); (2) exploit the derived model to provide a microfluidic assay (similar to the biocircuit-regulatory scanning (BRS) assay presented in Chapter 7) that scans various configurations of the DNA memory. Constraints related to state coverage, state reachability, control/transition duration, and reagent usage are introduced.

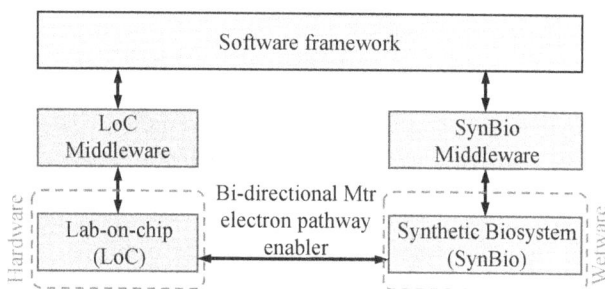

FIGURE 11.1
The hardware/wetware/software co-design environment and the inter-communication channels.

Utilizing Cyber-Physical Microfluidics in the Presence of Sensor Spoofing Attacks

Cyber-physical (CP) microfluidic biochips use an array of optical sensors (e.g., fluorescence sensors or a camera) to carry out quantitative analysis (Chapter 2). These sensors generate data streams that can be used to reconfigure protocol conditions and more importantly report analysis results. Similar to other sensing technologies, e.g., sonar, wireless sensor networks, etc [59, 70], microfluidic sensors and data acquisition/processing algorithms are vulnerable to spoofing attacks that may bias or nullify obtained results. Hence, the transition of CP-integration capabilities of microfluidic biochips to the marketplace relies on whether sensor data streams are collected, processed, and transferred in a trustworthy manner.

Microfluidic sensors are a good target for input spoofing; an unsophisticated adversary can interact with the camera simply by obscuring the ground plane image detected by the camera. Furthermore, if an array of fluorescence sensors is used, an adversary with control over the fluorescence setup can exercise an implicit control over the excitation and emission filters to tamper with the detection wavelength (Figure 1.2). A malicious activity can also be launched through an intent hijacking of the data-acquisition/processing algorithms. Such a hijacking attack has been successfully demonstrated using an unmanned aerial vehicle (UAV) that uses a classical data-processing algorithm, known as Lucas-Kanade algorithm [59]. This adversarial capability can be transferred to CP microfluidic biochips that uses similar processing techniques [161].

Therefore, comprehensive security assessment of microfluidic sensing is imperative in order to counter the above threats. Effective defense mechanisms are needed to secure biochips against such a type of attacks.

DNA Barcoding-Based Side-Channel Fingerprinting for Piracy Prevention

As described in Section 9.1, an undesirable side-effect of the current manufacturing model of microfluidic systems is the potential for untrusted third-parties, who in the course of performing their intended duties, also steal bioassay (intellectual property) or alter designs to modify the functionality of the end product. To secure microfluidic biochips against such attacks, a few defense mechanisms have been introduced including microfluidic encryption (Section 10.1) and bioassay locking [251]. A major drawback of these methods is that their effectiveness relies on the scale of the implemented bioassay—a bioassay with a limited number of fluidic operations may not be effectively secured against brute-force theft attacks. For example, if microfluidic encryption is used *without* randomizing the produced 8-bit keys (Section 10.1), an attacker can break the 8-bit key simply by buying 256 DMFBs to recover the secret key. Hence, the authentication of a microfluidic bioassay design must be strengthened using side-channel fingerprinting, which allows inherent (and secret) microfluidic or biological characteristics of a specific biochip to be exploited in providing a unique security key. This is similar to side-channel fingerprinting in integrated circuits where power, area, and delay characteristics can be used in deriving the keys [9].

A simple, yet effective, method for side-channel fingerprinting can be developed using DNA barcoding, as demonstrated in Section 10.5. This method can be integrated with microfluidic encryption or bioassay locking. Such a compound defense mechanism can be effective in securing biochips against different forms of piracy attacks even if the constituent bioassay consists of a limited number of fluidic operations. Each biochip will be fabricated and shipped with a unique set of DNA barcodes that are stored on the chip. Once the end user injects the DNA samples, an automated procedure is executed to mix the samples with these barcodes. Next, a DNA amplification and quantification process is executed to identify these barcodes. If a positive signal is received, the system generates a "corresponding key" or actuation sequence to "unlock" the remaining parts of the bioassay. On the other hand, if no signal is received (indicating that an attacker attempts to steal the proprietary bioassay), then the biochip will be locked and no biochemical operations will be executed. Note that a positive signal can be received only if the right type of primers, which are tightly coupled with the designed barcode, are used during DNA amplification. Hence, these primers can be shipped separately so that they can be loaded by an authenticated user only. Figure 11.2 illustrates

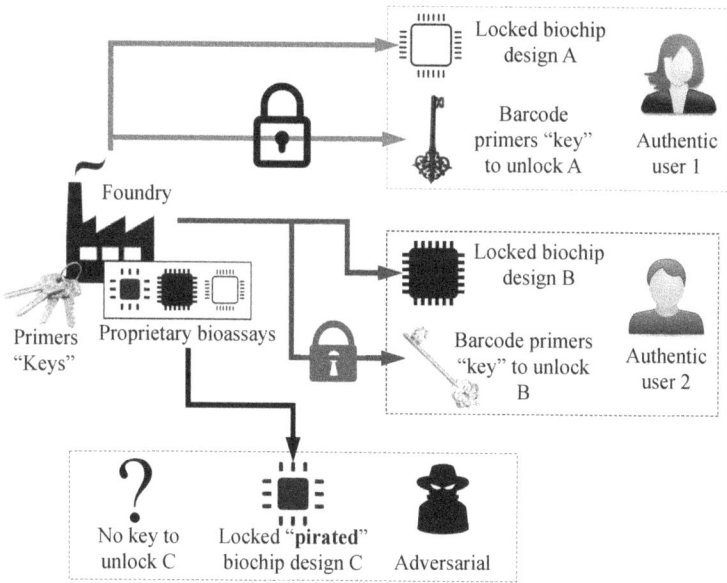

FIGURE 11.2
The use of DNA barcoding and related primers to secure proprietary bioassays.

the above fingerprinting scheme. Also, note that an authenticated user does not need to be aware of the barcode sequence nor the chemicals used in this fingerprinting process since the "unlocking" mechanism is automated.

Clearly, the above defense mechanism can be effective in securing biochip operations. However, potential threat models and assumptions related to the "safety zones," e.g., secure unlocking, must be thoroughly examined.

A

Proof of Theorem 5.1: A Fully Connected Routing Crossbar

Our objective here is to provide an inductive proof for Theorem 5.1 in Chapter 5. We aim to design an n_b-to-m_b routing crossbar, as shown in Figure A.1(a). The associated graph model $\mathcal{F}_{(n_b \times m_b,\ G_b)}$ (defined in Section 5.2) is depicted in Figure A.1(b); n_b is the number of inputs, m_b is the number of outputs, and G_b is a directed acyclic graph that represents the crossbar transposers and their interconnections. It is required to design a routing crossbar such that it is fully connected; the definition of full connectivity is given below.

Definition A.1 *An n_b-to-m_b routing crossbar is fully connected if and only if we can construct $\mathcal{F}_{(n_b \times m_b,\ G_b)}$ such that a droplet at any input $i_k \in \{i_1, i_2, ..., i_{n_b}\}$ can reach any output $o_j \in \{o_1, o_2, ..., o_{m_b}\}$.*

More specifically, the problem statement is described as follows:

"What constraint must be enforced on q such that a droplet at i_1 can reach o_{m_b}, and a droplet at i_{n_b} can reach o_1 through $\mathcal{F}_{(n_b \times m_b,\ G_b)}$, given that n_b and m_b are even integers?"

To simplify notation, we denote $\mathcal{F}_{(n_b \times m_b,\ G_b)}$ by $\mathcal{F}_{n_b \times m_b}$ throughout this discussion. Assume that $\mathcal{F}_{n_b \times m_b}$ is placed in a two-dimensional space with x-y coordinates; the origin is located at the top-left corner (Figure A.1(b)). Let $B(x, y)$ be a bounding box applied to $\mathcal{F}_{n_b \times m_b}$ such that it covers x horizontal levels and y vertical levels starting from the origin. Clearly, the crossbar consists of interconnected transposers that are topologically equivalent; thus we view the model $\mathcal{F}_{n_b \times m_b}$ as it is constructed by iteration, as illustrated in Figure A.2. This feature intuitively inspires us to use proof by induction to investigate crossbar full connectivity.

The Base Case: We start with the smallest scale of a routing crossbar, which is a single transposer (defined in Section III). At this scale, the inputs and outputs of the associated graph $\mathcal{F}_{2 \times 2}$ are directly connected. The graph model $\mathcal{F}_{2 \times 2}$ of the top left-most transposer is encapsulated in the bounding box $B(2, 2)$, as shown in Figure A.1(b). Based on the reconfigurable design of the transposer (Section 5.1), we introduce the following lemma:

Lemma A.1 *A full (half) transposer is a 2-to-2 (2-to-1) fully connected crossbar.*

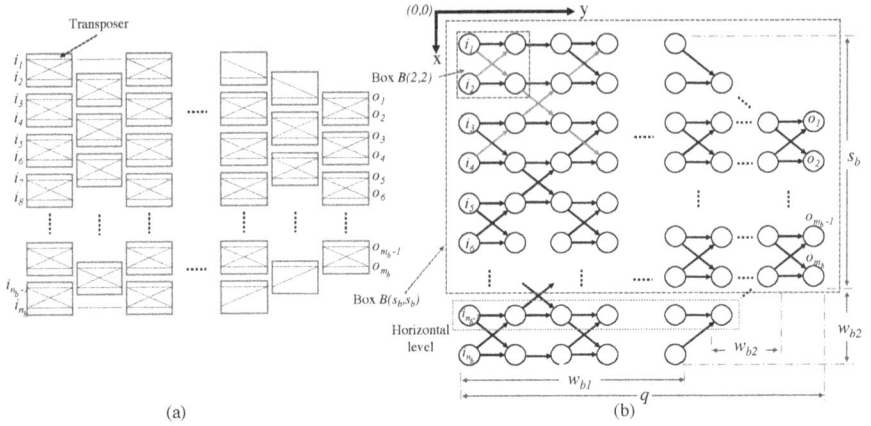

FIGURE A.1
(a) A schematic view of an n_b-to-m_b routing crossbar; (b) the associated graph model $\mathcal{F}_{(n_b \times m_b,\ G_b)}$.

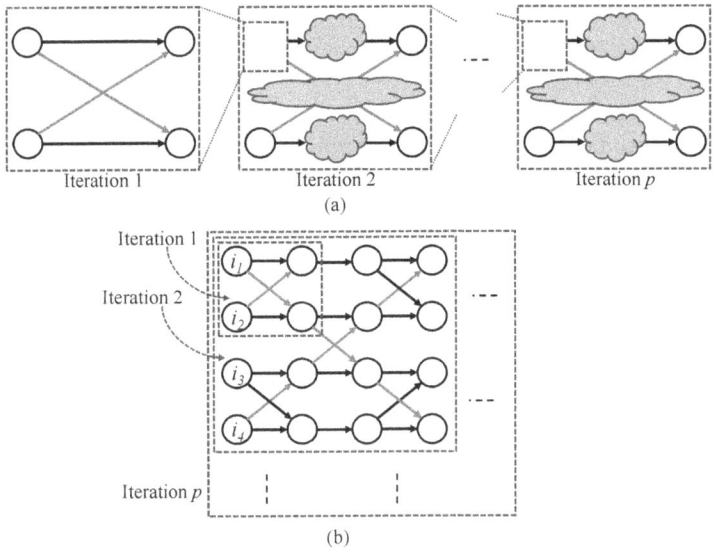

FIGURE A.2
Iterative construction of an $\mathcal{F}_{n_b \times m_b}$.

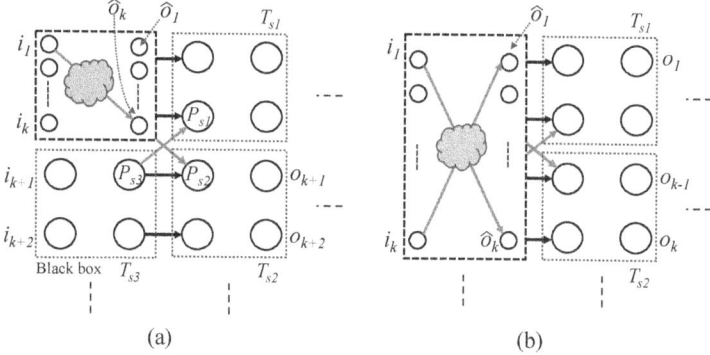

FIGURE A.3

Expanding the graph model $\mathcal{F}_{k \times k}$ to (a) $\mathcal{F}_{(k+2) \times (k+2)}$, (b) $\mathcal{F}_{k \times (k+2)}$.

Proof *We present the proof by contradiction. Suppose that a full transposer is not fully connected. This implies that in the associated graph $\mathcal{F}_{2 \times 2}$, there must be at least one input $i_k \in \{i_1, i_2\}$ that is not directly connected to one of the outputs $o_j \in \{o_1, o_2\}$. However, since only input and output nodes exist in $\mathcal{F}_{2 \times 2}$, i.e., there are no intermediate vertical levels between the inputs and outputs, then there are no alternative paths between i_k and o_j. This contradicts the original assumption about reconfigurability of transposers; that is, there is always a path between any input and any output. As a result, a full transposer must be fully connected to preserve reconfigurability. The proof is similar for half transposers.*

The Induction Step: Next, we study a generic model $\mathcal{F}_{k \times k}$, where k is an even integer, and $\mathcal{F}_{k \times k}$ is bounded by $B(k, k)$. Suppose that $\mathcal{F}_{k \times k}$ represents a crossbar that is fully connected. Then, the input i_1 is connected to the output $\hat{o}_k$. Similarly, i_k is connected to $\hat{o}_1$; see Figure A.3(a). By expanding the graph model to $\mathcal{F}_{(k+2) \times (k+2)}$ (bounded by $B(k + 2, k + 2)$), we obtain the following result.

Lemma A.2 *If a graph $\mathcal{F}_{k \times k}$ represents a fully connected crossbar, then the expanded graph $\mathcal{F}_{(k+2) \times (k+2)}$ also models a fully connected crossbar.*

Proof *By expanding $\mathcal{F}_{k \times k}$ to $\mathcal{F}_{(k+2) \times (k+2)}$, we connect three transposers T_{s1}, T_{s2}, and T_{s3} to $\mathcal{F}_{k \times k}$, as shown in Figure A.3(a). Since each transposer is fully connected (Lemma A.1), we guarantee that the inputs of T_{s3} (i.e., $i_{(k+1)}$ and $i_{(k+2)}$) are connected to P_{s3}. Likewise, P_{s2} in T_{s2} is connected to the two outputs $o_{(k+1)}$ and $o_{(k+2)}$, and P_{s1} is connected to the outputs of T_{s1}. According to the graph model topology, directed edges are constructed from P_{s3} to P_{s1} and P_{s2}, and from $\hat{o}_k$ to P_{s1} and P_{s2}. This implies that there is always a path from any input in $\mathcal{F}_{k \times k}$ and T_{s3} to all the outputs of T_{s1} and T_{s2}. As a* result, $\mathcal{F}_{(k+2)\times(k+2)}$ *represents a fully connected crossbar. We reach the same conclusion if* T_{s1}, T_{s2}, *or both are replaced with half transposers.*

Furthermore, consider expanding $\mathcal{F}_{k\times k}$ to $\mathcal{F}_{k\times(k+2)}$ following the same approach; see Figure A.3(b). Intuitively, $\mathcal{F}_{k\times(k+2)}$ also represents a fully connected crossbar, since i_1 is connected to o_k, and i_k is connected to o_1. Therefore, we introduce the following lemma.

Lemma A.3 *If a graph* $\mathcal{F}_{k\times k}$ *represents a fully connected crossbar, then the expanded graph* $\mathcal{F}_{k\times(k+2)}$ *also models a fully connected crossbar.*

Proof *Similar to the proof of Lemma A.2.*

Using Lemma A.1 and Lemma A.2, and by induction, we obtain the following theorem.

Theorem A.1 *A graph* $\mathcal{F}_{k\times k}$ *represents a fully connected crossbar if k is an even integer and* $\mathcal{F}_{k\times k}$ *is bounded by* $B(k,k)$.

Consider Figure A.1(b), the graph model $\mathcal{F}_{s_b\times s_b}$ bounded by $B(s_b, s_b)$ is fully connected if s_b is even, according to Theorem A.1. To test this condition, the value of s_b can be calculated as follows: $s_b = m_b + \frac{n_b - m_b}{2} = \frac{n_b + m_b}{2}$. It is obvious that if m_b and n_b are even integers (given assumption), then s_b is even. Therefore, $\mathcal{F}_{s_b\times s_b}$ is fully connected. Furthermore, due to the square shape of $B(s_b, s_b)$, we conclude that $q = s_b = \frac{n_b + m_b}{2}$. Based on Lemma A.3, the previous formula can be generalized to the following inequality: $q \geq \frac{n_b + m_b}{2}$, where n_b and m_b are even.

The same proof can also be applied if we move the origin of the space to the bottom left of Figure A.1(b) and repeat the calculation of q based on a bounding box expanding from the bottom part of the graph $\mathcal{F}_{n_b\times m_b}$. We conclude that o_1 is connected to i_{n_b}. Therefore, we obtain the required result as follows.

Theorem 5.1 *An n_b-to-m_b, q-level valve-based crossbar is a fully connected fabric if n_b and m_b are even integers, and $q \geq \frac{m_b + n_b}{2}$.*

Note that this theorem provides a sufficient condition, which requires m_b and n_b to be even. A necessary condition remains to be derived as part of future work.

B

Modeling a Fully Connected Routing Crossbar

Using Figure A.1(b), we can calculate the parameters w_{b1} and w_{b2}, given m_b, n_b, and q. Note that q is set to the lower bound, i.e., $q = \frac{m_b + n_b}{2}$.

$$w_{b2} = \frac{n_b - m_b}{2} \tag{B.1}$$

$$
\begin{aligned}
w_{b1} &= q - w_{b2} - 1 \\
&= \frac{m_b + n_b}{2} - \frac{n_b - m_b}{2} - 1 \\
&= m_b - 1
\end{aligned}
\tag{B.2}
$$

Algorithm B.1 is implemented to generate the graph model $\mathcal{F}_{n_b \times m_b}$ of a fully connected routing crossbar. For the sake of illustration, we apply the steps of the algorithm to generate the graph $\mathcal{F}_{10 \times 6}$; see Figure B.1. The algorithm begins by calculating the parameters w_{b1} and w_{b2} (Line 2) using Equations (B.1)-(B.2); $w_{b1} = 5$ and $w_{b2} = 2$ for $\mathcal{F}_{10 \times 6}$. Then the algorithm generates the input nodes which reside at the first vertical level of $\mathcal{F}_{10 \times 6}$ (Line 3). Next, the subsequent $w_{b1} - 1$ vertical levels are generated (Lines 5-13); note that these levels incorporate decision nodes for only full transposers. In each level (Lines 7-12), n_b nodes are generated and connected with the appropriate nodes located in the previous level.

After completing the generation of the first w_{b1} vertical levels, the following (w_{b2}) levels contain decision nodes that represent half transposers. Therefore, the variable s_h (skipped horizontal levels) is used to account for the change in the number of nodes generated at each vertical level (Line 16). The steps used for generating the nodes and the associated edges (Lines 15-24) are similar to the previous part. Finally, the algorithm generates the output nodes and the edges that connect them to the previous level (Lines 26-31).

The worst-case computational complexity of this algorithm is $O(n_b{}^2 \cdot m_b)$.

Algorithm B.1: Graph Model $\mathcal{F}_{n_b \times m_b}$ Generator

---

Input: m_b, n_b, q

Output: Set of nodes $\mathcal{D}_{n_b \times m_b}$ and set of edges $\mathcal{Y}_{n_b \times m_b}$

1: $\mathcal{D}_{n_b \times m_b} \leftarrow \emptyset$; $\mathcal{Y}_{n_b \times m_b} \leftarrow \emptyset$;

2: $w_{b1}, w_{b2} \leftarrow \text{INITIALIZE}(n_b, m_b, q)$;

3: $\mathcal{D}_{n_b \times m_b} \leftarrow \text{GENERATEINPUTNODES}(n_b)$;   ▷ 1st vertical level

4:   ▷ Generate the nodes and the directed edges for the next $w_{b1} - 1$ vertical levels

5: **for** $v_l = 2$ to w_{b1} **do**

6:   ▷ v_l: vertical level

7:     **for** $h_l = 1$ to n_b **do**

8:         $nd \leftarrow \text{GENERATENODE}(v_l, h_l)$;

9:         $ed \leftarrow \text{GENEDGESFROMPREVIOUSLEVELTOCURRENTNODE}(v_l - 1)$;

10:         $\mathcal{D}_{n_b \times m_b} \leftarrow \mathcal{D}_{n_b \times m_b} \cup nd$;

11:         $\mathcal{S}_{n_b \times m_b} \leftarrow \mathcal{S}_{n_b \times m_b} \cup ed$;

12:     ▷ Generate the nodes and the directed edges for the next w_{b2} vertical levels (including half transposers)

13: **for** $v_l = w_{b1} + 1$ to $w_{b1} + w_{b2}$ **do**

14:     $s_h = 1$;   ▷ Number of skipped horizontal levels

15:     **for** $h_l = 1 + s_h$ to $n_b - s_h$ **do**

16:         $nd \leftarrow \text{GENERATENODE}(v_l, h_l)$;

17:         $ed \leftarrow \text{GENEDGESFROMPREVIOUSLEVELTOCURRENTNODE}(v_l - 1)$;

18:         $\mathcal{D}_{n_b \times m_b} \leftarrow \mathcal{D}_{n_b \times m_b} \cup nd$;

19:         $\mathcal{S}_{n_b \times m_b} \leftarrow \mathcal{S}_{n_b \times m_b} \cup ed$;

20:     $s_h = s_h + 1$;

21:     ▷ Generate the nodes and the edges for the last (output) vertical level

22: **for** $h_l = w_{b2}$ to $w_{b2} + m_b$ **do**

23:     $nd \leftarrow \text{GENERATEOUTPUTNODES}(h_l)$;

24:     $ed \leftarrow \text{GENEDGESFROMPREVIOUSLEVELTOCURRENTNODE}()$; ▷ Last vertical level

25:     $\mathcal{D}_{n_b \times m_b} \leftarrow \mathcal{D}_{n_b \times m_b} \cup nd$;

26:     $\mathcal{S}_{n_b \times m_b} \leftarrow \mathcal{S}_{n_b \times m_b} \cup ed$;

    return $\mathcal{D}_{n_b \times m_b}, \mathcal{S}_{n_b \times m_b}$;

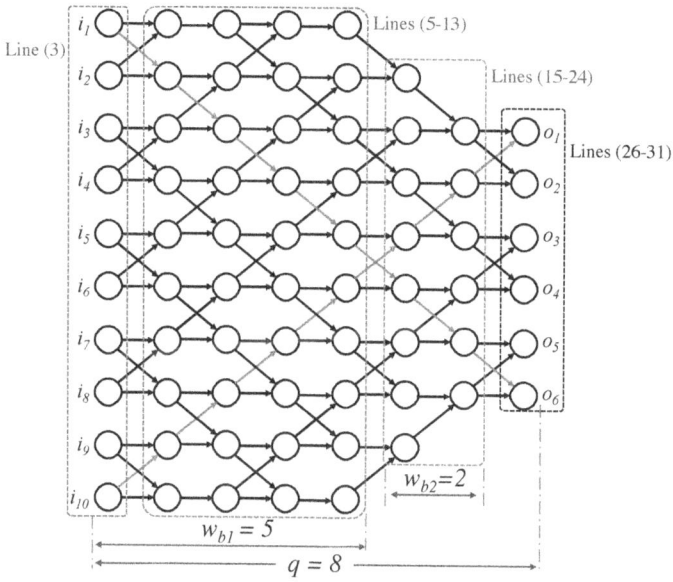

FIGURE B.1
The application of Algorithm B.1 on a graph $\mathcal{F}_{10\times6}$.

C

Proof of Lemma 6.1: Derivation of Control Delay Vector Ψ

Our objective here is to provide a proof by induction for Lemma 6.1, considering a reconfigurable flow-based biochip (RFB) $\theta_{sc}(x_v, y_v)$, where x_v is the number of flow valves and y_v is the number of control pins. We first derive the control-delay value b_x of a flow valve L_x. Second, we combine all the delay values $\{b_1, b_2, ..., b_{x_v}\}$ associated with all biochip valves in a vector $\Psi \in \mathbb{R}^{x_v}$, which is known as the control delay vector for $\theta_{sc}(x_v, y_v)$.

Recall that the control of flow valves in an RFB requires two types of control pins: (1) primary pins $\{C_1^a, C_2^a, ..., C_{a_v}^a\}$, where a_v is the number of primary pins; (2) demultiplexing pins $\{C_1^g, C_2^g, ..., C_{g_v}^g\}$, where g_v is the number of demultiplexing pins. Hence, a generic control configuration of a valve L_x in $\theta_{sc}(x_v, y_v)$ can be performed using m_p primary pins and n_p demultiplexing pins, where $1 \leq m_p \leq a_v, 0 \leq n_p \leq g_v$. This generic configuration is therefore denoted by (m_p, n_p). For example, Figure C.1 shows six possible configurations (m_p, n_p), where $1 \leq m_p \leq 2, 0 \leq n_p \leq 2$.

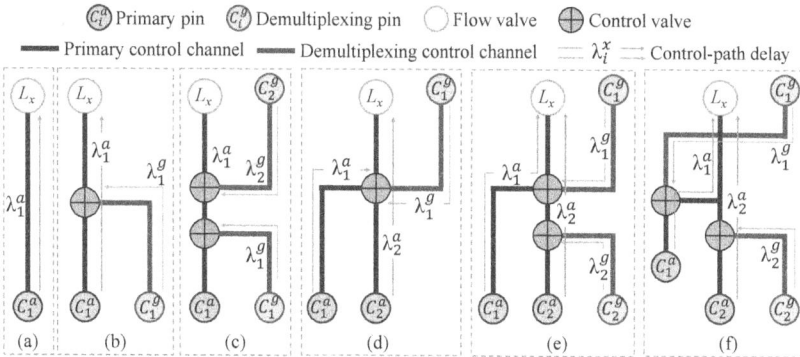

FIGURE C.1

Control configurations for a valve L_x: (a) (1,0) configuration. (b) (1,1) configuration. (c) (1,2) configuration. (d) (2,1) configuration. (e) (2,2) configuration. (f) (2,2) configuration. The parameters λ_i^a and λ_i^g refer to the valve-control delays associated with a primary pin C_i^a and a demultiplexing pin C_i^g, respectively.

Based on the number of primary pins m_p used in actuating L_x, we introduce two types of valve control through the following definition.

Definition C.1 *A valve L_x in an RFB $\theta_{sc}(x_v, y_v)$ can be actuated using two different control schemes: (1) Multiplexed-control (MC) scheme, where the actuation of L_x is performed by using a single primary pin, i.e., $m_p = 1$, and n_p demultiplexing pins; $n_p \geq 0$. (2) Networked-control (NC), where the control of L_x is performed by using an m_p primary pins and n_p demultiplexing pins, where $m_p > 1$, $n_p \geq 1$.*

For example, the configurations in Figure C.1(a-c) are MC, whereas the configurations in Figure C.1(d-f) are NC. Since our synthesis methodology is focused on MC-driven RFBs, we focus on deriving b_x for a valve L_x that has MC and keep the derivation for the NC-driven valve for future work.

Note that adopting MC to actuate a valve L_x, i.e., by using a single primary pin C_y^a, indicates that there is only a single control path which connects C_y^a to L_x; see Figure C.1(b). In other words, we do not consider the case where duplicate paths are used for connection. Intuitively, if a primary pin C_y^a is unable to actuate a valve L_x, then C_y^a must be busy actuating another valve L_z through a different control path, and hence having a duplicate path between L_x and C_y^a is meaningless.

Note that prior work in the synthesis of flow-based biochips considers a single control pin ($m_p = 1$) for actuating a valve L_x, even with employing pin-sharing among multiple valves. Hence, the control configuration considered in the previous work is also labeled as MC.

Recall that primary pins are used to provide the pressure (or vacuum) to actuate the flow valves, whereas demultiplexing pins are used to direct the pressure-driven flow through control channels from a primary pin to a particular valve. These two functions cannot be overlapping in any control configuration (MC or NC), because the control channels need to be setup first before a pressure (or vacuum) pulse is provided by a primary pin. Analytically, this non-overlapping condition implies that the control delay b_x of a valve L_x can be interpreted as $b_x = b_x^a + b_x^g$, where b_x^a and b_x^g are the delay associated with primary pins and demultiplexing pins, respectively. Throughout the proof, we also refer to b_x^g as the delay associated with control-path activation.

To derive the relation for b_x^a and b_x^g, we first construct a canonical MC topology for a valve L_x, as shown in Figure C.2. A set of k demultiplexing pins $\{C_0^g, C_1^g, ..., C_{k-1}^g\}$ are used to actuate the control path which connects L_x to a primary pin C_1^a. The control path is activated if all the control valves $\{l_1^1, l_1^2, l_2^2, ..., l_k^k\}$ are opened. An MC topology is canonical if each demultiplexing pin is connected to a different number of control valves. For example, the pin C_0^g is not connected to any valve, the pin C_1^g is connected to a single control valve l_1^1, C_2^g is connected to two control valves $\{l_1^1, l_2^2\}$, and C_k^g is connected to k control valves. This design can capture all possible control settings within the control path for an MC-driven valve, and therefore offers a generic MC platform that can be used to derive the relations for b_x^a and b_x^g.

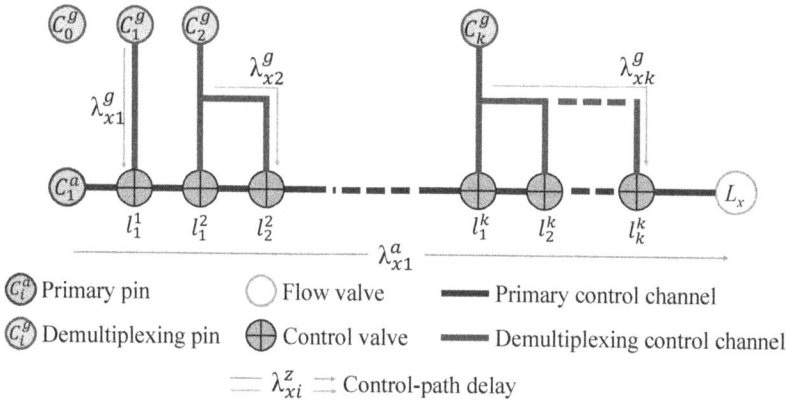

FIGURE C.2
Canonical MC configuration for a valve L_x.

Using the canonical MC design in Figure C.2, the control path, denoted by CP, that connects the primary pin C_1^a with L_x can be activated only after all the control valves in the set $\widetilde{\mathcal{L}} = \{l_1^1, l_1^2, l_2^2, ..., l_k^k\}$ are actuated. In other words, if a demultiplexing pin C_y^g is connected to a set of control valves $\widetilde{\mathcal{L}}^y \subset \widetilde{\mathcal{L}}$, then CP cannot be activated before a fluid injected via C_y^g reaches all the valves in the set $\widetilde{\mathcal{L}}^y = \{l_1^y, l_2^y, ..., l_y^y\}$. Consider a value dv_i^y that describes the delay of fluidic transport from C_y^g to a control valve $l_i^y \in \widetilde{\mathcal{L}}^y$. Based on the above observation, we compute the actuation delay λ_{xy}^g for a demultiplexing pin C_y^g as follows.

Definition C.2 *Consider a demultiplexing pin C_y^g that is used in canonical MC of a valve L_x. Also, consider a set of control valves $\widetilde{\mathcal{L}}^y \subset \widetilde{\mathcal{L}}$ that is connected to C_y^g, where $\widetilde{\mathcal{L}}^y = \{l_1^y, l_2^y, ..., l_y^y\}$, and the fluidic delays associated with these connections are presented in a vector $DV^y = [dv_1^y \; dv_2^y \; ... \; dv_y^y]^\mathsf{T}$. The actuation delay λ_{xy}^g of C_y^g is defined as the minimum duration through which a fluid stream injected from C_y^g reaches all control valves in $\widetilde{\mathcal{L}}^y$, and therefore the actuation delay λ_{xy}^g is estimated using the relation $\lambda_{xy}^g = \max\{DV^y\}$.*

Figure C.2 demonstrates the actuation delays λ_{x1}^g, λ_{x2}^g, and λ_{xk}^g associated with demultiplexing pins C_1^g, C_2^g, and C_k^g, respectively.

Recall from Section 6.2 that the connectivity vector of a valve L_x is $\phi_x = [\lambda_{x1}^a \; \lambda_{x2}^a \; ... \; \lambda_{xa_v}^a | \lambda_{x1}^g \; \lambda_{x2}^g \; ... \; \lambda_{xg_v}^g] = [\phi_x^a | \phi_x^g]$, where $\lambda_{xy}^a > 0$ indicates that a primary pin C_y^a is used to actuate L_x with actuation delay λ_{xy}^a, $\lambda_{xy}^g > 0$ indicates that a demultiplexing pin C_y^g is used to actuate L_x with actuation delay λ_{xy}^g, a_v is the number of primary pins, g_v is the number of demultiplexing pins, ϕ_x^a is the connectivity associated with the primary pins, and ϕ_x^g is the connectivity associated with the demultiplexing pins. By using the canonical MC scheme to actuate L_x, the connectivity vector ϕ_x becomes $\phi_x = [1|0\;1\;1\;1\;...\;1]$, where $\phi_x \in \mathbb{R}^{k+1}$. In the following discussion, we redefine ϕ_x as $\phi_{\{x,k\}}$ to consider the number of demultiplexing pins connected to the control path for L_x. Similarly, b_x^a and b_x^g are redefined as $b_{\{x,k\}}^a$ and $b_{\{x,k\}}^g$, respectively.

To derive the relation for $b_{\{x,k\}}^a$ (using proof by induction), we introduce the base cases for a canonical MC scheme.

Base Case: We investigate the simplest three MC configurations for a valve L_x; see Figure C.3. The simplest case (Figure C.3(a)) is where no demultiplexing pin is connected; $k = 0$. In this case, the connectivity vector associated with demultiplexing pins $\phi_{\{x,0\}}^g$ is empty, and there is no actuation delay by any demultiplexing pin. In other words, $b_{\{x,0\}}^g = 0$ for this base case.

In the second base case, shown in Figure C.3(b), a single demultiplexing pin C_1^g is connected to the control path, i.e., $k = 1$. Hence, the connectivity vector $\phi_{\{x,1\}}^g$ associated with the demultiplexing pins is simply $[\lambda_{x1}^g]$, and therefore the control delay associated with activating the control path is $b_{\{x,1\}}^g = \phi_{\{x,1\}}^g = \lambda_{x1}^g$. Note that this case also applies to an MC configuration where a single demultiplexing pin is used to actuate any number of control valves located on the control path; the connectivity vector $\phi_{\{x,1\}}^g$ does not include information about the individual control valves, but λ_{xy}^g captures the longest fluidic delay among control-valve connections.

The third base case is shown in Figure C.3(c). Two demultiplexing pins $\{C_1^g, C_2^g\}$ are connected to the control path, i.e., $k = 2$. The connectivity vector associated with this case is $\phi_{\{x,2\}}^g = [\lambda_{x1}^g \ \lambda_{x2}^g]$. To obtain the value of $b_{\{x,2\}}^g$ by using $\phi_{\{x,2\}}^g$, we investigate the scenarios through which the control path for L_x can be activated. We observe two control scenarios: (1) Serial actuation of $\{C_1^g, C_2^g\}$, in which the biochip controller pressurizes one demultiplexing

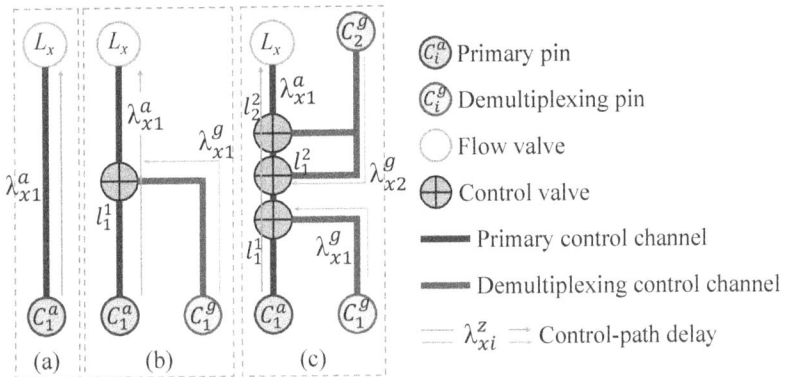

FIGURE C.3
The base MC configurations for a valve L_x: (a) No demultiplexing pins are connected; (b) A single demultiplexing pin is connected; (c) Two demultiplexing pins are connected.

pin at a time. By using this control mechanism, $b^g_{\{x,2\}}$ can be computed as follows: $b^g_{\{x,2\}} = \lambda^g_{x1} + \lambda^g_{x2} = \mathrm{sum}\{\phi^g_{\{x,2\}}\}$. (2) Parallel actuation of $\{C^g_1, C^g_2\}$, in which the controller pressurizes the two pins concurrently. By using this control mechanism, $b^g_{\{x,2\}}$ is computed as follows: $b^g_{\{x,2\}} = \max\{\lambda^g_{x1}, \lambda^g_{x2}\} = \max\{\phi^g_{\{x,2\}}\}$. Since the demultiplexing pins $\{C^g_1, C^g_2\}$ are used to activate only the control path for L_x, then parallel control of $\{C^g_1, C^g_2\}$ is considered, and therefore $b^g_{\{x,2\}} = \max\{\phi^g_{\{x,2\}}\}$.

Induction Hypothesis: Next, we study a generic MC configuration, where $k \geq 3$ demultiplexing pins are connected to the control path for L_x. By using a biochip controller that pressurizes all the k demultiplexing pins concurrently, then the value of $b^g_{\{x,k\}}$ can be computed similar to the base case as follows:

$$b^g_{\{x,k\}} = \max\{\lambda^g_{x1}, \lambda^g_{x2}, ..., \lambda^g_{xk}\} = \max\{\phi^g_{\{x,k\}}\} \qquad (\text{C.1})$$

Inductive Step: By expanding the control path to connect an additional demultiplexing pin C^g_{k+1}, we obtain the following result.

Lemma C.1 *Consider a valve L_x that is actuated using MC, and k demultiplexing pins are connected to the control path. If the control delay associated with activating the control path is $b^g_{\{x,k\}}$, then the delay of control-path activation when $k+1$ demultiplexing pins are connected is $\max\{b^g_{\{x,k\}}, \lambda^g_{x(k+1)}\}$.*

    Proof *By expanding the control path, an additional set of control valves $\tilde{\mathcal{L}}^{k+1} = \{l^{k+1}_1, l^{k+1}_2, ..., l^{k+1}_{k+1}\}$ are added, and these valves can be actuated using a demultiplexing pin C^g_{k+1}. By using the assumption that a biochip controller can pressurizes all the $k+1$ demultiplexing pins concurrently, then the activation of the control path is associated with a delay $b^g_{\{x,k+1\}} = \max\{\lambda^g_{x1}, \lambda^g_{x2}, ..., \lambda^g_{xk}, \lambda^g_{x(k+1)}\} = \max\{\max\{\lambda^g_{x1}, \lambda^g_{x2}, ..., \lambda^g_{xk}\}, \lambda^g_{x(k+1)}\}$. By using Equation (C.1), we reach the following result: $b^g_{\{x,k+1\}} = \max\{b^g_{\{x,k\}}, \lambda^g_{x(k+1)}\}$.*

    Note that in all cases of MC for a valve L_x, only a single primary pin C^a_x is used to actuate L_x after the control path is activated. As a result, the actuation delay associated with the primary pin is $b^a_{\{x,1\}} = b^a_{\{x,2\}} = b^a_{\{x,k\}} = \lambda^a_{x1}$. If the other $m_p - 1$ primary pin are used to actuate other valves using MC, then the connectivity vector $\phi^a_{\{x,k\}}$ associated with L_x is defined as $[\lambda^a_{x1} \; 0 \; 0 \; ... \; 0]$, where $\phi^a_{\{x,k\}} \in \mathbb{R}^{m_p}$. Hence, the actuation delay $b^a_{\{x,k\}}$ is estimated using the following relation:

$$b^a_{\{x,k\}} = \lambda^a_{x1} = \max\{\phi^a_{\{x,k\}}\} \qquad (\text{C.2})$$

Using Lemma C.1 and Equation (C.2), and by induction, we obtain the required result for $b_{\{x,k\}}$ (or simply b_x) and Ψ.

Lemma 6.1 *The control delay of a flow valve L_x is a value $b_x = b^a_x + b^g_x = \max\{\phi^a_x\} + \max\{\phi^g_x\}$, and the control delay vector for all the valves in $\theta_{sc}(x_v, y_v)$ is $\Psi = [b_1 \; b_2 \; ... \; b_{x_v}]^{\mathsf{T}}$.*

D

Proof of Theorem 6.1: Derivation of Control Latency α_i

Our objective here is to provide a proof by induction for Theorem 6.1. This theorem extends the result obtained by Lemma 6.1 for a biochip that has a set of x_v valves $\mathcal{L} = \{L_1, L_2, ..., L_{x_v}\}$, and a subset of these valves ($\mathcal{L}_{subs} \subset \mathcal{L}$) are connected within a "flow" path[1] that can be used to route a sample droplet. Hence, we aim to derive the control latency α_i, which describes the duration for actuating all the valves in $\mathcal{L}_{subs}$ and thus activating the flow path. We construct the proof by considering all possible configurations of control pins.

Recall that a *control vector* $\vartheta_i : \{i \in \mathbb{N}, t \leq CV\}$ is a vector of length x_v such that $\vartheta_i[x] = 1$ indicates that a flow valve L_x must be actuated as a part of the control vector, and $\vartheta_i[x] = 0$ otherwise. For example, Figure D.1 shows an RFB with 6 flow valves, and the flow path from **S** to **E** can be activated if only the valves $\{L_1, L_2, L_5, L_6\}$ are actuated. Hence, the control vector ϑ_i associated with this flow path is defined as $\vartheta_i = [1\ 1\ 0\ 0\ 1\ 1]$. We henceforth use the terms "control vector" and "flow path" interchangeably. Throughout this proof, we focus on a single control vector, i.e, $CV = 1$ to derive the relation for α_i. Afterwards, we consider a generic case, where $CV > 1$, to obtain the control latency vector Θ associated with all control vectors in $\theta_{sc}(x_v, y_v)$.

Consider a biochip $\theta_{sc}(x_v, y_v)$ that includes a set of flow valves $\mathcal{L} = \{L_1, L_2, ..., L_{x_v}\}$. To derive the relation for α_i, we construct a canonical MC scheme for a control vector ϑ_i. Assume without loss of generality that $\vartheta_i = [0\ 1\ 1\ ...\ 1]$. This indicates that only one valve (L_1) does not need to be actuated for ϑ_i (flow-path activation). This assumption is used for illustrative purposes and it does not impact the procedure of the proof. To avoid confusion, we henceforth refer to L_1, the valve that is not a part of ϑ_i, as L'.

A simple approach to construct a canonical MC scheme for ϑ_i is to replicate the canonical scheme from Figure C.2 for every valve L_x. However, recall that our previous analysis considers that every MC-driven valve L_x is actuated using a single primary pin and an independent set of demultiplexing pins. Hence, by replicating the canonical MC for a valve L_x, we unnecessarily double the number of control pins needed for activating ϑ_i.

---

[1]Flow path is not to be confused with control path. A flow path is fluidic path that is used to route a sample through flow channels, whereas a control path describes the fluidic path for actuating a valve L_x.

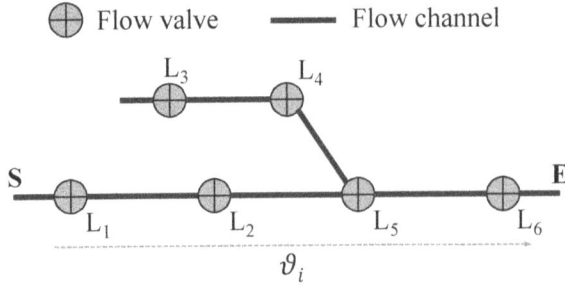

FIGURE D.1
Demonstration of the flow path associated with a control vector ϑ_i.

To overcome this problem, we need a canonical MC scheme for ϑ_i that allows multiple valves to share a primary pin. For this purpose, we construct a canonical MC topology that connects m_p primary pins $\{C_1^a, C_2^a, ..., C_{m_p}^a\}$ to x_v flow valves $\{L', L_1^1, L_1^2, L_2^2, ..., L_{m_p}^{m_p}\}$; see Figure D.2. Recall that an MC scheme for an individual valve L_x is subject to the condition that L_x is actuated using a single primary pin; this condition must be fulfilled by a canonical MC scheme for ϑ_i as well. Also, note that the control latency for ϑ_i is not affected by which primary pin actuates the valve L', and therefore it is left disconnected in Figure D.2.

An MC topology associated with ϑ_i is canonical if each primary pin is connected to a different number of flow valves in the set $\{L_1^1, L_1^2, L_2^2, ..., L_{m_p}^{m_p}\}$. For example, as shown in Figure D.2, the primary pin C_1^a is connected to a single valve L_1^1, C_2^a is connected to two valves $\{L_1^2, L_2^2\}$, and $C_{m_p}^a$ is connected

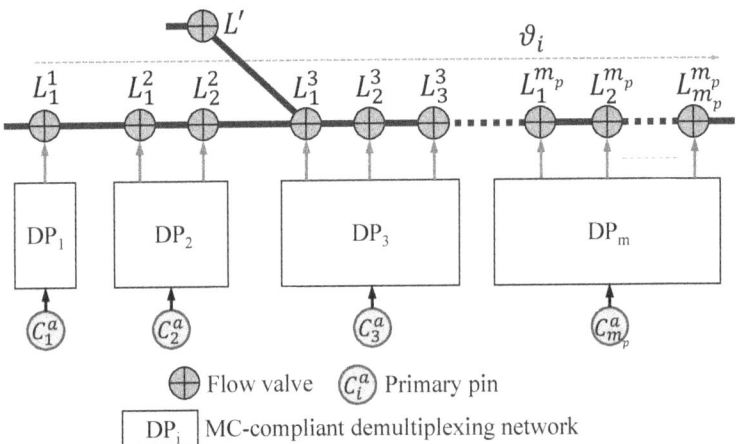

FIGURE D.2
Canonical MC configuration associated with a control vector ϑ_i.

to m_p flow valves $\{L_1^{m_p}, L_2^{m_p}, ..., L_{m_p}^{m_p}\}$. This design captures all possible control settings needed for the activation of ϑ_i, and therefore offers a generic MC platform that can be used to derive the relation for α_i.

By using the canonical MC design in Figure D.2, ϑ_i can be activated only after all the valves in $\{L_1^1, L_1^2, L_2^2, ..., L_{m_p}^{m_p}\}$ are actuated. Since these valves are actuated by different primary pins, we introduce the connectivity Ω^{m_p} below. We also present a constraint on Ω^{m_p} for an MC-compliant pin-sharing design.

Definition D.1 *Consider a biochip* $\theta_{sc}(x_v, y_v)$ *that has a set of flow valves* $\mathcal{L} = \{L', L_1^1, \ L_1^2, L_2^2, ..., L_{m_p}^{m_p}\}$ *that is actuated using MC using* m_p *primary pins. We define the connectivity vector of a primary pin* C_y^a *as* $\rho_y = [SN_{1y}^a \ SN_{2y}^a \ ... \ SN_{x_vy}^a]^\intercal \in \mathbb{R}^{x_v}, y \in [1, a_v]$. *In addition, we define a binary vector* $SN_y = sign(\rho_y)$, *where* $SN_y[x] = 1$ *indicates that a valve* $L_x \in \mathcal{L}$ *is actuated using a primary pin* C_y^a, *and* $SN_y[x] = 0$ *otherwise. Furthermore, we define the connectivity matrix* Ω^{m_p} *as follows:* $\Omega^{m_p} = [SN_1 \ SN_2 \ ... \ SN_{m_p}]^\intercal \in \mathbb{R}^{m_p \times x_v}$. *To provide an MC scheme using primary pins, the following condition must be satisfied for each valve* L_x.

$$\sum_{i=1}^{m_p} \Omega^{m_p}[i, x] = 1 \tag{D.1}$$

This condition implies that a valve L_x can be actuated only by a single primary pin. While a primary pin can flexibly be shared among multiple valves, sharing of a demultiplexing pin must also be compliant with the MC scheme (see Section 6.2). The following definition introduces an MC-related constraint that regulates sharing of demultiplexing pins.

Definition D.2 *Let* RD_{kj} *be a binary variable, where* $RD_{kj} = 1$ *if a demultiplexing pin* C_k^g *is used in the activation of a control path that is pressurized using a primary pin* C_j^a, *and* $RD_{kj} = 0$ *otherwise. An MC-compliant scheme that constitutes sharing of demultiplexing pins must satisfy the following condition:*

$$\sum_{j=1}^{A} RD_{kj} = 1 \tag{D.2}$$

By fulfilling the condition in Equation (D.2), we ensure that any demultiplexing pin can be used only in the activation of control paths that are connected to a single primary pin. For example, in Figure D.2, demultiplexing pins used in DP_1 to actuate L_1^1 cannot be used in DP_2 or DP_{m_p}. This condition enables us to make the assumption that valves connected to different primary pins, i.e., valves in different primary-pin groups, can be actuated concurrently, if the biochip controller supports parallel actuation. In other words, there are no control dependencies between valves in different primary-pin groups.

Recall that $\vartheta_i = [0 \ 1 \ 1 \ ... \ 1]$, $\Omega^{m_p} = [SN_1 \ SN_2 \ ... \ SN_{m_p}]^\intercal$, and $\Psi = [b_1 \ b_2 \ ... \ b_{x_v}]^\intercal$ (Lemma 6.1). Also, consider the vector operator $\circ$ that

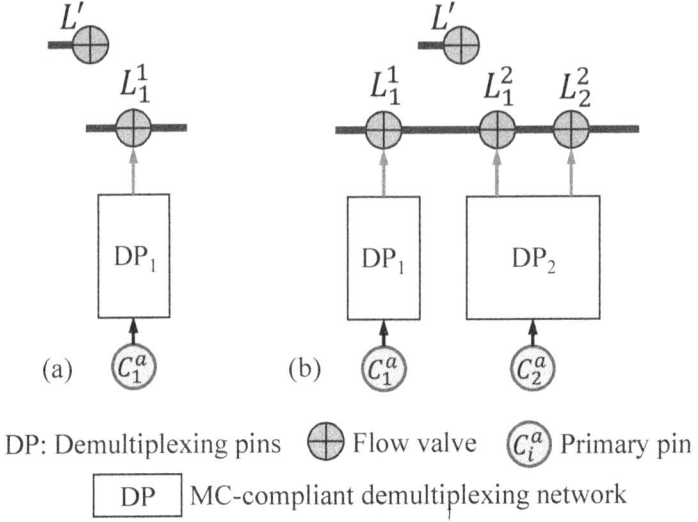

FIGURE D.3

The base MC configurations associated with ϑ_i: (a) a single primary pin is connected to a single valve; (b) a single primary pin connected to two valves; (c) two primary pins are connected to three valves.

provides the element-wise product of two vectors, i.e., $E \circ B = [e_1 \ e_2 \ ... \ e_n] \circ [b_1 \ b_2 \ ... \ b_n] = [e_1 b_1 \ e_2 b_2 \ ... \ e_n b_n]$. By using the canonical MC scheme (Figure D.2), we derive the relation for α_i (using proof by induction). In the following discussion, we redefine α_i as $\alpha_{\{i,m_p\}}$ to consider the number of primary pins connected to $\mathcal{L}$. Similarly, ϑ_i is redefined as $\vartheta_{\{i,m_p\}}$. We start the proof by introducing two base cases for a canonical MC scheme.

The Base Case: We investigate the simplest two MC configuration associated with a control vector $\vartheta_{\{i,m_p\}}$; see Figure D.3. The simplest configuration (Figure D.3(a)) is where only a single primary pin is used to actuate the valves in the set $\mathcal{L} = \{L', L_1^1\}$, i.e., $m_p = 1$. In this case, the connectivity matrix Ω^1 includes only a single row/vector SN_1, which is defined as $SN_1 = [1 \ 1]$. Since only L_1^1 is located on the flow path, i.e., $\vartheta_{\{i,1\}} = [0 \ 1]$, then activating this control path requires only activating the valve L_1^1 using the primary pin C_1^a. Hence, the relation for the control latency $\alpha_{\{i,1\}}$ associated with $\vartheta_{\{i,1\}}$ is derived as follows[2]: $\alpha_{\{i,1\}} = b_1 = [0 \ 1] \cdot ([1 \ 1] \circ [b_0 \ b_1]^\mathsf{T}) = \vartheta_{\{i,1\}} \cdot (SN_1 \circ \Psi)$.

In the second base case, shown in Figure D.3(b), two primary pins are used to actuate the valves in the set $\mathcal{L} = \{L', L_1^1, L_1^2, L_2^2\}$, i.e., $m_p = 2$. In this case, the connectivity matrix Ω^2 includes two vectors $\{SN_1, SN_2\}$, which can be defined as $SN_1 = [1 \ 1 \ 0 \ 0]$, $SN_2 = [0 \ 0 \ 1 \ 1]$. Also, $\vartheta_{\{i,2\}} = [0 \ 1 \ 1 \ 1]$. By using

---

[2]To avoid confusion, we define the control delay value associated with the valve L' to be b_0.

two primary pins and by considering an MC-compliant demultiplexing network, a biochip controller can actuate two groups of valves $\{L', L_1^1\}$, $\{L_1^2, L_2^2\}$ concurrently. Note however that the valves $\{L_1^2, L_2^2\}$ need to be actuated serially since they are connected to the same primary pin. Also, note that L' does not have to be actuated to enable $\vartheta_{\{i,2\}}$. As a result, the control latency $\alpha_{\{i,2\}}$ is associated with the longest control latency among the two primary-pin groups. In other words, $\alpha_{\{i,2\}} = \max\{\alpha_{\{i,2\}}^1, \alpha_{\{i,2\}}^2\}$, where $\alpha_{\{i,m_p\}}^i$ is the latency associated with the actuation of a subset of valves $\{L_1^i, L_2^i, ..., L_i^i\}$, i.e., the valve group of primary a pin C_i^a. Both $\alpha_{\{i,2\}}^1$ and $\alpha_{\{i,2\}}^2$ can be computed as follows: $\alpha_{\{i,2\}}^1 = b_1 = \vartheta_{\{i,2\}} \cdot (SN_1 \circ \Psi)$ and $\alpha_{\{i,2\}}^2 = b_2 + b_3 = \vartheta_{\{i,2\}} \cdot (SN_2 \circ \Psi)$. Therefore, $\alpha_{\{i,2\}} = \max\{\vartheta_{\{i,2\}} \cdot (SN_1 \circ \Psi), \vartheta_{\{i,2\}} \cdot (SN_2 \circ \Psi)\}$.

Induction Hypothesis: Next, we study a generic MC configuration, where $m_p \geq 3$ primary pins are connected to the valves in the set $\mathcal{L}$. By considering an MC-compliant demultiplexing network, the value of $\alpha_{\{i,m_p\}}$ can be computed similar to the base case as follows:

$$\alpha_{\{i,m_p\}} = \max\{\alpha_{\{i,m_p\}}^1, \alpha_{\{i,m_p\}}^2, ..., \alpha_{\{i,m_p\}}^{m_p}\}$$
$$= \max_{1 \leq y \leq m_p} \{\vartheta_{\{i,m_p\}} \cdot (SN_y \circ \Psi)\} \tag{D.3}$$

Inductive Step:

By expanding the MC to connect an additional primary pin $C_{m_p+1}^a$, we obtain the following result.

Lemma D.1 *Consider an MC-driven RFB* $\theta_{sc}(x_v, y_v)$ *that has a set of valves* $\{L', L_1^1, L_1^2, L_2^2, ..., L_{m_p}^{m_p}\}$ *and* m_p *primary pins. Also, consider a control vector* $\vartheta_{\{i,m_p\}}$ *that is used to route a sample droplet. If the control latency associated with activating* $\vartheta_{\{i,m_p\}}$ *is* $\alpha_{\{i,m_p\}}$*, then the latency for a control vector* $\vartheta_{\{i,m_p+1\}}$ *that is activated using* $m_p + 1$ *primary pins is* $\max\{\alpha_{\{i,m_p\}}, \vartheta_{\{i,m_p+1\}} \cdot (SN_{m_p+1} \circ \Psi)\}$.

Proof *By expanding the control vector* ϑ_i*, the following changes occur in the MC scheme: (1) an additional set of flow valves* $\mathcal{L}^{m_p+1} = \{L_1^{m_p+1}, L_2^{m_p+1}, ..., L_{m_p+1}^{m_p+1}\}$ *are added; (2) a new primary pin* $C_{m_p+1}^a$ *is connected to these valves and used for pressurization; (3) a new set of demultiplexing pins are connected to the control paths of these valves. Hence, by using the assumption that a biochip controller can pressurize all demultiplexing pins concurrently, then the actuation of flow valves in the set* $\bigcup_{i=1}^{m_p+1} \mathcal{L}^i$ *is associated with a latency* $\alpha_{\{i,m_p+1\}} = \max\{\vartheta_{\{i,m_p+1\}} \cdot (SN_1 \circ \Psi), \vartheta_{\{i,m_p+1\}} \cdot (SN_2 \circ \Psi), ..., \vartheta_{\{i,m_p+1\}} \cdot (SN_m \circ \Psi), \vartheta_{\{i,m_p+1\}} \cdot (SN_{m_p+1} \circ \Psi)\}$. *Note that* $\vartheta_{\{i,m_p\}} \cdot (SN_i \circ \Psi) = \vartheta_{\{i,m_p+1\}} \cdot (SN_i \circ \Psi)$ *if* $i \leq m$*, although the vectors* $\vartheta_{\{i,m_p\}}$ *and* $\vartheta_{\{i,m_p+1\}}$ *have different lengths. When* $i \leq m$*, all the additional entries in* $\vartheta_{\{i,m_p+1\}}$ *will be cancelled by the dot product. Therefore,* $\alpha_{\{i,m_p+1\}} = \max\{\max\{\vartheta_{\{i,m_p\}} \cdot (SN_1 \circ \Psi), \vartheta_{\{i,m_p\}} \cdot (SN_2 \circ \Psi), ..., \vartheta_{\{i,m_p\}} \cdot (SN_m \circ \Psi)\}, \vartheta_{\{i,m_p+1\}} \cdot (SN_{m_p+1} \circ \Psi)\}$. *By using Equation* (D.3)*, we obtain the following result:* $\alpha_{\{i,m_p+1\}} = \max\{\alpha_{\{i,m_p\}}, \vartheta_{\{i,m_p+1\}} \cdot (SN_{m_p+1} \circ \Psi)\}$.

By Using Lemma D.1 and Equation (D.3), and by induction, we obtain the required result for $\alpha_{\{i,m_p\}}$ (or simply α_i) and Θ.

Theorem 6.1 *If* α_i, *where* $\alpha_i \in \mathbb{R}, i \in \mathbb{N}, i \leq CV$, *is the control latency value associated with a control vector* ϑ_i *in a chip* $\theta_{sc}(x_v, y_v)$, *then* $\alpha_i = \max\limits_{1 \leq y \leq a_v} \{\vartheta_i \cdot (SN_y \circ \Psi)\} = \max\limits_{1 \leq y \leq a_v} \{\vartheta_i \cdot (sign(\rho_y) \circ \Psi)\}$, *where* $\circ$ *is the element-wise product. In addition, the cumulative control latency vector for all control vectors in* $\theta_{sc}(x_v, y_v)$ *is* $\Theta = \max\limits_{1 \leq y \leq a_v} \{\Gamma \cdot (sign(\rho_y) \circ \Psi)\}$.

E

Proof of Lemma 7.1: Properties of Aliquot-Generation Trees

Our objective here is to provide a proof for Lemma 7.1 in Chapter 7. This lemma highlights two important properties that are inherent in aliquot-generation trees; those are, the *overlapping-subproblems* and *optimal-substructure* properties.

Consider an aliquot-generation tree $\mathcal{Z}^g_{[o_1...o_N]}$, which contains N leaf nodes (aliquots), as shown in Figure E.1. The tree $\mathcal{Z}^g_{[o_1...o_N]}$ consists of a set of $W^g_{[1...N]}$ nodes $\{\mathcal{Z}^g_{([o_1...o_N],1)}, \mathcal{Z}^g_{([o_1...o_N],2)}, ...\}$. Each node $\mathcal{Z}^g_{([o_1...o_N],j)}$ is defined as $\mathcal{Z}^g_{([o_1...o_N],j)} = \{\psi^g_{([o_1...o_N],j)}, ct^g_{([o_1...o_N],j)}, ht^g_{([o_1...o_N],j)}\}$, where $\psi^g_{([o_1...o_N],j)}$ is the operation type, $ct^g_{([o_1...o_N],j)}$ is the operation cost, and $ht^g_{([o_1...o_N],j)}$ is the height of the node. We define the overall cost $CT^g_{[o_1...o_N]}$ of $\mathcal{Z}^g_{[o_1...o_N]}$ as follows.

Definition E.1 *The overall cost* $CT^g_{[o_1...o_N]}$ *of an aliquot-generation tree* $\mathcal{Z}^g_{[o_1...o_N]} = \{\mathcal{Z}^g_{([o_1...o_N],1)}, \mathcal{Z}^g_{([o_1...o_N],2)}, ...\}$*, where* $\mathcal{Z}^g_{([o_1...o_N],j)} = \{\psi^g_{([o_1...o_N],j)}, ct^g_{([o_1...o_N],j)}, ht^g_{([o_1...o_N],j)}\}$*, is defined as* $CT^g_{[o_1...o_N]} = \sum_j (ct^g_{([o_1...o_N],j)} \cdot ht^g_{([o_1...o_N],j)})$.

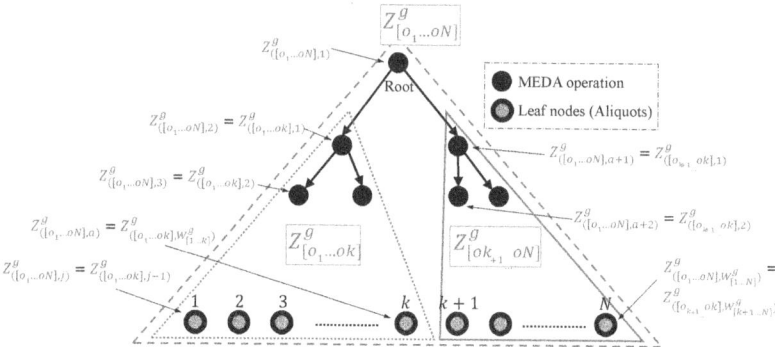

FIGURE E.1
An aliquot-generation tree $\mathcal{Z}^g_{[o_1...o_N]}$ and its subtrees $\mathcal{Z}^g_{[o_1...o_k]}$ and $\mathcal{Z}^g_{[o_{k+1}...o_N]}$.

Based on the above definition, we first present a proof for the overlapping-subproblems property. Next, we prove that $\mathcal{Z}^g_{[o_1...o_N]}$ exhibits the optimal-substructure property.

E.1   Overlapping-Subproblems Property

Based on Figure E.1, our goal is to examine whether the cost $CT^g_{[o_1...o_N]}$ of $\mathcal{Z}^g_{[o_1...o_N]}$ can be computed by breaking $\mathcal{Z}^g_{[o_1...o_N]}$ into two subtrees $\mathcal{Z}^g_{[o_1...o_k]}$ and $\mathcal{Z}^g_{[o_{k+1}...o_N]}$ and computing their relative costs $CT^g_{[o_1...o_k]}$ and $CT^g_{[o_{k+1}...o_N]}$, respectively. If $CT^g_{[o_1...o_N]}$ can be computed in terms of $CT^g_{[o_1...o_k]}$ and $CT^g_{[o_{k+1}...o_N]}$, then the tree $\mathcal{Z}^g_{[o_1...o_N]}$ exhibits the overlapping-subproblems property.

We present a proof by construction. If $CT^g_{[o_1...o_k]}$ is the overall cost of the left subtree (bounded by a dotted-line triangle in Figure E.1), then based on Definition E.1, $CT^g_{[o_1...o_k]}$ can be computed as follows.

$$CT^g_{[o_1...o_k]} = ct^g_{([o_1...o_k],1)} \cdot ht^g_{([o_1...o_k],1)} + ... + ct^g_{([o_1...o_k],W^g_{[1...k]})} \cdot ht^g_{([o_1...o_k],W^g_{[1...k]})} \tag{E.1}$$

Since $\mathcal{Z}^g_{[o_1...o_k]}$ is a subtree of $\mathcal{Z}^g_{[o_1...o_N]}$, then the tree nodes $\mathcal{Z}^g_{([o_1...o_k],j)}$ also belong to the tree $\mathcal{Z}^g_{[o_1...o_N]}$ (Figure E.1), therefore Equation E.1 can be rewritten as follows.

$$CT^g_{[o_1...o_k]} = ct^g_{([o_1...o_N],2)} \cdot ht^g_{([o_1...o_N],2)} + ... + ct^g_{([o_1...o_N],a)} \cdot ht^g_{([o_1...o_N],a)} \tag{E.2}$$

Similarly, if $CT^g_{[o_{k+1}...o_N]}$ is the overall cost of the right subtree (bounded by a solid-line triangle in Figure E.1), then based on Definition E.1, $CT^g_{[o_{k+1}...o_N]}$ can be computed as follows.

$$CT^g_{[o_{k+1}...o_N]} = ct^g_{([o_{k+1}...o_N],1)} \cdot ht^g_{([o_{k+1}...o_N],1)} + ... + \\ ct^g_{([o_{k+1}...o_N],W^g_{[k+1...N]})} \cdot ht^g_{([o_{k+1}...o_N],W^g_{[k+1...N]})} \tag{E.3}$$

Since $\mathcal{Z}^g_{[o_{k+1}...o_N]}$ is a subtree of $\mathcal{Z}^g_{[o_1...o_N]}$, then the tree nodes $\mathcal{Z}^g_{([o_{k+1}...o_N],j)}$ also belong to the tree $\mathcal{Z}^g_{[o_1...o_N]}$ (Figure E.1), therefore Equation E.3 can be rewritten as follows.

$$CT^g_{[o_{k+1}...o_N]} = ct^g_{([o_1...o_N],a+1)} \cdot ht^g_{([o_1...o_N],a+1)} + ... + \\ ct^g_{([o_1...o_N],W^g_{[1...N]})} \cdot ht^g_{([o_1...o_N],W^g_{[1...N]})} \tag{E.4}$$

Next, based on Definition E.1, we compute the overall cost $CT^g_{[o_1...o_N]}$ of the tree $\mathcal{Z}^g_{[o_1...o_N]}$ (bounded by a dashed-line triangle in Figure E.1) as follows.

$$CT^g_{[o_1...o_N]} = ct^g_{([o_1...o_N],1)} \cdot ht^g_{([o_1...o_N],1)}$$
$$+ ct^g_{([o_1...o_N],2)} \cdot ht^g_{([o_1...o_N],2)}$$
$$+ ...$$
$$+ ct^g_{([o_1...o_N],a)} \cdot ht^g_{([o_1...o_N],a)} \qquad \text{(E.5)}$$
$$+ ct^g_{([o_1...o_N],a+1)} \cdot ht^g_{([o_1...o_N],a+1)}$$
$$+ ...$$
$$+ ct^g_{([o_1...o_N],W^g_{[1...N]})} \cdot ht^g_{([o_1...o_N],W^g_{[1...N]})}$$

By substituting Equation E.2 and Equation E.4 into Equation E.5, we reach the following result.

$$CT^g_{[o_1...o_N]} = ct^g_{([o_1...o_N],1)} \cdot ht^g_{([o_1...o_N],1)}$$
$$+ CT^g_{[o_1...o_k]} + CT^g_{[o_{k+1}...o_N]} \qquad \text{(E.6)}$$

Hence, we observe that $CT^g_{[o_1...o_N]}$ can be computed in terms of $CT^g_{[o_1...o_k]}$ and $CT^g_{[o_{k+1}...o_N]}$. Therefore, we obtain the following conclusion regarding the overlapping-subproblems property.

Lemma E.1 *An aliquot-generation tree $\mathcal{Z}^g_{[o_1...o_N]}$ with a root $\mathcal{Z}^g_{([o_1...o_N],RT)}$ exhibits overlapping subproblems, therefore the cost function $CT^g_{[o_1...o_N]}$ can be computed as follows: $CT^g_{[o_1...o_N]} = ct^g_{([o_1...o_N],RT)} \cdot ht^g_{([o_1...o_N],RT)} + CT^g_{[o_1...o_k]} + CT^g_{[o_{k+1}...o_N]}; 1 \leq k < N$, where $CT^g_{[o_1...o_k]}$ and $CT^g_{[o_{k+1}...o_N]}$ are the cost values associated with the left and right subtrees, respectively.*

E.2 Optimal-Substructure Property

Next, suppose that the tree $\mathcal{Z}^g_{[o_1...o_N]}$ is associated with an optimal cost $CT^g_{[o_1...o_N]}$, i.e., the minimum cost. Based on Lemma E.1, we aim to examine whether the values of the cost functions $CT^g_{[o_1...o_k]}$ and $CT^g_{[o_{k+1}...o_N]}$ are optimal when $CT^g_{[o_1...o_N]}$ is optimal. If $CT^g_{[o_1...o_k]}$ and $CT^g_{[o_{k+1}...o_N]}$ are proven to be optimal, then the tree $\mathcal{Z}^g_{[o_1...o_N]}$ exhibits the optimal-substructure property.

We present a proof by contradiction using the generic form in Figure E.1. Suppose that the tree $\mathcal{Z}^g_{[o_1...o_N]}$ is optimal since it is associated with an optimal cost $CT^g_{[o_1...o_N]}$. Also, the tree $\mathcal{Z}^g_{[o_1...o_N]}$ consists of two subtrees $\mathcal{Z}^g_{[o_1...o_k]}$ and $\mathcal{Z}^g_{[o_{k+1}...o_N]}$. Hence, our proof is presented considering two cases.

Case 1: Suppose for contradiction that the left subtree $\mathcal{Z}^g_{[o_1...o_k]}$ is not optimal. This assumption means that there is another subtree structure, denoted by

$\widetilde{\mathcal{Z}}^g_{[o_1...o_k]}$, that has a lower cost value $\widetilde{CT}^g_{[o_1...o_k]}$, where $\widetilde{CT}^g_{[o_1...o_k]} < CT^g_{[o_1...o_k]}$. Based on this assumption, we construct a new tree $\widetilde{\mathcal{Z}}^g_{[o_1...o_N]}$ from $\mathcal{Z}^g_{[o_1...o_N]}$ by replacing the subtree $\mathcal{Z}^g_{[o_1...o_k]}$ with the optimal subtree $\widetilde{\mathcal{Z}}^g_{[o_1...o_k]}$. By using Lemma E.1, we compute the overall cost $\widetilde{CT}^g_{[o_1...o_N]}$ of the new tree $\widetilde{\mathcal{Z}}^g_{[o_1...o_k]}$ as follows:

$$\begin{aligned}
\widetilde{CT}^g_{[o_1...o_N]} = {}& ct^g_{([o_1...o_N],1)} \cdot ht^g_{([o_1...o_N],1)} \\
& + \widetilde{CT}^g_{[o_1...o_k]} + CT^g_{[o_{k+1}...o_N]}
\end{aligned} \tag{E.7}$$

By comparing Equation E.6 and Equation E.7, we observe that the assumption ($\widetilde{CT}^g_{[o_1...o_k]} < CT^g_{[o_1...o_k]}$) implies that $\widetilde{CT}^g_{[o_1...o_N]} < CT^g_{[o_1...o_N]}$. This result contradicts with the optimality of $\mathcal{Z}^g_{[o_1...o_N]}$ that was given at the beginning. As a result, if $\mathcal{Z}^g_{[o_1...o_N]}$ is optimal, then the subtree $\mathcal{Z}^g_{[o_1...o_k]}$ must also be optimal.

Case 2: A similar proof to Case 1 can be derived, while considering the right subtree $\mathcal{Z}^g_{[o_{k+1}...o_N]}$, to show that the subtree $\mathcal{Z}^g_{[o_{k+1}...o_N]}$ must be optimal if $\mathcal{Z}^g_{[o_1...o_N]}$ is optimal.

Based on the presented proof, we reach the following result.

Lemma E.2 *Consider an aliquot-generation tree $\mathcal{Z}^g_{[o_1...o_N]}$ that contains two subtrees $\mathcal{Z}^g_{[o_1...o_k]}$ and $\mathcal{Z}^g_{[o_{k+1}...o_N]}$. The tree $\mathcal{Z}^g_{[o_1...o_N]}$ exhibits optimal substructure, therefore $\mathcal{Z}^g_{[o_1...o_N]}$ is optimal if both subtrees $\mathcal{Z}^g_{[o_1...o_k]}$ and $\mathcal{Z}^g_{[o_{k+1}...o_N]}$ are optimal.*

Using Lemma E.1 and Lemma E.2, we obtain the required result as follows.

Lemma 7.1 *An aliquot-generation tree $\mathcal{Z}^g_{[o_1...o_N]}$ exhibits the "overlapping-subproblems" and "optimal-substructure" properties.*

F

Proof of Theorem 7.1: Recursion in Aliquot-Generation Trees

Our objective here is to provide a proof by induction for Theorem 7.1. This theorem builds on the result obtained from Lemma 7.1, which highlights two important properties of aliquot-generation trees; these properties are the overlapping subproblems and optimal substructure. Our goal is to show that, by using these properties, an *optimal* aliquot-generation tree $\mathcal{Z}^g_{[o_1 \ldots o_N]}$ can be constructed using dynamic programming.

Recall that an aliquot-generation tree $\mathcal{Z}^g_{[o_1 \ldots o_N]}$ consists of a set of $W^g_{[1 \ldots N]}$ nodes $\{\mathcal{Z}^g_{([o_1 \ldots o_N], 1)}, \mathcal{Z}^g_{([o_1 \ldots o_N], 2)}, \ldots\}$. Each node $\mathcal{Z}^g_{([o_1 \ldots o_N], j)}$ is defined as $\mathcal{Z}^g_{([o_1 \ldots o_N], j)} = \{\psi^g_{([o_1 \ldots o_N], j)}, ct^g_{([o_1 \ldots o_N], j)}, ht^g_{([o_1 \ldots o_N], j)}\}$, where $\psi^g_{([o_1 \ldots o_N], j)}$ is the operation type, $ct^g_{([o_1 \ldots o_N], j)}$ is the operation cost, and $ht^g_{([o_1 \ldots o_N], j)}$ is the height of the node. The tree $\mathcal{Z}^g_{[o_1 \ldots o_N]}$ contains a root node, denoted by $\mathcal{Z}^g_{([o_1 \ldots o_N], RT)}$, and N leaf nodes, which represent the aliquots. Also, recall that the cost of a splitting operation is lower than the cost of an aliquoting operation (Section 7.6), i.e., $cost(\text{split}) < cost(\text{aliquot})$. Finally, as described in Appendix E, the overall cost of $\mathcal{Z}^g_{[o_1 \ldots o_N]}$ is $CT^g_{[o_1 \ldots o_N]} = \sum_j ct^g_{([o_1 \ldots o_N], j)} \cdot ht^g_{([o_1 \ldots o_N], j)}$.

The Base Case: We start with the smallest scale of an aliquot-generation tree, which contains only two leaf nodes, i.e., $\mathcal{Z}^g_{[o_1 \ o_2]}$, as shown in Figure F.1. At this scale, a single MEDA operation is needed to generate the aliquots, on condition that the aliquoting constraints (Section 7.6) can be satisfied. If the two aliquots have the same volume, then a splitting operation is performed Figure F.1(a). However, if the two aliquots have different volumes, then an aliquoting operation is executed Figure F.1(b).

On the other hand, if the aliquoting constraints cannot be satisfied directly, then two consecutive aliquoting operations are performed to generate the target aliquot volume; these steps are referred to as a *fragment* operation (Figure F.1(c))—the optimality of the fragment operation based on these steps is discussed as part of the presented proof. For convenience, we consider the steps of a fragment operation as a single operation $\mathcal{Z}^g_{([o_1 \ o_2], j)}$ and its associated cost is $ct^g_{([o_1 \ o_2], j)} > cost(\text{aliquot})$. Based on the above discussion, we introduce the following lemma.

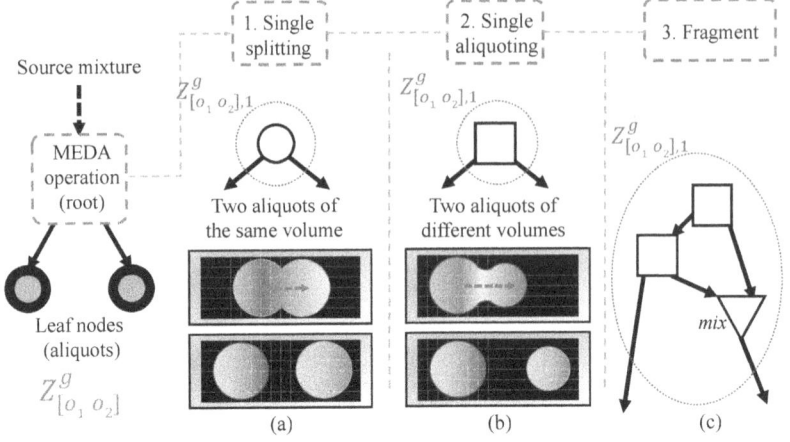

FIGURE F.1
Possible MEDA operations to generate $\mathcal{Z}^g_{[o_1\ o_2]}$: (a) a single splitting operation;
(b) a single aliquoting operation; (c) a fragment operation.

Lemma F.1 *An aliquot-generation tree $\mathcal{Z}^g_{[o_1\ o_2]}$ with root $\mathcal{Z}^g_{([o_1\ o_2],RT)}$ and two leaf nodes $\{\mathcal{Z}^g_{([o_1\ o_2],2)},\ \mathcal{Z}^g_{([o_1\ o_2],3)}\}$ is optimal, and the optimal cost $CT^g_{[o_1\ o_2]}$ is computed as follows: $CT^g_{[o_1\ o_2]} = \min\{CT^g_{[o_1]} + CT^g_{[o_2]}\} + ct^g_{([o_1\ o_2],RT)} \cdot ht^g_{([o_1\ o_2],RT)}$.*

　　Proof *We present the proof by contradiction. Suppose that the tree $\mathcal{Z}^g_{[o_1\ o_2]}$ is not optimal. This implies that there is another tree $\widetilde{\mathcal{Z}}^g_{[o_1\ o_2]}$ that has a lower cost, i.e., $\widetilde{CT}^g_{[o_1\ o_N]} < CT^g_{[o_1\ o_N]}$. Since the leaf nodes are the same in both trees, then this implies that $\widetilde{ct}^g_{([o_1\ o_2],RT)} < ct^g_{([o_1\ o_2],RT)}$. This inequality is studied based on three cases related to aliquots' volumes:*
Case 1: Suppose that the two leaf nodes represent two aliquots that have the same volume, meaning that $ct^g_{([o_1\ o_2],RT)} = cost(split)$. Since $cost(split) < cost(aliquot) < cost(fragment)$, then $ct^g_{([o_1\ o_2],RT)}$ exhibits the lowest value. This contradicts the original assumption that $\widetilde{ct}^g_{([o_1\ o_2],RT)} < ct^g_{([o_1\ o_2],RT)}$. As a result, $\mathcal{Z}^g_{[o_1\ o_2]}$ must be optimal if $\mathcal{Z}^g_{([o_1\ o_2],RT)}$ represents a splitting operation.
　　Since $\mathcal{Z}^g_{[o_1\ o_2]}$ is optimal and using Lemma E.2, the subtrees $\mathcal{Z}^g_{[o_1]}$ and $\mathcal{Z}^g_{[o_2]}$ are also optimal. Note that $CT^g_{[o_1]} = ct^g_{([o_1\ o_2],2)}$ and $CT^g_{[o_2]} = ct^g_{([o_1\ o_2],3)}$. Hence, the cost $CT^g_{[o_1\ o_2]}$ is computed as follows: $CT^g_{[o_1\ o_2]} = ct^g_{([o_1\ o_2],2)} + ct^g_{([o_1\ o_2],3)} + ct^g_{([o_1\ o_2],RT)} = CT^g_{[o_1]} + CT^g_{[o_2]} + ct^g_{([o_1\ o_2],RT)} = \min\{CT^g_{[o_1]} + CT^g_{[o_2]}\} + ct^g_{([o_1\ o_2],RT)} \cdot ht^g_{([o_1\ o_2],RT)}$, which is the required result.
Case 2: Suppose that the two leaf nodes $\{\mathcal{Z}^g_{([o_1\ o_2,2)},\ \mathcal{Z}^g_{([o_1\ o_2,3)}\}$ represent aliquots of different volumes, and that the aliquoting constraints are satisfied.

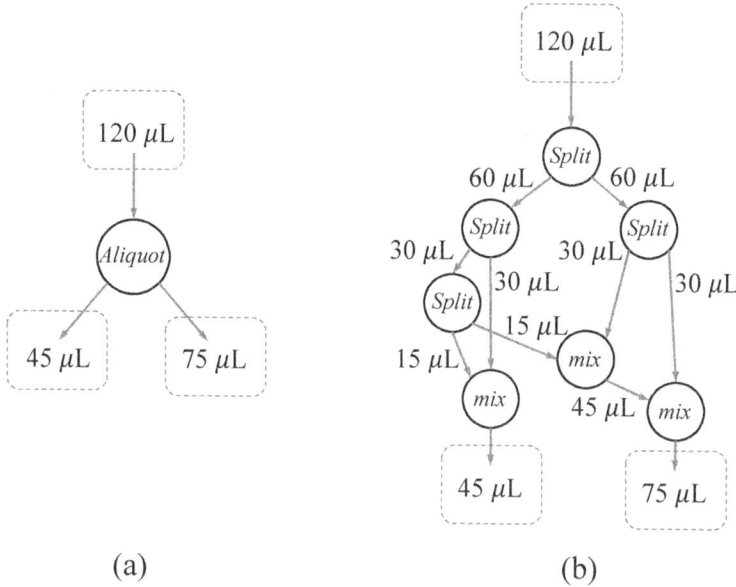

(a)   (b)

FIGURE F.2
Generation of two aliquots with different volumes using: (a) a single MEDA-enabled aliquoting operation; (b) serial splitting and mixing operations.

This implies that $ct^g_{([o_1 \; o_2], RT)} = cost(aliquot)$. Hence, the original assumption that $\widetilde{ct}^g_{([o_1 \; o_2], RT)} < ct^g_{([o_1 \; o_2], RT)}$ cannot be true unless $\widetilde{ct}^g_{([o_1 \; o_2], RT)} = cost(split)$. However, since the aliquots' volumes are different, then these aliquots cannot be generated using a single splitting operation. The only possible method to generate two aliquots of different volumes using splitting is to perform serial splitting followed by mixing steps to attain the right volumes[1]. Figure F.2 compares the aliquoting and the serial splitting mechanisms. The best case scenario for the latter mechanism is achieved when only three splitting and two mixing operations are performed. Clearly, this still exhibits a higher cost $\widetilde{ct}^g_{([o_1 \; o_2], RT)}$ compared to the aliquoting cost $ct^g_{([o_1 \; o_2], RT)}$; this contradicts the original assumption that $\widetilde{ct}^g_{([o_1 \; o_2], RT)} < ct^g_{([o_1 \; o_2], RT)}$. As a result, $\mathcal{Z}^g_{[o_1 \; o_2]}$ must be optimal if $\mathcal{Z}^g_{([o_1 \; o_2], RT)}$ represents an aliquoting operation.

Also, similar to Case 1, the cost $CT^g_{[o_1 \; o_2]}$ can now be computed as follows: $CT^g_{[o_1 \; o_2]} = ct^g_{([o_1 \; o_2], 2)} + ct^g_{([o_1 \; o_2], 3)} + ct^g_{([o_1 \; o_2], RT)} = CT^g_{[o_1]} + CT^g_{[o_2]} + ct^g_{([o_1 \; o_2], RT)} = \min\{CT^g_{[o_1]} + CT^g_{[o_2]}\} + ct^g_{([o_1 \; o_2], RT)} \cdot ht^g_{([o_1 \; o_2], RT)}$, which is the required result.

---

[1]This is equivalent to solving the sample-preparation problem using conventional DMFBs.

Case 3: *Suppose that the two leaf nodes* $\{\mathcal{Z}^g_{([o_1 \ o_2],2)}, \ \mathcal{Z}^g_{([o_1 \ o_2],3)}\}$ *represent aliquots of different volumes, and that the aliquoting constraints are not satisfied based on these volumes. This implies that* $ct^g_{([o_1 \ o_2],RT)} = cost(fragment)$. *Hence, the original assumption that* $\widetilde{ct}^g_{([o_1 \ o_2],RT)} < ct^g_{([o_1 \ o_2],RT)}$ *cannot be true unless* $\widetilde{ct}^g_{([o_1 \ o_2],RT)} = cost(split)$ *or* $\widetilde{ct}^g_{([o_1 \ o_2],RT)} = cost(aliquot)$.

Case 3.A: *We first examine the proposition that* $\widetilde{ct}^g_{([o_1 \ o_2],RT)} = cost(split)$. *Since the aliquots' volumes are different, then these aliquots cannot be generated using a single splitting operation. Hence, this proposition is not valid.*

Case 3.B: *We now examine the proposition that* $\widetilde{ct}^g_{([o_1 \ o_2],RT)} = cost(aliquot)$. *Since the aliquoting conditions are not satisfied, then these droplets cannot also be generated using a single aliquoting operation. Hence, this proposition is also not valid.*

As a result, using a single primitive operation to generate the target aliquot is not possible[2]; we need to implement more than one primitive operation to obtain the required volume. Below we present a proof for the optimal sequence of primitive operations that can be used to implement the steps of a fragment operation.

Suppose that the source droplet, e.g., the 120 μL droplet in Figure F.2(a), has a principal radius of curvature W_{src}, *the bigger droplet, e.g., the 75 μL droplet in Figure F.2(a), has a principal radius of curvature* W_1, *and the target aliquot, e.g., the 45 μL droplet in Figure F.2(a), has a principal radius of curvature* W_2. *There are only three possible implementations of the fragment operation to obtain the target aliquot (Figure F.3(a-c)):*

Case 3.C.I (Splitting and Mixing): *We use consecutive splitting operations to reduce* W_{src} *until we reach the target* W_2; *see Figure F.3(a). This approach is similar to Case 2, which proves that using splitting only is more costly compared with using aliquoting operations. This implies that this implementation does not lead to the optimal cost for fragment.*

Case 3.C.II (Splitting, Aliquoting, and Mixing): *We use consecutive splitting operations to reduce* W_{src} *until we obtain a droplet with radius* $\frac{1}{k}W_{src}$, *which can be used to generate a droplet with radius* W_2 *through an aliquoting process; see Figure F.3(b). This is also similar to Case 2 as multiple splitting operations may be needed. Hence, this implementation is also costly and it may not lead to the optimal cost for fragment. Note that this proof is also applicable to the case where an aliquoting operation is used first followed by consecutive splitting operations. More specifically, the aliquoting operation generates an intermediate aliquot with radius* $W_2^* > W_2$, *which satisfies the aliquoting constraints. Next, this aliquot is processed through consecutive splitting operations.*

Case 3.C.III (Aliquoting and Mixing): *We use a single aliquoting operation to generate an intermediate aliquot with radius* W_2^*, *where* $W_2 < W_2^* < W_{2.max}$. *This droplet in turn is used through a second aliquoting operation to*

---

[2]A primitive operation is either splitting or aliquoting. We use the keyword "primitive" to distinguish between these operations and the fragment operation.

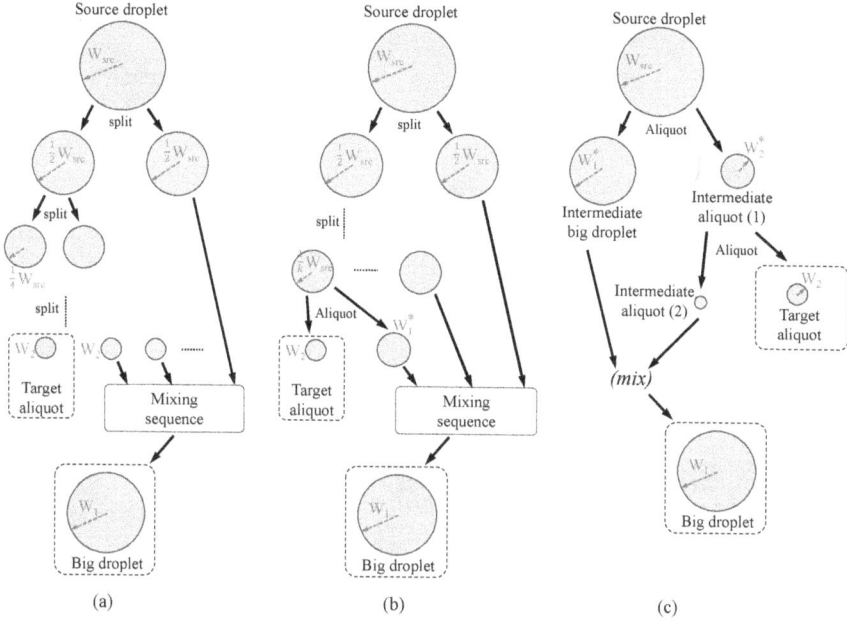

FIGURE F.3
Possible implementations of the fragment operation: (a) using splitting and mixing operations; (b) using splitting, aliquoting, and mixing operations; (c) using aliquoting and mixing operations.

generate the target aliquot that has a radius W_2; see Figure F.3(c). Note that only two aliquoting operations are executed, and this assumption is based on the following facts: (1) an intermediate aliquot can always be generated when W_2^* lies in the interval $(W_2, W_{2.max})$; (2) the difference between W_2^* and W_2 is significantly small compared to the scale of the bigger droplet W_1, therefore the aliquoting constraints are always satisfied for the second aliquoting operation.

Based on the above discussion, the approach described in Case 3.C.III, which is guaranteed to use only two aliquoting steps, is the optimal method to perform the fragment operation. The cost $CT^g_{[o_1\ o_2]}$ can now be computed as follows: $CT^g_{[o_1\ o_2]} = ct^g_{([o_1\ o_2],2)} + ct^g_{([o_1\ o_2],3)} + ct^g_{([o_1\ o_2],RT)} = CT^g_{[o_1]} + CT^g_{[o_2]} + ct^g_{([o_1\ o_2],RT)} = \min\{CT^g_{[o_1]} + CT^g_{[o_2]}\} + ct^g_{([o_1\ o_2],RT)} \cdot ht^g_{([o_1\ o_2],RT)},$ *where $\mathcal{Z}^g_{([o_1\ o_2],RT)}$ is the fragment operation.*

Inductive Step: Next, we study a generic structure of $\mathcal{Z}^g_{[o_1\ o_N]}$. Suppose that $\mathcal{Z}^g_{[o_1\ o_N]}$ is an optimal aliquot-generation tree. Then, by splitting the tree into two subtrees, $\mathcal{Z}^g_{[o_1\ o_k]}$ and $\mathcal{Z}^g_{[o_{k+1}\ o_N]}$ (Figure E.1), we obtain the following result.

Lemma F.2 *If an aliquot-generation tree $\mathcal{Z}^g_{[o_1...o_N]}$ with root $\mathcal{Z}^g_{([o_1...o_N],RT)}$ is optimal, then:*

(1) its subtrees $\mathcal{Z}^g_{[o_1...o_k]}$ and $\mathcal{Z}^g_{[o_{k+1}...o_N]}$ are also optimal, and

(2) the parameter k is specified as follows: $k = \operatorname{argmin}_{1 \le k < N}\{CT^g_{[o_1...o_k]} + CT^g_{[o_{k+1}...o_N]}\}$

(3) the cost $CT^g_{[o_1...o_N]}$ is computed as follows:

$$CT^g_{[o_1...o_N]} = \min_{1 \le j < N}\{CT^g_{[o_1...o_j]} + CT^g_{[o_{j+1}...o_N]}\}$$
$$+ ct^g_{([o_1...o_N],RT)} \cdot ht^g_{([o_1...o_N],RT)}$$

Proof *The proof for (1) is similar to the proof of Lemma E.2. The proof for (2) and (3) is presented using construction as follows. Since both subtrees $\mathcal{Z}^g_{[o_1...o_k]}$ and $\mathcal{Z}^g_{[o_{k+1}...o_N]}$ are optimal, then there is no other subtrees $\mathcal{Z}^g_{[o_1...o_j]}$ and $\mathcal{Z}^g_{[o_{j+1}...o_N]}$ (where $j \ne k$) that can have lower cost values. In other words, the following conditions must be satisfied:*

$$CT^g_{[o_1...o_k]} + CT^g_{[o_{k+1}...o_N]} > CT^g_{[o_1]} + CT^g_{[o_2...o_N]}$$
$$CT^g_{[o_1...o_k]} + CT^g_{[o_{k+1}...o_N]} > CT^g_{[o_1\ o_2]} + CT^g_{[o_3...o_N]}$$
$$CT^g_{[o_1...o_k]} + CT^g_{[o_{k+1}...o_N]} > CT^g_{[o_1...o_j]} + CT^g_{[o_{j+1}...o_N]}; j > 2; j \ne k$$
$$CT^g_{[o_1...o_k]} + CT^g_{[o_{k+1}...o_N]} > CT^g_{[o_1...o_{N-1}]} + CT^g_{[o_N]}$$

By combining the above constraints, we obtain the required result for k as follows.

$$k = \operatorname*{argmin}_{1 \le k < N}\{CT^g_{[o_1...o_k]} + CT^g_{[o_{k+1}...o_N]}\}$$

Based on the above result and by using Lemma E.1, we obtain the required result for $CT^g_{[o_1...o_N]}$ as follows.

$$CT^g_{[o_1...o_N]} = \min_{1 \le j < N}\{CT^g_{[o_1...o_j]} + CT^g_{[o_{j+1}...o_N]}\}$$
$$+ ct^g_{([o_1...o_N],RT)} \cdot ht^g_{([o_1...o_N],RT)} \tag{F.1}$$

Using Lemma F.1 and Lemma F.2, and by induction, we obtain the required result as follows.

Theorem 7.1 *An optimal aliquot-generation tree $\mathcal{Z}^g_{[o_1...o_N]}$ with root $\mathcal{Z}^g_{([p_1...p_n],RT)}$ can be constructed using dynamic programming, where the recurrence relation is defined as follows:*

$$CT^g_{[o_1...o_N]} = \begin{cases} \min_{1 \le j < N}\{CT^g_{[o_1...o_j]} + CT^g_{[o_{j+1}...o_N]}\} \\ \quad + (ct^g_{([o_1...o_N],RT)} \cdot ht^g_{([o_1...o_N],RT)}) & \text{if } N \ge 1 \\ \\ 0 & \text{if } N < 1 \end{cases}$$

Bibliography

[1] ANDE: Rapid DNA for a safer world. https://www.ande.com/.

[2] COMSOL Multiphysics Modeling Software. http://www.comsol.com/.

[3] Fluidigm Corporation. [Online] https://www.fluidigm.com/systems. (Date last accessed June 9, 2019).

[4] IntegenX. https://www.integenx.com/.

[5] Lockheed Martin. https://www.lockheedmartin.com/us.html.

[6] Seeker platform for newborn screening. [Online] http://baebies.com/products/seeker/. (Date last accessed June 9, 2019).

[7] Videos for error recovery in integrated cyberphysical digital-microfluidic platform. [Online] https://goo.gl/jp1FYy.

[8] Rasmus Adler, Ina Schaefer, Mario Trapp, and Arnd Poetzsch-Heffter. Component-based modeling and verification of dynamic adaptation in safety-critical embedded systems. *ACM Transactions on Embedded Computing Systems (TECS)*, 10(2):20, 2010.

[9] Dakshi Agrawal, Selcuk Baktir, Deniz Karakoyunlu, Pankaj Rohatgi, and Berk Sunar. Trojan detection using IC fingerprinting. In *IEEE Symposium on Security and Privacy (SP)*, pages 296–310, 2007.

[10] Sk Subidh Ali, Mohamed Ibrahim, Jeyavijayan Rajendran, Ozgur Sinanoglu, and Krishnendu Chakrabarty. Supply-chain security of digital microfluidic biochips. *Computer*, 49(8):36–43, 2016.

[11] Sk Subidh Ali, Mohamed Ibrahim, Ozgur Sinanoglu, Krishnendu Chakrabarty, and Ramesh Karri. Security implications of cyberphysical digital microfluidic biochips. In *Proceedings International Conference on Computer Design (ICCD)*, pages 483–486, 2015.

[12] Sk Subidh Ali, Mohamed Ibrahim, Ozgur Sinanoglu, Krishnendu Chakrabarty, and Ramesh Karri. Microfluidic encryption of on-chip biochemical assays. In *Proceedings IEEE Biomedical Circuits and Systems (BioCAS) Conference*, pages 152–155, 2016.

[13] Sk Subidh Ali, Mohamed Ibrahim, Ozgur Sinanoglu, Krishnendu Chakrabarty, and Ramesh Karri. Security assessment of cyberphysical digital microfluidic biochips. *IEEE/ACM Transactions on Computational Biology and Bioinformatics (TCBB)*, 13(3):445–458, 2016.

[14] Mirela Alistar, Paul Pop, and Jan Madsen. Redundancy optimization for error recovery in digital microfluidic biochips. *Design Automation for Embedded Systems*, 19(1-2):129–159, 2015.

[15] Yousra M. Alkabani and Farinaz Koushanfar. Active hardware metering for intellectual property protection and security. In *USENIX Security Symposium*, pages 20:1–20:16, 2007.

[16] Carlos L Araya, Celia Payen, Maitreya J Dunham, and Stanley Fields. Whole-genome sequencing of a laboratory-evolved yeast strain. *BMC Genomics*, 11:88–88, 2010.

[17] Laura Arrigoni, Andreas S Richter, Emily Betancourt, Kerstin Bruder, Sarah Diehl, Thomas Manke, and Ulrike Bönisch. Standardizing chromatin research: a simple and universal method for ChIP-seq. *Nucleic acids research*, 44(7):e67–e67, 2015.

[18] Christel Baier, Joost-Pieter Katoen, and Kim Guldstrand Larsen. *Principles of model checking*. 2008.

[19] Jean-Christophe Baret, Oliver J Miller, Valerie Taly, Michaël Ryckelynck, Abdeslam El-Harrak, Lucas Frenz, Christian Rick, Michael L Samuels, J Brian Hutchison, Jeremy J Agresti, et al. Fluorescence-activated droplet sorting (FADS): efficient microfluidic cell sorting based on enzymatic activity. *Lab on a Chip*, 9(13):1850–1858, 2009.

[20] Sanjoy Baruah, Marko Bertogna, and Giorgio Buttazzo. *Multiprocessor Scheduling for Real-Time Systems*. Springer, 2015.

[21] Luca Benini, Alessandro Bogliolo, Giuseppe A Paleologo, and Giovanni De Micheli. Policy optimization for dynamic power management. *IEEE Transactions on Computer-Aided Design of Integrated Circuits and Systems (TCAD)*, 18(6):813–833, 1999.

[22] Michel Berkelaar, Kjell Eikland, and Peter Notebaert. lpsolve, version 5.5. http://lpsolve.sourceforge.net/5.5/, 2004.

[23] Swapnil Bhatia, Craig LaBoda, Vanessa Yanez, Traci Haddock-Angelli, and Douglas Densmore. Permutation machines. *ACS Synthetic Biology*, 5(8):827–834, 2016.

[24] Sukanta Bhattacharjee, Yi-Ling Chen, Juinn-Dar Huang, and Bhargab B Bhattacharya. Concentration-resilient mixture preparation with digital microfluidic lab-on-chip. *ACM Transactions on Embedded Computing Systems (TECS)*, 17(2):49, 2018.

[25] Thomas E Bihari and Karsten Schwan. Dynamic adaptation of real-time software. *ACM Transactions on Computer Systems (TOCS)*, 9(2):143–174, 1991.

[26] Bio-Rad. Real-Time PCR Applications Guide. `http://www.bio-rad.com/webroot/web/pdf/lsr/literature/Bulletin_5279.pdf`. (Date last accessed June 9, 2019).

[27] Bioconductor. Open source software for bioinformatics. [Online] https://www.bioconductor.org/.

[28] BioNumbers. Number of cell types in human body. `http://bionumbers.hms.harvard.edu/bionumber.aspx?&id=103626`.

[29] Adrian Bird. Perceptions of epigenetics. *Nature*, 447(7143):396–398, 2007.

[30] Dario Bogojevic, M Dean Chamberlain, Irena Barbulovic-Nad, and Aaron R Wheeler. A digital microfluidic method for multiplexed cell-based apoptosis assays. *Lab on a Chip*, 12:627–634, 2012.

[31] Deborah J Boles, Jonathan L Benton, Germaine J Siew, Miriam H Levy, Prasanna K Thwar, Melissa A Sandahl, Jeremy L Rouse, Lisa C Perkins, Arjun P Sudarsan, Roxana Jalili, et al. Droplet-based pyrosequencing using digital microfluidics. *Analytical Chemistry*, 83(22):8439–8447, 2011.

[32] Vincenzo Bonifaci, Alberto Marchetti-Spaccamela, Sebastian Stiller, and Andreas Wiese. Feasibility analysis in the sporadic dag task model. In *Proceedings IEEE Euromicro Conference on Real-Time Systems (ECRTS)*, pages 225–233, 2013.

[33] Pamela D Schoppee Bortz and Brian R Wamhoff. Chromatin immunoprecipitation (ChIP): revisiting the efficacy of sample preparation, sonication, quantification of sheared DNA, and analysis via PCR. *PLoS One*, 6(10):e26015, 2011.

[34] John F Brothers, Kahkeshan Hijazi, Celine Mascaux, Randa A El-Zein, Margaret R Spitz, and Avrum Spira. Bridging the clinical gaps: genetic, epigenetic and transcriptomic biomarkers for the early detection of lung cancer in the post-national lung screening trial era. *BMC Medicine*, 11(1):1–15, 2013.

[35] Eric Brouzes, Martina Medkova, Neal Savenelli, Dave Marran, Mariusz Twardowski, J Brian Hutchison, Jonathan M Rothberg, Darren R Link, Norbert Perrimon, and Michael L Samuels. Droplet microfluidic technology for single-cell high-throughput screening. *Proceedings of the National Academy of Sciences*, 106(34):14195–14200, 2009.

[36] Christopher R Brown, Changhui Mao, Elena Falkovskaia, Melissa S Jurica, and Hinrich Boeger. Linking stochastic fluctuations in chromatin structure and gene expression. *PLoS Biol*, 11(8):e1001621, 2013.

[37] Brigitte Bruijns, Arian van Asten, Roald Tiggelaar, and Han Gardeniers. Microfluidic devices for forensic DNA analysis: A review. *Biosensors*, 6(3):41, 2016.

[38] Stephen A. Bustin. *A-Z of Quantitative PCR, 1st Edition*. International University Line, San Diego, CA, 2004.

[39] Colin F Camerer, Teck-Hua Ho, and Juin Kuan Chong. Behavioural game theory: Thinking, learning and teaching. In *Advances in Understanding Strategic Behaviour*, pages 120–180. 2004.

[40] Zhenning Cao, Changya Chen, Bing He, Kai Tan, and Chang Lu. A microfluidic device for epigenomic profiling using 100 cells. *Nature Methods*, 12(10):959–962, 2015.

[41] Krishnendu Chakrabarty and Tao Xu. *Digital Microfluidic Biochips: Design Automation and Optimization*. CRC Press, 2010.

[42] Rajat Subhra Chakraborty and Swarup Bhunia. Harpoon: an obfuscation-based soc design methodology for hardware protection. *IEEE Transactions on Computer-Aided Design of Integrated Circuits and Systems (TCAD)*, 28(10):1493–1502, 2009.

[43] Huili Chen, Seetal Potluri, and Farinaz Koushanfar. BioChipWork: Reverse engineering of microfluidic biochips. In *Proceedings International Conference on Computer Design (ICCD)*, pages 9–16, 2017.

[44] Jian Chen, Chengcheng Xue, Yang Zhao, Deyong Chen, Min-Hsien Wu, and Junbo Wang. Microfluidic impedance flow cytometry enabling high-throughput single-cell electrical property characterization. *International Journal of Molecular Sciences*, 16(5):9804–9830, 2015.

[45] L Chen and RB Fair. Digital microfluidics chip with integrated intra-droplet magnetic bead manipulation. *Microfluidics and Nanofluidics*, 19(6):1349–1361, 2015.

[46] Liji Chen. *Sparse Sample Detection Using Magnetic Bead Manipulation on a Digital Microfluidic Device*. PhD thesis, Duke University, 2016.

[47] Ming-Tang Chen and Ron Weiss. Artificial cell-cell communication in yeast Saccharomyces cerevisiae using signaling elements from Arabidopsis thaliana. *Nature Biotechnology*, 23(12):1551, 2005.

[48] Taolue Chen, Vojtěch Forejt, Marta Kwiatkowska, David Parker, and Aistis Simaitis. Automatic verification of competitive stochastic systems. *Formal Methods in System Design*, 43(1):61–92, 2013.

[49] Curtis D Chin, Tassaneewan Laksanasopin, Yuk Kee Cheung, David Steinmiller, Vincent Linder, Hesam Parsa, Jennifer Wang, Hannah Moore, Robert Rouse, Gisele Umviligihozo, et al. Microfluidics-based diagnostics of infectious diseases in the developing world. *Nature Medicine*, 17(8):1015–1019, 2011.

[50] Daniel T Chiu, Dino Di Carlo, Patrick S Doyle, Carl Hansen, Richard M Maceiczyk, Robert CR Wootton, et al. Small but perfectly formed? successes, challenges, and opportunities for microfluidics in the chemical and biological sciences. *Chem*, 2(2):201–223, 2017.

[51] Minsik Cho and David Z Pan. A high-performance droplet routing algorithm for digital microfluidic biochips. *IEEE Transactions on Computer-Aided Design of Integrated Circuits and Systems (TCAD)*, 27(10):1714–1724, 2008.

[52] Sung Kwon Cho, Hyejin Moon, and Chang-Jin Kim. Creating, transporting, cutting, and merging liquid droplets by electrowetting-based actuation for digital microfluidic circuits. *Journal of Microelectromechanical Systems*, 12(1):70–80, 2003.

[53] Kyungyong Choi, Jee-Yeon Kim, Jae-Hyuk Ahn, Ji-Min Choi, Maesoon Im, and Yang-Kyu Choi. Integration of field effect transistor-based biosensors with a digital microfluidic device for a lab-on-a-chip application. *Lab on a Chip*, 12(8):1533–1539, 2012.

[54] Thomas M Cioppa and Thomas W Lucas. Efficient nearly orthogonal and space-filling Latin hypercubes. *Technometrics*, 49(1):45–55, 2007.

[55] Federal Trade Commission. LabMD, inc. v. Federal Trade Commission. [Online] https://www.ftc.gov/enforcement/cases-proceedings/102-3099/labmd-inc-matter, March 2016.

[56] Jason Cong, Lei He, Cheng-Kok Koh, and Patrick H Madden. Performance optimization of VLSI interconnect layout. *Integration, the VLSI Journal*, 21(1-2):1–94, 1996.

[57] Jose C Contreras-Naranjo, Qingshan Wei, and Aydogan Ozcan. Mobile phone-based microscopy, sensing, and diagnostics. *IEEE Journal of Selected Topics in Quantum Electronics*, 22(3):1–14, 2016.

[58] Thomas H Cormen, Charles E Leiserson, Ronald L Rivest, and Clifford Stein. *Introduction to Algorithms*. MIT Press Cambridge, 2001.

[59] Drew Davidson, Hao Wu, Robert Jellinek, Thomas Ristenpart, and Vikas Singh. Controlling UAVs with sensor input spoofing attacks. In *USENIX Conference on Offensive Technologies*, pages 221–231, 2016.

[60] Douglas Densmore and J Christopher Anderson. Combinational logic design in synthetic biology. In *IEEE International Symposium on Circuits and Systems (ISCAS)*, pages 301–304, 2009.

[61] Trung Anh Dinh, Shigeru Yamashita, and Tsung-Yi Ho. A network-flow-based optimal sample preparation algorithm for digital microfluidic biochips. In *Design Automation Conference (ASP-DAC), 2014 19th Asia and South Pacific*, pages 225–230, 2014.

[62] Cheng Dong, Tianlan Chen, Jie Gao, Yanwei Jia, Pui-In Mak, Mang-I Vai, and Rui P Martins. On the droplet velocity and electrode lifetime of digital microfluidics: voltage actuation techniques and comparison. *Microfluidics and Nanofluidics*, 18(4):673–683, 2015.

[63] David C Duffy et al. Rapid prototyping of microfluidic systems in poly (dimethylsiloxane). *Analytical Chemistry*, 70(23):4974–4984, 1998.

[64] David Easley and Arpita Ghosh. Incentives, gamification, and game theory: an economic approach to badge design. In *Proceedings ACM Conference on Electronic Commerce (EC)*, pages 359–376, 2013.

[65] Emanuel Elizalde, Raúl Urteaga, and Claudio LA Berli. Rational design of capillary-driven flows for paper-based microfluidics. *Lab on a Chip*, 15(10):2173–2180, 2015.

[66] Ksenia Ermoshina, Francesca Musiani, and Harry Halpin. End-to-end encrypted messaging protocols: An overview. In *International Conference on Internet Science*, pages 244–254. Springer, 2016.

[67] Irwin A Eydelnant, Uvaraj Uddayasankar, Meng Wen Liao, Aaron R Wheeler, et al. Virtual microwells for digital microfluidic reagent dispensing and cell culture. *Lab on a Chip*, 12(4):750–757, 2012.

[68] Richard B Fair. Digital microfluidics: is a true lab-on-a-chip possible? *Microfluidics and Nanofluidics*, 3(3):245–281, 2007.

[69] Richard B Fair, Andrey Khlystov, Tina D Tailor, Vladislav Ivanov, Randall D Evans, Vijay Srinivasan, Vamsee K Pamula, Michael G Pollack, Peter B Griffin, and Jack Zhou. Chemical and biological applications of digital-microfluidic devices. *IEEE Design & Test*, 24(1), 2007.

[70] Yawen Fan, Zhenghao Zhang, Matthew Trinkle, Aleksandar D Dimitrovski, Ju Bin Song, and Husheng Li. A cross-layer defense mechanism against GPS spoofing attacks on PMUs in smart grids. *IEEE Transactions on Smart Grid*, 6(6):2659–2668, 2015.

[71] Luis M Fidalgo and Sebastian J Maerkl. A software-programmable microfluidic device for automated biology. *Lab on a Chip*, 11(9):1612–1619, 2011.

[72] Farshad Firouzi, Bahar Farahani, Mohamed Ibrahim, and Krishnendu Chakrabarty. Keynote paper: From EDA to IoT ehealth: Promises, challenges, and solutions. *IEEE Transactions on Computer-Aided Design of Integrated Circuits and Systems (TCAD)*, 37(12):2965–2978, 2018.

[73] Marco Fondi and Pietro Liò. Multi-omics and metabolic modelling pipelines: challenges and tools for systems microbiology. *Microbiological Research*, 171:52–64, 2015.

[74] Ane Elida Fonneløp, Helen Johannessen, Thore Egeland, and Peter Gill. Contamination during criminal investigation: detecting police contamination and secondary DNA transfer from evidence bags. *Forensic Science International: Genetics*, 23:121–129, 2016.

[75] Centers for Disease Control and Prevention. U.s. diabetes surveillance system. [Online] https://gis.cdc.gov/grasp/diabetes/DiabetesAtlas.html.

[76] The Apache Software Foundation. Apache spark. [Online] http://spark.apache.org/.

[77] The Apache Software Foundation. Cassandra spark. [Online] http://cassandra.apache.org/.

[78] Terrence S Furey. ChIP-seq and beyond: new and improved methodologies to detect and characterize protein-DNA interactions. *Nature Reviews Genetics*, 13(12):840, 2012.

[79] Jie Gao, Xianming Liu, Tianlan Chen, Pui-In Mak, Yuguang Du, Mang-I Vai, Bingcheng Lin, and Rui P Martins. An intelligent digital microfluidic system with fuzzy-enhanced feedback for multi-droplet manipulation. *Lab on a Chip*, 13(3):443–451, 2013.

[80] GenMark Dx. GenMark ePlex System. [Online] https://www.genmarkdx.com/solutions/systems/eplex-system/. (Date last accessed June 9, 2019).

[81] Sergio Gómez, Alexandre Arenas, J Borge-Holthoefer, Sandro Meloni, and Yamir Moreno. Discrete-time Markov chain approach to contact-based disease spreading in complex networks. *EPL (Europhysics Letters)*, 89(3):38009, 2010.

[82] David Gomez-Cabrero, Imad Abugessaisa, Dieter Maier, Andrew Teschendorff, Matthias Merkenschlager, Andreas Gisel, Esteban Ballestar, Erik Bongcam-Rudloff, Ana Conesa, and Jesper Tegnér. Data integration in the era of omics: current and future challenges. *BMC Systems Biology*, 8(2):1, 2014.

[83] J Gong and CJ Kim. All-electronic droplet generation on-chip with real-time feedback control for EWOD digital microfluidics. *Lab on a Chip*, 8(6):898–906, 2008.

[84] Mel Greaves. Evolutionary determinants of cancer. *Cancer Discovery*, 5(8):806–820, 2015.

[85] Daniel Grissom and Philip Brisk. Fast online synthesis of generally programmable digital microfluidic biochips. In *Proceedings IEEE/ACM/IFIP International Conference on Hardware/Software Codesign and System Synthesis (CODES+ISSS)*, pages 413–422, 2012.

[86] Daniel Grissom and Philip Brisk. A field-programmable pin-constrained digital microfluidic biochip. In *Proceedings ACM/IEEE Design Automation Conference (DAC)*, page 46, 2013.

[87] William H Grover, Robin HC Ivester, Erik C Jensen, and Richard A Mathies. Development and multiplexed control of latching pneumatic valves using microfluidic logical structures. *Lab on a Chip*, 6(5):623–631, 2006.

[88] Hao Gu, Michel HG Duits, and Frieder Mugele. Droplets formation and merging in two-phase flow microfluidics. *International Journal of Molecular Sciences*, 12(4):2572–2597, 2011.

[89] Ujjwal Guin, Daniel DiMase, and Mohammad Tehranipoor. Counterfeit integrated circuits: detection, avoidance, and the challenges ahead. *Journal of Electronic Testing*, 30(1):9–23, 2014.

[90] James W Haefner. *Modeling Biological Systems:: Principles and Applications*. Springer Science & Business Media, 2005.

[91] Brian S Hardy, Kawika Uechi, Janet Zhen, and H Pirouz Kavehpour. The deformation of flexible PDMS microchannels under a pressure driven flow. *Lab on a Chip*, 9(7):935–938, 2009.

[92] Paul DN Hebert, Alina Cywinska, Shelley L Ball, et al. Biological identifications through dna barcodes. *Proceedings of the Royal Society of London B: Biological Sciences*, 270(1512):313–321, 2003.

[93] Chee Meng Benjamin Ho, Sum Huan Ng, King Ho Holden Li, and Yong-Jin Yoon. 3D printed microfluidics for biological applications. *Lab on a Chip*, 15(18):3627–3637, 2015.

[94] Yutaka Hori, Chaitanya Kantak, Richard M Murray, and Adam R Abate. Cell-free extract based optimization of biomolecular circuits with droplet microfluidics. *Lab on a Chip*, 17(18):3037–3042, 2017.

[95] Sanjin Hosic, Shashi K Murthy, and Abigail N Koppes. Microfluidic sample preparation for single cell analysis. *Analytical Chemistry*, 88(1):354–380, 2015.

[96] Walter Houser. Could what happened to Sony happen to us? *IT Professional*, 17(2):54–57, 2015.

[97] Ching-Wei Hsieh, Zipeng Li, and Tsung-Yi Ho. Piracy prevention of digital microfluidic biochips. In *Proceedings IEEE Asia South Pacific Design Automation Conference (ASP-DAC)*, pages 512–517, 2017.

[98] Yi-Ling Hsieh, Tsung-Yi Ho, and Krishnendu Chakrabarty. Biochip synthesis and dynamic error recovery for sample preparation using digital microfluidics. *IEEE Transactions on Computer-Aided Design of Integrated Circuits and Systems (TCAD)*, 33(2):183–196, 2014.

[99] Bang-Ning Hsu. *Accelerated Sepsis Diagnosis by Seamless Integration of Nucleic Acid Purification and Detection*. PhD thesis, Duke University, 2014.

[100] Kai Hu, Krishnendu Chakrabarty, and Tsung-Yi Ho. *Computer-Aided Design of Microfluidic Very Large Scale Integration (mVLSI) Biochips*. Springer, 2017.

[101] Kai Hu, Trung Anh Dinh, Tsung-Yi Ho, and Krishnendu Chakrabarty. Control-layer routing and control-pin minimization for flow-based microfluidic biochips. *IEEE Transactions on Computer-Aided Design of Integrated Circuits and Systems (TCAD)*, 36(1):55–68, 2017.

[102] Kai Hu, Bang-Ning Hsu, Andrew Madison, Krishnendu Chakrabarty, and Richard Fair. Fault detection, real-time error recovery, and experimental demonstration for digital microfluidic biochips. In *Proceedings Design, Automation, and Test in Europe (DATE)*, pages 559–564, 2013.

[103] Kai Hu, Mohamed Ibrahim, Liji Chen, Zipeng Li, Krishnendu Chakrabarty, and Richard Fair. Experimental demonstration of error recovery in an integrated cyberphysical digital-microfluidic platform. In *Proceedings IEEE Biomedical Circuits and Systems (BioCAS) Conference*, pages 1–4, 2015.

[104] Zhishan Hua, Jeremy L Rouse, Allen E Eckhardt, Vijay Srinivasan, Vamsee K Pamula, Wiley A Schell, Jonathan L Benton, Thomas G Mitchell, and Michael G Pollack. Multiplexed real-time polymerase chain reaction on a digital microfluidic platform. *Analytical Chemistry*, 82(6):2310–2316, 2010.

[105] Tsung-Wei Huang, Chun-Hsien Lin, and Tsung-Yi Ho. A contamination aware droplet routing algorithm for the synthesis of digital microfluidic biochips. *IEEE Transactions on Computer-Aided Design of Integrated Circuits and Systems (TCAD)*, 29(11):1682–1695, 2010.

[106] Mohamed Ibrahim, Bhargab B Bhattacharya, and Krishnendu Chakrabarty. Bioscan: Parameter-space exploration of synthetic biocircuits using MEDA biochips. In *Proceedings Design, Automation, and Test in Europe (DATE)*, pages 1519–1524, 2019.

[107] Mohamed Ibrahim, Craig Boswell, Krishnendu Chakrabarty, Kristin Scott, and Miroslav Pajic. A real-time digital-microfluidic platform for epigenetics. In *Proceedings IEEE/ACM International Conference on Compilers, Architectures and Synthesis of Embedded Systems (CASES)*, pages 10:1–10, 2016.

[108] Mohamed Ibrahim and Krishnendu Chakrabarty. Efficient error recovery in cyberphysical digital-microfluidic biochips. *IEEE Transactions on Multi-Scale Computing Systems (TMSCS)*, 1(1):46–58, 2015.

[109] Mohamed Ibrahim and Krishnendu Chakrabarty. Error recovery in digital microfluidics for personalized medicine. In *Proceedings Design, Automation, and Test in Europe (DATE)*, pages 247–252, 2015.

[110] Mohamed Ibrahim and Krishnendu Chakrabarty. Cyberphysical adaptation in digital-microfluidic biochips. In *Proceedings IEEE Biomedical Circuits and Systems (BioCAS) Conference*, pages 444–447, 2016.

[111] Mohamed Ibrahim and Krishnendu Chakrabarty. Digital-microfluidic biochips for quantitative analysis: Bridging the gap between microfluidics and microbiology. In *Proceedings Design, Automation, and Test in Europe (DATE)*, pages 1791–1796, 2017.

[112] Mohamed Ibrahim and Krishnendu Chakrabarty. Cyber-physical digital-microfluidic biochips: Bridging the gap between microfluidics and microbiology. *Proceedings of the IEEE*, 106(9):1717–1743, 2018.

[113] Mohamed Ibrahim, Krishnendu Chakrabarty, and Ulf Schlichtmann. CoSyn: Efficient single-cell analysis using a hybrid microfluidic platform. In *Proceedings Design, Automation, and Test in Europe (DATE)*, pages 1677–1682, 2017.

[114] Mohamed Ibrahim, Krishnendu Chakrabarty, and Ulf Schlichtmann. Synthesis of a cyberphysical hybrid microfluidic platform for single-cell analysis. *IEEE Transactions on Computer-Aided Design of Integrated Circuits and Systems (TCAD)*, 38(7):1237–1250, 2019.

[115] Mohamed Ibrahim, Krishnendu Chakrabarty, and Kristin Scott. Integrated and real-time quantitative analysis using cyberphysical digital-microfluidic biochips. In *Proceedings Design, Automation, and Test in Europe (DATE)*, pages 630–635, 2016.

[116] Mohamed Ibrahim, Krishnendu Chakrabarty, and Kristin Scott. Synthesis of cyberphysical digital-microfluidic biochips for real-time quantitative analysis. *IEEE Transactions on Computer-Aided Design of Integrated Circuits and Systems (TCAD)*, 36(5):733–746, 2017.

[117] Mohamed Ibrahim, Krishnendu Chakrabarty, and Jun Zeng. BioCyBig: A cyberphysical system for integrative microfluidics-driven analysis of genomic association studies. *IEEE Transactions on Big Data (TBD)*, 2016.

[118] Mohamed Ibrahim, Zipeng Li, and Krishnendu Chakrabarty. Advances in design automation techniques for digital-microfluidic biochips. In *Formal Modeling and Verification of Cyber-Physical Systems*, pages 190–223. Springer, 2015.

[119] Mohamed Ibrahim, Aditya Sridhar, Krishnendu Chakrabarty, and Ulf Schlichtmann. Sortex: Efficient timing-driven synthesis of reconfigurable flow-based biochips for scalable single-cell screening. In *Proceedings International Conference on Computer-Aided Design (ICCAD)*, pages 623–630, 2017.

[120] Mohamed Ibrahim, Aditya Sridhar, Krishnendu Chakrabarty, and Ulf Schlichtmann. Synthesis of reconfigurable flow-based biochips for scalable single-cell screening. *IEEE Transactions on Computer-Aided Design of Integrated Circuits and Systems (TCAD)*, 2018.

[121] Illumina. Illumina NeoPrep Library Prep System. [Online] http://www.illumina.com/systems/neoprep-library-system.html/. (Date last accessed June 9, 2019).

[122] Christopher Jaress, Philip Brisk, and Daniel Grissom. Rapid online fault recovery for cyber-physical digital microfluidic biochips. In *Proceedings IEEE VLSI Test Symposium (VTS)*, pages 1–6, 2015.

[123] Mais J Jebrail, Ronald F Renzi, Anupama Sinha, Jim Van De Vreugde, Carmen Gondhalekar, Cesar Ambriz, Robert J Meagher, and Steven S Branda. A solvent replenishment solution for managing evaporation of biochemical reactions in air-matrix digital microfluidics devices. *Lab on a Chip*, 15(1):151–158, 2015.

[124] Mais J Jebrail, Anupama Sinha, Samantha Vellucci, Ronald F Renzi, Cesar Ambriz, Carmen Gondhalekar, Joseph S Schoeniger, Kamlesh D Patel, and Steven S Branda. World-to-digital-microfluidic interface enabling extraction and purification of RNA from human whole blood. *Analytical Chemistry*, 86(8):3856–3862, 2014.

[125] Heather M Jensen, Aaron E Albers, Konstantin R Malley, Yuri Y Londer, Bruce E Cohen, Brett A Helms, Peter Weigele, Jay T Groves,

and Caroline M Ajo-Franklin. Engineering of a synthetic electron conduit in living cells. *Proceedings of the National Academy of Sciences*, 107(45):19213–19218, 2010.

[126] Timothy J Johnson, David Ross, and Laurie E Locascio. Rapid microfluidic mixing. *Analytical chemistry*, 74(1):45–51, 2002.

[127] The Wall Street Journal. Theranos Results Could Throw Off Medical Decisions, Study Finds. [Online] https://www.wsj.com/articles/theranos-results-could-throw-off-medical-decisions-study-finds-1459196177, March 2016.

[128] Martin Kantlehner, Roland Kirchner, Petra Hartmann, Joachim W Ellwart, Marianna Alunni-Fabbroni, and Axel Schumacher. A high-throughput DNA methylation analysis of a single cell. *Nucleic Acids Research*, 39(7):1–9, 2011.

[129] Marc Karle, Sandeep Kumar Vashist, Roland Zengerle, and Felix von Stetten. Microfluidic solutions enabling continuous processing and monitoring of biological samples: A review. *Analytica chimica acta*, 929:1–22, 2016.

[130] Ramesh Karri, Jeyavijayan Rajendran, Kurt Rosenfeld, and Mohammad Tehranipoor. Trustworthy hardware: Identifying and classifying hardware Trojans. *IEEE Computer*, 43(10):39–46, 2010.

[131] Albert J Keung and Ahmad S Khalil. A unifying model of epigenetic regulation. *Science*, 351(6274):661–662, 2016.

[132] Peter V Kharchenko, Lev Silberstein, and David T Scadden. Bayesian approach to single-cell differential expression analysis. *Nature methods*, 11(7):740–742, 2014.

[133] Dongshin Kim, Naomi C Chesler, and David J Beebe. A method for dynamic system characterization using hydraulic series resistance. *Lab on a Chip*, 6(5):639–644, 2006.

[134] Jungkyu Kim, Amanda M Stockton, Erik C Jensen, and Richard A Mathies. Pneumatically actuated microvalve circuits for programmable automation of chemical and biochemical analysis. *Lab on a Chip*, 16(5):812–819, 2016.

[135] B. Kirby. *Micro-and Nanoscale Fluid Mechanics: Transport in Microfluidic Devices*. Cambridge University Press, 2010.

[136] Allon M Klein, Linas Mazutis, Ilke Akartuna, Naren Tallapragada, Adrian Veres, Victor Li, Leonid Peshkin, David A Weitz, and Marc W Kirschner. Droplet barcoding for single-cell transcriptomics applied to embryonic stem cells. *Cell*, pages 1187–1201, 2015.

[137] Satendra Kumar, Ankur Gupta, Sudip Roy, and Bhargab B Bhattacharya. Design automation of multiple-demand mixture preparation using a K-array rotary mixer on digital microfluidic biochips. In *Proceedings International Conference on Computer Design (ICCD)*, pages 273–280, 2016.

[138] Bambang Kuswandi, Jurriaan Huskens, Willem Verboom, et al. Optical sensing systems for microfluidic devices: a review. *Analytica chimica acta*, 601(2):141–155, 2007.

[139] Marta Kwiatkowska. Quantitative verification: Models, techniques and tools. In *Proceedings 6th Joint Meeting on European Software Engineering Conference and the ACM SIGSOFT Symposium on the Foundations of Software Engineering (ESEC/FSE)*, pages 449–458, 2007.

[140] Marta Kwiatkowska, Gethin Norman, and David Parker. PRISM 4.0: Verification of probabilistic real-time systems. In *Proceedings International Conference on Computer aided Verification (CAV)*, pages 585–591, 2011.

[141] Marta Kwiatkowska and David Parker. Automated verification and strategy synthesis for probabilistic systems. In *Proceedings International Symposium on Automated Technology for Verification and Analysis*, pages 5–22, 2013.

[142] The Emerald Cloud Lab. The ecl: Built by scientists for scientists. [Online] https://www.transcriptic.com.

[143] Nelson M Lafreniere, Jared M Mudrik, Alphonsus HC Ng, Brendon Seale, Neil Spooner, and Aaron R Wheeler. Attractive design: An elution solvent optimization platform for magnetic-bead-based fractionation using digital microfluidics and design of experiments. *Analytical chemistry*, 87(7):3902–3910, 2015.

[144] Morteza Lahijanian, Joseph Wasniewski, Sean B Andersson, and Calin Belta. Motion planning and control from temporal logic specifications with probabilistic satisfaction guarantees. In *Proceedings IEEE International Conference on Robotics and Automation (ICRA*, pages 3227–3232, 2010.

[145] Kelvin Yi-Tse Lai, Ming-Feng Shiu, Yi-Wen Lu, Yingchieh Ho, Yu-Chi Kao, Yu-Tao Yang, Gary Wang, Keng-Ming Liu, Hsie-Chia Chang, and Chen-Yi Lee. A field-programmable lab-on-a-chip with built-in self-test circuit and low-power sensor-fusion solution in 0.35 μm standard CMOS process. In *Proceedings IEEE A-SSCC*, pages 1–4, 2015.

[146] Pei Yun Lee, John Costumbrado, Chih-Yuan Hsu, and Yong Hoon Kim. Agarose gel electrophoresis for the separation of DNA fragments. *Journal of Visualized Experiments: JoVE*, (62), 2012.

[147] Chunxiao Li, Anand Raghunathan, and Niraj K Jha. Hijacking an insulin pump: Security attacks and defenses for a diabetes therapy system. In *Proc. IEEE International Conference on e-Health Networking Applications and Services (Healthcom)*, pages 150–156, 2011.

[148] Zipeng Li, Kelvin Yi-Tse Lai, Krishnendu Chakrabarty, Tsung-Yi Ho, and Chen-Yi Lee. Droplet size-aware and error-correcting sample preparation using micro-electrode-dot-array digital microfluidic biochips. *IEEE Transactions on Biomedical Circuits and Systems (TBioCAS)*, 2017.

[149] Chen Liao and Shiyan Hu. Physical-level synthesis for digital lab-on-a-chip considering variation, contamination, and defect. *IEEE Transactions on NanoBioscience (TNB)*, 13(1):3–11, 2014.

[150] YC Lim, AZ Kouzani, and W Duan. Lab-on-a-chip: a component view. *Microsystem Technologies*, 16(12):1995–2015, 2010.

[151] Cliff Chiung-Yu Lin and Yao-Wen Chang. Cross-contamination aware design methodology for pin-constrained digital microfluidic biochips. *IEEE Transactions on Computer-Aided Design of Integrated Circuits and Systems (TCAD)*, 30(6):817–828, 2011.

[152] Chunfeng Liu, Bing Li, Bhargab B Bhattacharya, Krishnendu Chakrabarty, Tsung-Yi Ho, and Ulf Schlichtmann. Testing microfluidic fully programmable valve arrays (FPVAs). In *Proceedings Design, Automation, and Test in Europe (DATE)*, pages 91–96, 2017.

[153] Kenneth J Livak and Thomas D Schmittgen. Analysis of relative gene expression data using real-time quantitative PCR and the $2^{-\Delta\Delta C_T}$ method. *Methods*, 25(4):402–408, 2001.

[154] Kenneth J Livak, Quin F Wills, Alex J Tipping, Krishnalekha Datta, Rowena Mittal, Andrew J Goldson, Darren W Sexton, and Chris C Holmes. Methods for qPCR gene expression profiling applied to 1440 lymphoblastoid single cells. *Methods*, 59(1):71–79, 2013.

[155] K Loens, D Ursi, H Goossens, and M Ieven. Molecular diagnosis of Mycoplasma pneumoniae respiratory tract infections. *Journal of Clinical Microbiology*, 41(11):4915–4923, 2003.

[156] Vlado A Lubarda and Kurt A Talke. Analysis of the equilibrium droplet shape based on an ellipsoidal droplet model. *Langmuir*, 27(17):10705–10713, 2011.

[157] Yan Luo, Bhargab B Bhattacharya, Tsung-Yi Ho, and Krishnendu Chakrabarty. Design and optimization of a cyberphysical digital-microfluidic biochip for the polymerase chain reaction. *IEEE Transactions on Computer-Aided Design of Integrated Circuits and Systems (TCAD)*, 34(1):29–42, 2015.

[158] Yan Luo, Krishnendu Chakrabarty, and Tsung-Yi Ho. A cyberphysical synthesis approach for error recovery in digital microfluidic biochips. In *Proceedings Design, Automation, and Test in Europe (DATE)*, pages 1239–1244, 2012.

[159] Yan Luo, Krishnendu Chakrabarty, and Tsung-Yi Ho. Error recovery in cyberphysical digital microfluidic biochips. *IEEE Transactions on Computer-Aided Design of Integrated Circuits and Systems (TCAD)*, 32:59–72, 2013.

[160] Yan Luo, Krishnendu Chakrabarty, and Tsung-Yi Ho. Real-time error recovery in cyberphysical digital-microfluidic biochips using a compact dictionary. *IEEE Transactions on Computer-Aided Design of Integrated Circuits and Systems (TCAD)*, 32(12):1839–1852, 2013.

[161] Van Luu-The, Nathalie Paquet, Ezequiel Calvo, and Jean Cumps. Improved real-time RT-PCR method for high-throughput measurements using second derivative calculation and double correction. *Biotechniques*, 38(2):287–296, 2005.

[162] Laurens van der Maaten and Geoffrey Hinton. Visualizing data using t-SNE. *Journal of Machine Learning Research*, 9(Nov):2579–2605, 2008.

[163] Andrew C Madison. *Scalable Genome Engineering in Electrowetting on Dielectric Digital Microfluidic Systems*. PhD thesis, Duke University, 2015.

[164] M. Maier. DirichletReg: Dirichlet regression for compositional data in R. 2014.

[165] Lidija Malic, Daniel Brassard, Teodor Veres, and Maryam Tabrizian. Integration and detection of biochemical assays in digital microfluidic LOC devices. *Lab on a Chip*, 10(4):418–431, 2010.

[166] Joshua S Marcus, W French Anderson, and Stephen R Quake. Parallel picoliter RT-PCR assays using microfluidics. *Analytical Chemistry*, 78(3):956–958, 2006.

[167] R Timothy Marler and Jasbir S Arora. The weighted sum method for multi-objective optimization: New insights. *Structural and Multidisciplinary Optimization*, 41(6):853–862, 2010.

[168] Vivien Marx. Cell culture: a better brew. *Nature*, 496(7444):253–258, 2013.

[169] Jiri Matousek. *Geometric discrepancy: An illustrated guide*, volume 18. Springer Science & Business Media, 2009.

[170] Linas Mazutis, John Gilbert, W Lloyd Ung, David A Weitz, Andrew D Griffiths, and John A Heyman. Single-cell analysis and sorting using droplet-based microfluidics. *Nature Protocols*, 8(5):870–891, 2013.

[171] Wajid Hassan Minhass, Jeffrey McDaniel, Michael Raagaard, Philip Brisk, Paul Pop, and Jan Madsen. Scheduling and fluid routing for flow-based microfluidic laboratories-on-a-chip. *IEEE Transactions on Computer-Aided Design of Integrated Circuits and Systems (TCAD)*, 2017.

[172] Wajid Hassan Minhass, Paul Pop, Jan Madsen, and Felician Stefan Blaga. Architectural synthesis of flow-based microfluidic large-scale integration biochips. *Proceedings IEEE/ACM International Conference on Compilers, Architectures and Synthesis of Embedded Systems (CASES)*, 2012.

[173] Wajid Hassan Minhass, Paul Pop, Jan Madsen, and Tsung-Yi Ho. Control synthesis for the flow-based microfluidic large-scale integration biochips. *Proceedings IEEE Asia South Pacific Design Automation Conference (ASP-DAC)*, 2013.

[174] Debasis Mitra, Sarmishtha Ghoshal, Hafizur Rahaman, Krishnendu Chakrabarty, and Bhargab B Bhattacharya. Offline washing schemes for residue removal in digital microfluidic biochips. *ACM Transactions on Design Automation of Electronic Systems*, 21(1):17, 2015.

[175] Debasis Mitra, Sudip Roy, Sukanta Bhattacharjee, Krishnendu Chakrabarty, and Bhargab B Bhattacharya. On-chip sample preparation for multiple targets using digital microfluidics. *IEEE Transactions on Computer-Aided Design of Integrated Circuits and Systems (TCAD)*, 33(8):1131–1144, 2014.

[176] Victoria Moignard, Steven Woodhouse, Laleh Haghverdi, Andrew J Lilly, Yosuke Tanaka, Adam C Wilkinson, Florian Buettner, Iain C Macaulay, Wajid Jawaid, Evangelia Diamanti, et al. Decoding the regulatory network of early blood development from single-cell gene expression measurements. *Nature Biotechnology*, 33(3):269–276, 2015.

[177] Aloysius K Mok. *Fundamental design problems of distributed systems for the hard-real-time environment*. PhD thesis, Massachusetts Institute of Technology, 1983.

[178] Amy Mongersun, Ian Smeenk, Guillem Pratx, Prashanth Asuri, and Paul Abbyad. Droplet microfluidic platform for the determination of single-cell lactate release. *Analytical Chemistry*, 88(6):3257–3263, 2016.

[179] Frieder Mugele and Jean-Christophe Baret. Electrowetting: from basics to applications. *Journal of Physics: Condensed Matter*, 17(28):R705, 2005.

[180] Miguel Angel Murran and Homayoun Najjaran. Capacitance-based droplet position estimator for digital microfluidic devices. *Lab on a Chip*, 12(11):2053–2059, 2012.

[181] Mark A Musen, Blackford Middleton, and Robert A Greenes. Clinical decision-support systems. In *Biomedical informatics*, pages 643–674. 2014.

[182] Healthcare IT News. Games for health. [Online] http://www.healthcareitnews.com/directory/games-health.

[183] Alphonsus HC Ng, M Dean Chamberlain, Haozhong Situ, Victor Lee, and Aaron R Wheeler. Digital microfluidic immunocytochemistry in single cells. *Nature communications*, 6:7513–7513, 2014.

[184] Alphonsus HC Ng, Kihwan Choi, Robert P Luoma, John M Robinson, and Aaron R Wheeler. Digital microfluidic magnetic separation for particle-based immunoassays. *Analytical Chemistry*, 84(20):8805–8812, 2012.

[185] Alphonsus HC Ng and Aaron R Wheeler. Next-generation microfluidic point-of-care diagnostics. *Clinical Chemistry*, 61(10):1233–1234, 2015.

[186] Harald Niederreiter. *Random number generation and quasi-Monte Carlo methods*, volume 63. SIAM, 1992.

[187] Haig Norian, Ryan M Field, Ioannis Kymissis, and Kenneth L Shepard. An integrated CMOS quantitative-polymerase-chain-reaction lab-on-chip for point-of-care diagnostics. *Lab on a Chip*, 14(20):4076–4084, 2014.

[188] Aisling O'Driscoll, Jurate Daugelaite, and Roy D Sleator. 'big data', Hadoop and cloud computing in genomics. *Journal of Biomedical Informatics*, 46(5):774–781, 2013.

[189] Kwang W Oh, Kangsun Lee, Byungwook Ahn, and Edward P Furlani. Design of pressure-driven microfluidic networks using electric circuit analogy. *Lab on a Chip*, 12(3):515–545, 2012.

[190] Kwang W Oh, Chinsung Park, Kak Namkoong, Jintae Kim, Kyeong-Sik Ock, Suhyeon Kim, Young-A Kim, Yoon-Kyoung Cho, and Christopher Ko. World-to-chip microfluidic interface with built-in valves for multi-chamber chip-based pcr assays. *Lab on a Chip*, 5(8):845–850, 2005.

[191] Kevin Oishi and Eric Klavins. Framework for engineering finite state machines in gene regulatory networks. *ACS Synthetic Biology*, 3(9):652–665, 2014.

[192] Lale Özbakir, Adil Baykasoğlu, and Pınar Tapkan. Bees algorithm for generalized assignment problem. *Applied Mathematics and Computation*, 215(11):3782–3795, 2010.

[193] Aydogan Ozcan and Euan McLeod. Lensless imaging and sensing. *Annual Review of Biomedical Engineering*, 18(1):77–102, 2016.

[194] Phil Paik, Vamsee K Pamula, and Richard B Fair. Rapid droplet mixers for digital microfluidic systems. *Lab on a Chip*, 3(4):253–259, 2003.

[195] AG Papathanasiou and AG Boudouvis. Manifestation of the connection between dielectric breakdown strength and contact angle saturation in electrowetting. *Applied Physics Letters*, 86(16):164102, 2005.

[196] Michael W Pfaffl. Quantification strategies in real-time PCR. In Stephen A Bustin, editor, *AZ of quantitative PCR*. International University Line, 2004.

[197] Linh TX Phan, Insup Lee, and Oleg Sokolsky. Compositional analysis of multi-mode systems. In *Proceedings IEEE Euromicro Conference on Real-Time Systems (ECRTS)*, pages 197–206, 2010.

[198] Ines Pickrahn, Gabriele Kreindl, Eva Müller, Bettina Dunkelmann, Waltraud Zahrer, Jan Cemper-Kiesslich, and Franz Neuhuber. Contamination incidents in the pre-analytical phase of forensic DNA analysis in austri—statistics of 17 years. *Forensic Science International: Genetics*, 31:12–18, 2017.

[199] Silvia Pineda, Francisco X Real, Manolis Kogevinas, Alfredo Carrato, Stephen J Chanock, Núria Malats, and Kristel Van Steen. Integration analysis of three omics data using penalized regression methods: An application to bladder cancer. *PLoS Genetics*, 11(12):1–22, 2015.

[200] Sudip Poddar, Sukanta Bhattarcharjee, Subhas C Nandy, Krishnendu Chakrabarty, and Bhargab B Bhattacharya. Optimization of multi-target sample preparation on-demand with digital microfluidic biochips. *IEEE Transactions on Computer-Aided Design of Integrated Circuits and Systems (TCAD)*, 2018.

[201] Sudip Poddar, Sarmishtha Ghoshal, Krishnendu Chakrabarty, and Bhargab B Bhattacharya. Error-correcting sample preparation with cyberphysical digital microfluidic lab-on-chip. *ACM Transactions on Design Automation of Electronic Systems*, 22(1):2:1–29, 2016.

[202] MG Pollack, AD Shenderov, and RB Fair. Electrowetting-based actuation of droplets for integrated microfluidics. *Lab on a Chip*, 2(2):96–101, 2002.

[203] Paul Pop, Mirela Alistar, Elena Stuart, and Jan Madsen. Design methodology for digital microfluidic biochips. In *Fault-Tolerant Digital Microfluidic Biochips*, pages 13–28. Springer, 2016.

[204] Paul Pop, Petru Eles, and Zebo Peng. *Analysis and Synthesis of Distributed Real-Time Embedded Systems*. New York, NY, USA: Springer, 2013.

[205] Daniel J Power. Decision support systems: a historical overview. In *Handbook on Decision Support Systems 1*, pages 121–140. 2008.

[206] Matthew Rabin. Incorporating fairness into game theory and economics. *The American Economic Review*, pages 1281–1302, 1993.

[207] Jeyavijayan Rajendran, Aman Ali, Ozgur Sinanoglu, and Ramesh Karri. Belling the CAD: Toward security-centric electronic system design. *IEEE Transactions on Computer-Aided Design of Integrated Circuits and Systems (TCAD)*, 34(11):1756–1769, 2015.

[208] Daniel Regalado, Shon Harris, Allen Harper, Chris Eagle, Jonathan Ness, Branko Spasojevic, Ryan Linn, and Stephen Sims. *Gray Hat Hacking The Ethical Hacker's Handbook*. McGraw-Hill Education Group, 2015.

[209] Hong Ren, Richard B Fair, and Micheal G Pollack. Automated on-chip droplet dispensing with volume control by electro-wetting actuation and capacitance metering. *Sensors and Actuators B: Chemical*, 98(2):319–327, 2004.

[210] Lidice Garcia Rios et al. Big data infrastructure for analyzing data generated by wireless sensor networks. In *Proceedings IEEE International Congress on Big Data*, pages 816–823, 2014.

[211] Marylyn D Ritchie, Emily R Holzinger, Ruowang Li, Sarah A Pendergrass, and Dokyoon Kim. Methods of integrating data to uncover genotype-phenotype interactions. *Nature Reviews Genetics*, 16(2):85–97, 2015.

[212] A Rival, D Jary, C Delattre, Y Fouillet, G Castellan, A Bellemin-Comte, and X Gidrol. An EWOD-based microfluidic chip for single-cell isolation, mRNA purification and subsequent multiplex qPCR. *Lab on a Chip*, 14(19):3739–3749, 2014.

[213] Sudip Roy, Bhargab B Bhattacharya, and Krishnendu Chakrabarty. Optimization of dilution and mixing of biochemical samples using digital microfluidic biochips. *IEEE Transactions on Computer-Aided Design of Integrated Circuits and Systems (TCAD)*, 29(11):1696–1708, 2010.

[214] Sudip Roy, Bhargab B Bhattacharya, Sarmishtha Ghoshal, and Krishnendu Chakrabarty. Theory and analysis of generalized mixing and dilution of biochemical fluids using digital microfluidic biochips. *ACM Journal on Emerging Technologies in Computing Systems*, 11(1):2:1–33, 2014.

[215] Sudip Roy, Partha P Chakrabarti, Srijan Kumar, Krishnendu Chakrabarty, and Bhargab B Bhattacharya. Layout-aware mixture preparation of biochemical fluids on application-specific digital microfluidic biochips. *ACM Transactions on Design Automation of Electronic Systems (TODAES)*, 20(3):45, 2015.

[216] Laura A Rozo Duque, Jose M Monsalve Diaz, and Chengmo Yang. Improving MPSoC reliability through adapting runtime task schedule based on time-correlated fault behavior. In *Proceedings Design, Automation, and Test in Europe (DATE)*, pages 818–823, 2015.

[217] Robert G Rutledge and Don Stewart. A kinetic-based sigmoidal model for the polymerase chain reaction and its application to high-capacity absolute quantitative real-time PCR. *BMC biotechnology*, 8(1):47, 2008.

[218] Assieh Saadatpour, Shujing Lai, Guoji Guo, and Guo-Cheng Yuan. Single-cell analysis in cancer genomics. *Trends in Genetics*, 31(10):576–586, 2015.

[219] Eric K Sackmann, Anna L Fulton, and David J Beebe. The present and future role of microfluidics in biomedical research. *Nature*, 507(7491):181–189, 2014.

[220] Abusayeed Saifullah, David Ferry, Jing Li, Kunal Agrawal, Chenyang Lu, and Christopher D Gill. Parallel real-time scheduling of dags. *IEEE Transactions on Parallel and Distributed Systems (TPDS)*, 25(12):3242–3252, 2014.

[221] Ehsan Samiei, Maryam Tabrizian, and Mina Hoorfar. A review of digital microfluidics as portable platforms for lab-on a-chip applications. *Lab on a Chip*, 2016.

[222] Philip Schneider and David H Eberly. *Geometric tools for computer graphics*. Elsevier, 2002.

[223] Thermo Fisher Scientific. NanoDrop Spectrophotometers–Assessment of Nucleic Acid Purity. [Online] http://www.nanodrop.com/Library/T042-NanoDrop-Spectrophotometers-Nucleic-Acid-Purity-Ratios.pdf. (Date last accessed June 9, 2019).

[224] Yu Fen Samantha Seah, Hongxing Hu, and Christoph A Merten. Microfluidic single-cell technology in immunology and antibody screening. *Molecular Aspects of Medicine*, 59:47–61, 2018.

[225] Stephen D Senturia. CAD challenges for microsensors, microactuators, and microsystems. *Proceedings of the IEEE*, 86(8):1611–1626, 1998.

[226] Gaurav J Shah, Aaron T Ohta, Eric P-Y Chiou, Ming C Wu, et al. EWOD-driven droplet microfluidic device integrated with optoelectronic tweezers as an automated platform for cellular isolation and analysis. *Lab on a Chip*, 9(12):1732–1739, 2009.

[227] Nachiket Shembekar, Chawaree Chaipan, Ramesh Utharala, and Christoph A Merten. Droplet-based microfluidics in drug discovery, transcriptomics and high-throughput molecular genetics. *Lab on a Chip*, 16(8):1314–1331, 2016.

[228] Hsien-Hua Shen, Shih-Kang Fan, Chang-Jin Kim, and Da-Jeng Yao. EWOD microfluidic systems for biomedical applications. *Microfluidics and Nanofluidics*, 16(5):965–987, 2014.

[229] Steve CC Shih, Philip C Gach, Jess Sustarich, Blake A Simmons, Paul D Adams, Seema Singh, and Anup K Singh. A droplet-to-digital (D2D) microfluidic device for single cell assays. *Lab on a Chip*, 15(1):225–236, 2015.

[230] Steve CC Shih, Garima Goyal, Peter W Kim, Nicolas Koutsoubelis, Jay D Keasling, Paul D Adams, Nathan J Hillson, and Anup K Singh. A versatile microfluidic device for automating synthetic biology. *ACS Synthetic Biology*, 4(10):1151–1164, 2015.

[231] Yasser Shoukry, Paul Martin, Yair Yona, Suhas Diggavi, and Mani Srivastava. Pycra: Physical challenge-response authentication for active sensors under spoofing attacks. In *Proceedings ACM SIGSAC Conference on Computer and Communications Security (CCS)*, pages 1004–1015, 2015.

[232] Ryan Silva, Swapnil Bhatia, and Douglas Densmore. A reconfigurable continuous-flow fluidic routing fabric using a modular, scalable primitive. *Lab on a Chip*, 16(14):2730–2741, 2016.

[233] David Sims, Ian Sudbery, Nicholas E Ilott, Andreas Heger, and Chris P Ponting. Sequencing depth and coverage: key considerations in genomic analyses. *Nature Reviews Genetics*, 15(2):121–132, 2014.

[234] Ramakrishna Sista, Zhishan Hua, Prasanna Thwar, Arjun Sudarsan, Vijay Srinivasan, Allen Eckhardt, Michael Pollack, and Vamsee Pamula. Development of a digital microfluidic platform for point of care testing. *Lab on a Chip*, 8(12):2091–2104, 2008.

[235] Ramakrishna S Sista, Allen E Eckhardt, Vijay Srinivasan, Michael G Pollack, Srinivas Palanki, and Vamsee K Pamula. Heterogeneous immunoassays using magnetic beads on a digital microfluidic platform. *Lab on a Chip*, 8(12):2188–2196, 2008.

[236] Ramakrishna S Sista, Tong Wang, Ning Wu, Carrie Graham, Allen Eckhardt, Theodore Winger, Vijay Srinivasan, Deeksha Bali, David S Millington, and Vamsee K Pamula. Multiplex newborn screening for Pompe, Fabry, Hunter, Gaucher, and Hurler diseases using a digital microfluidic platform. *Clinica Chimica Acta*, 424:12–18, 2013.

[237] Mano Sivaganesan, Shawn Seifring, Manju Varma, Richard A Haugland, and Orin C Shanks. A Bayesian method for calculating real-time quantitative PCR calibration curves using absolute plasmid DNA standards. *BMC bioinformatics*, 9(1):120, 2008.

[238] Vijay Srinivasan, Vamsee K Pamula, and Richard B Fair. Droplet-based microfluidic lab-on-a-chip for glucose detection. *Analytica Chimica Acta*, 507(1):145–150, 2004.

[239] Vijay Srinivasan, Vamsee K Pamula, and Richard B Fair. An integrated digital microfluidic lab-on-a-chip for clinical diagnostics on human physiological fluids. *Lab on a Chip*, 4(4):310–315, 2004.

[240] Vijay Srinivasan, Vamsee K Pamula, Phil Paik, and Richard B Fair. Protein stamping for MALDI mass spectrometry using an electrowetting-based microfluidic platform. *Proceedings of SPIE*, 5591:26–33, 2004.

[241] William Stallings. *Cryptography and network security: principles and practice*. Pearson Education India, 2003.

[242] Graham Steel. Formal analysis of PIN block attacks. *Theoretical Computer Science*, 367(1-2):257–270, 2006.

[243] Oliver Stegle, Sarah A Teichmann, and John C Marioni. Computational and analytical challenges in single-cell transcriptomics. *Nature Reviews Genetics*, 16(3):133–145, 2015.

[244] Zachary D Stephens, Skylar Y Lee, Faraz Faghri, Roy H Campbell, Chengxiang Zhai, Miles J Efron, Ravishankar Iyer, Michael C Schatz, Saurabh Sinha, and Gene E Robinson. Big data: astronomical or genomical? *PLoS Biol*, 13(7):1–11, 2015.

[245] Fei Su and Krishnendu Chakrabarty. Architectural-level synthesis of digital microfluidics-based biochips. In *Proceedings International Conference on Computer-Aided Design (ICCAD)*, pages 223–228, 2004.

[246] Fei Su and Krishnendu Chakrabarty. Module placement for fault-tolerant microfluidics-based biochips. In *ACM Transactions on Design Automation of Electronic Systems*, volume 11, pages 682–710, 2004.

[247] Fei Su and Krishnendu Chakrabarty. Unified high-level synthesis and module placement for defect-tolerant microfluidic biochips. In *Proceedings ACM/IEEE Design Automation Conference (DAC)*, pages 825–830, 2005.

[248] Fei Su and Krishnendu Chakrabarty. High-level synthesis of digital microfluidic biochips. *ACM Journal on Emerging Technologies in Computing Systems*, 3(4):1, 2008.

[249] Fei Su, William Hwang, and Krishnendu Chakrabarty. Droplet routing in the synthesis of digital microfluidic biochips. In *Proceedings Design, Automation, and Test in Europe (DATE)*, pages 1–6, 2006.

[250] Fei Su, Sule Ozev, and Krishnendu Chakrabarty. Ensuring the operational health of droplet-based microelectrofluidic biosensor systems. *IEEE Sensors Journal*, 5(4):763–773, 2005.

[251] Bhattacharjee Sukanta, Jack Tang, Mohamed Ibrahim, Krishnendu Chakrabarty, and Ramesh Karri. Locking of biochemical assays for digital microfluidic biochips. In *IEEE European Test Symposium (ETS)*, 2018.

[252] P. Tabeling. *Introduction to Microfluidics*. Oxford University Press, 2005.

[253] Boxin Tang. Orthogonal array-based Latin hypercubes. *Journal of the American statistical association*, 88(424):1392–1397, 1993.

[254] Jack Tang, Mohamed Ibrahim, and Krishnendu Chakrabarty. Randomized checkpoints: A practical defense for cyberphysical microfluidic systems. *IEEE Design & Test*, 2018.

[255] Jack Tang, Mohamed Ibrahim, Krishnendu Chakrabarty, and Ramesh Karri. Secure randomized checkpointing for digital microfluidic biochips. *IEEE Transactions on Computer-Aided Design of Integrated Circuits and Systems (TCAD)*, 2017.

[256] Jack Tang, Mohamed Ibrahim, Krishnendu Chakrabarty, and Ramesh Karri. Security implications of cyberphysical flow-based microfluidic biochips. In *IEEE Asian Test Symposium (ATS)*, pages 115–120, 2017.

[257] Jack Tang, Mohamed Ibrahim, Krishnendu Chakrabarty, and Ramesh Karri. Security trade-offs in microfluidic routing fabrics. In *IEEE International Conference on Computer Design (ICCD)*, pages 25–32, 2017.

[258] Jack Tang, Mohamed Ibrahim, Krishnendu Chakrabarty, and Ramesh Karri. Tamper-resistent pin-constrained digital microfluidic biochips. In *Proceedings ACM/IEEE Design Automation Conference (DAC)*, 2018.

[259] Jack Tang, Mohamed Ibrahim, Krishnendu Chakrabarty, and Ramesh Karri. Toward secure and trustworthy cyberphysical microfluidic biochips. *IEEE Transactions on Computer-Aided Design of Integrated Circuits and Systems (TCAD)*, 38(4):589–603, 2018.

[260] Jack Tang, Mohamed Ibrahim, Krishnendu Chakrabarty, and Ramesh Karri. Towards secure and trustworthy cyberphysical microfluidic biochips. *IEEE Transactions on Computer-Aided Design of Integrated Circuits and Systems (TCAD)*, 2018.

[261] Jack Tang, Mohamed Ibrahim, Krishnendu Chakrabarty, and Ramesh Karri. Analysis and design of tamper-mitigating microfluidic routing fabrics. *IEEE Transactions on Computer-Aided Design of Integrated Circuits and Systems (TCAD)*, 2019.

[262] Jack Tang, Mohamed Ibrahim, Krishnendu Chakrabarty, and Ramesh Karri. *Secure and Trustworthy Cyberphysical Microfluidic Biochips*. Springer, Cham, 2020.

[263] Jack Tang, Ramesh Karri, Mohamed Ibrahim, and Krishnendu Chakrabarty. Securing digital microfluidic biochips by randomizing checkpoints. In *IEEE International Test Conference (ITC)*, pages 1–8, 2016.

[264] Michaela A TerAvest, Tom J Zajdel, and Caroline M Ajo-Franklin. The Mtr pathway of shewanella oneidensis MR-1 couples substrate utilization to current production in escherichia coli. *ChemElectroChem*, 1(11):1874–1879, 2014.

[265] Todd Thorsen, Sebastian J Maerkl, and Stephen R Quake. Microfluidic large-scale integration. *Science*, 298(5593):580–584, 2002.

[266] Transcriptic. Automated cell and molecular biology laboratory. [Online] https://www.transcriptic.com.

[267] Stephen M Trimberger and Jason J Moore. FPGA security: Motivations, features, and applications. *Proceedings of the IEEE*, 102(8):1248–1265, 2014.

[268] Tsun-Ming Tseng, Bing Li, Mengchu Li, Tsung-Yi Ho, and Ulf Schlichtmann. Reliability-aware synthesis with dynamic device mapping and fluid routing for flow-based microfluidic biochips. *IEEE Transactions on Computer-Aided Design of Integrated Circuits and Systems (TCAD)*, 35(12):1981–1994, 2016.

[269] Tsun-Ming Tseng, Mengchu Li, Bing Li, Tsung-Yi Ho, and Ulf Schlichtmann. Columba: co-layout synthesis for continuous-flow microfluidic biochips. In *Proceedings ACM/IEEE Design Automation Conference (DAC)*, pages 147:1–6, 2016.

[270] HJJ Verheijen and MWJ Prins. Reversible electrowetting and trapping of charge: model and experiments. *Langmuir*, 15(20):6616–6620, 1999.

[271] Elisabeth Verpoorte and Nico F De Rooij. Microfluidics meets MEMS. *Proceedings of the IEEE*, 91(6):930–953, 2003.

[272] Timothy Quang Vu, Ricardo Miguel Bessa de Castro, and Lidong Qin. Bridging the gap: microfluidic devices for short and long distance cell–cell communication. *Lab on a Chip*, 17(6):1009–1023, 2017.

[273] Simon J Walsh. Legal perceptions of forensic DNA profiling: Part i: A review of the legal literature. *Forensic Science International*, 155(1):51–60, 2005.

[274] Gary Wang, Daniel Teng, and S-K Fan. Digital microfluidic operations on micro-electrode dot array architecture. *IET Nanobiotechnology*, 5(4):152–160, 2011.

[275] Qin Wang, Yue Xu, Shiliang Zuo, Hailong Yao, Tsung-Yi Ho, Bing Li, Ulf Schlichtmann, and Yici Cai. Pressure-aware control layer optimization for flow-based microfluidic biochips. *IEEE Transactions on Biomedical Circuits and Systems (TBioCAS)*, 2017.

[276] Qin Wang, Shiliang Zuo, Hailong Yao, Tsung-Yi Ho, Bing Li, Ulf Schlichtmann, and Yici Cai. Hamming-distance-based valve-switching optimization for control-layer multiplexing in flow-based microfluidic biochips. In *Proceedings IEEE Asia South Pacific Design Automation Conference (ASP-DAC)*, pages 524–529, 2017.

[277] Michael   Waskom.   Seaborn.   [Online] https://stanford.edu/~mwaskom/soft-ware/seaborn/.

[278] Wilfried Weber and Martin Fussenegger. Emerging biomedical applications of synthetic biology. *Nature Reviews Genetics*, 13(1):21, 2012.

[279] Bayly S Wheeler, Brandon T Ruderman, Huntington F Willard, and Kristin C Scott. Uncoupling of genomic and epigenetic signals in the maintenance and inheritance of heterochromatin domains in fission yeast. *Genetics*, 190(2):549–557, 2012.

[280] R. Wille, O. Keszocze, R. Drechsler, T. Boehnisch, and A. Kroker. Scalable one-pass synthesis for digital microfluidic biochips. *IEEE Design & Test*, 32(6):41–50, 2015.

[281] Angela R Wu, Joseph B Hiatt, Rong Lu, Joanne L Attema, Neethan A Lobo, Irving L Weissman, Michael F Clarke, and Stephen R Quake. Automated microfluidic chromatin immunoprecipitation from 2,000 cells. *Lab on a Chip*, 9(10):1365–1370, 2009.

[282] Angela R Wu, Tiara LA Kawahara, Nicole A Rapicavoli, Jan Van Riggelen, Emelyn H Shroff, Liwen Xu, Dean W Felsher, Howard Y Chang, and Stephen R Quake. High throughput automated chromatin immunoprecipitation as a platform for drug screening and antibody validation. *Lab on a Chip*, 12(12):2190–2198, 2012.

[283] Dong Wu, Bashir M Al-Hashimi, and Petru Eles. Scheduling and mapping of conditional task graph for the synthesis of low power embedded systems. *IEE Proceedings Computers and Digital Techniques*, 150(5):262–73, 2003.

[284] Yuan Xie and Wayne Wolf. Allocation and scheduling of conditional task graph in hardware/software co-synthesis. In *Proceedings Design, Automation, and Test in Europe (DATE)*, pages 620–625, 2001.

[285] Tao Xu and Krishnendu Chakrabarty. Integrated droplet routing in the synthesis of microfluidic biochips. In *Proceedings ACM/IEEE Design Automation Conference (DAC)*, pages 948–953, 2007.

[286] Tao Xu and Krishnendu Chakrabarty. Broadcast electrode-addressing for pin-constrained multi-functional digital microfluidic biochips. In *Proceedings ACM/IEEE Design Automation Conference (DAC)*, pages 173–178, 2008.

[287] Tao Xu, Krishnendu Chakrabarty, and Fei Su. Defect-aware high-level synthesis and module placement for microfluidic biochips. *IEEE Transactions on Biomedical Circuits and Systems (TBioCAS)*, 2(1):50–62, 2008.

[288] Tao Xu, William L Hwang, Fei Su, and Krishnendu Chakrabarty. Automated design of pin-constrained digital microfluidic biochips under droplet-interference constraints. *ACM Journal on Emerging Technologies in Computing Systems*, 3(3):14:1–23, 2007.

[289] Kaiping Xue, Shaohua Li, Jianan Hong, Yingjie Xue, Nenghai Yu, and Peilin Hong. Two-cloud secure database for numeric-related SQL range queries with privacy preserving. *IEEE Transactions on Information Forensics and Security (TIFS)*, 12(7):1596–1608, 2017.

[290] Hailong Yao, Tsung-Yi Ho, and Yici Cai. PACOR: Practical control-layer routing flow with length-matching constraint for flow-based microfluidic biochips. In *Proceedings ACM/IEEE Design Automation Conference (DAC)*, page 142, 2015.

[291] Yole Développment. Positive dynamics of microfluidic-based Point-of-Need. http://www.yole.fr/Microfluidics_PoN_IndustryEvolution.aspx. (Date last accessed June 9, 2019).

[292] Ping-Hung Yuh, Chia-Lin Yang, and Yao-Wen Chang. Placement of digital microfluidic biochips using the T-tree formulation. In *Proceedings ACM/IEEE Design Automation Conference (DAC)*, pages 931–934, 2006.

[293] Ping-Hung Yuh, Chia-Lin Yang, and Yao-Wen Chang. BioRoute: A network-flow based routing algorithm for digital microfluidic biochips.

In *Proceedings International Conference on Computer-Aided Design (ICCAD)*, pages 752–757, 2007.

[294] Yang Zhao and Krishnendu Chakrabarty. Digital microfluidic logic gates and their application to built-in self-test of lab-on-chip. *IEEE Transactions on Biomedical Circuits and Systems (TBioCAS)*, 4(4):250–262, 2010.

[295] Yang Zhao and Krishnendu Chakrabarty. Cross-contamination avoidance for droplet routing in digital microfluidic biochips. *IEEE Transactions on Computer-Aided Design of Integrated Circuits and Systems (TCAD)*, 31(6):817–830, 2012.

[296] Yang Zhao, Tao Xu, and Krishnendu Chakrabarty. Integrated control-path design and error recovery in the synthesis of digital microfluidic lab-on-chip. *ACM Journal on Emerging Technologies in Computing Systems (JETC)*, 6(3):11, 2010.

[297] Yuejun Zhao, Sang Kug Chung, Ui-Chong Yi, and Sung Kwon Cho. Droplet manipulation and microparticle sampling on perforated micro-filter membranes. *Journal of Micromechanics and Microengineering*, 18(2):025030, 2008.

[298] Zhanwei Zhong, Zipeng Li, and Krishnendu Chakrabarty. Adaptive error recovery in MEDA biochips based on droplet-aliquot operations and predictive analysis. In *Proceedings International Conference on Computer-Aided Design (ICCAD)*, pages 615–622, 2017.

[299] Jing Zhu, Chunmei Qiu, Mirkó Palla, ThaiHuu Nguyen, James J Russo, Jingyue Ju, and Qiao Lin. A microfluidic device for multiplex single-nucleotide polymorphism genotyping. *RSC advances*, 4(9):4269–4277, 2014.

[300] Rapolas Zilionis, Juozas Nainys, Adrian Veres, Virginia Savova, David Zemmour, Allon M Klein, and Linas Mazutis. Single-cell barcoding and sequencing using droplet microfluidics. *Nature Protocols*, 12(1):44–73, 2017.

Index

Note: Page numbers followed by "*n*" indicate footnote.

A

Adenine (A), 263
AD library, *see* Analysis and
 decision-making library
Agarose gel electrophoresis, 266
Aging reinforces DMFB security,
 256–257
 electrode actuations, 256
 thickness of dielectric layer,
  256–257
Algorithmic baseline (ReSyn),
 135, 136
Algorithmic models, mapping to, 120
 cell state machine, 122–123
 digital-microfluidic biochip
  modeling, 121–122
 protocol model, 122–123
 valve-based crossbar modeling,
  120–121
Aliquot-generation trees, 303
 properties of, 303–306
 recursion in, 307–312
Aliquots, 189, 201
Amplicon copy numbers using serial
 dilution, 272–273
Analysis and decision-making library
 (AD library), 49–50
Apache Cassandra, 91, 106
Apache Spark, 91
A priori, 43
Architectural baseline (ArcSyn), 135,
 136
ArcSyn, *see* Architectural baseline
Asymmetric cryptography, 264
Attacker, 226–227
Authenticated analyzers, 264
Authenticated collectors, 264

B

Barcoded DNA sample material, 264
Barcode propagation, 122
Barcoding-aware synthesis, 104–105
Barcoding, *see* Droplet indexing
Benchtop demonstration of DNA
 barcoding, 265
 DNA barcodes using traditional
  DNA-analysis flow, 266
 experiment procedure, 266–270
 interpretation of gel-electrohoresis
  results, 270–272
 specifications of DNA barcodes,
  266, 267
β-actin, 36
"Big Data" community goals, 94
Bioassay execution-time overhead,
 260–261
Bioassay level, 17
Biochemical-level defense mechanism,
 262–265
Biochip
 fabrication and chip specifications,
  219–221
 resources, 64, 65
Biocircuit-regulatory scanning (BRS),
 179, 180–181, 279
BioCyBig, 84, 88–89, 90
 application stack, 94–100
 "Big Data" community goals, 94
 cloud software, 92
 distributed-system architecture, 93
 human-interface system, 92
 plug-and-play control of
  microfluidic devices, 93–94
 software stack of, 92
 system architecture design, 90–91

Biomolecular analysis
    custom-built fluorescence sensor, 6
    sensing systems for, 5–7
Biomolecular assays, 249
Biomolecular mixture preparation,
    MEDA biochips for, 9–10
Biomolecular quantitative analysis,
    20–21
    security assessment of, 17–18
Biomolecular sample preparation, 21
BioRad iCycler, 38
BioScan imprecision, 189
BIST, *see* Built-in self-test
BRS, *see* Biocircuit-regulatory scanning
Brute-force attacks, protection against,
    259
Built-in self-test (BIST), 251–252

C

C^5 architecture, 119–120
    to error recovery, 211–212
CAD, *see* Computer-aided design
CanLib, 89–90, 96, 100, 101
    visual analytics, 100, 101
Cassandra data model, 106
Cassandra Query Language (CQL), 106
Causative exploration of biomolecules,
    see Down-stream analysis
CCD, *see* Charge-coupled device
CDC, *see* Centers for Disease Control
    and Prevention
cDNA, *see* Complementary DNA
Cell culture, 34
Cell differentiation, 111, 115
Cell encapsulation, 111
    and flow control, 114–115
Cell sorter using multiplexed control,
    see Sortex
Cell state machine, 122–123
Centers for Disease Control and
    Prevention (CDC), 232
CF, *see* Concentration factor
CFD, *see* Computational fluid dynamics
CFG, *see* Control flow graph
Charge-coupled device (CCD), 238
ChIP, *see* Chromatin
    immuneprecipitation
Chromatin immuneprecipitation
    (ChIP), 7, 103, 111

Cloud service, 95
Cloud software, 92
CMFBs, *see* Continuous-flow
    microfluidic biochips
CNV, *see* Copy-number variation
Co-Synthesis (CoSyn), 113, 276
    methodology, 123–125
Complementary DNA (cDNA), 35, 39
Composition graph, 195, 196
Computational fluid dynamics (CFD),
    3, 143
Computer-aided design (CAD), 215
Concentration factor (CF), 180
    analysis of CF sampling, 203–204
    sampling of CF space, 185–188
Continuous-flow microfluidic biochips
    (CMFBs), 1, 10; *see also*
    Digital microfluidic biochip
    (DMFB)
    flow-based components, 11
Continuous-flow microfluidics, 10–12
Control delay vector (ψ) derivation,
    291–295
Control flow graph (CFG), 42, 122
Control latency (α_i) derivation, 297–302
Copy-number variation (CNV), 84
CoSyn, *see* Co-Synthesis
CP, *see* Cyber-physical
CPS, *see* Cyber-physical system
CQL, *see* Cassandra Query Language
Crossbar architecture, 128–130
Cumulative system utilization, 76
Cyber-physical (CP), 280
    adaptation for quantitative
        analysis, 15–16
    DMFB, 239
    integration, 238–239
    microfluidic biochip, 13
    microfluidics in presence of sensor
        spoofing attacks, 280
Cyber-physical system (CPS), 15
Cyber space, 13
Cytosine (C), 263

D

DAG model, *see* Directed-acyclic graph
    model
DA method, *see* Direct-addressing
    method

Decision-making models, 99
Decision-support systems (DSS), 99
Degradation-aware resource sharing, 44–46
Degradation time, 255
Degree of concentration accuracy, 189
Deoxynucleotides (dNTPs), 242
Design-quality assessment, 138–139
Dictionary-based error recovery, 216–219
Dielectric layer, thickness of, 256–257
Digital microfluidic biochip (DMFB), 2, 4–5, 111, 125, 182
    cyber-physical integration of, 14
    design flow, 227
    DMFB-enabled encryption technique, 249
    implementation of immunoassay, 8
    magnetic-bead manipulation, 7–8
    MEDA biochips, 9–10
    modeling, 121–122
    on-chip heaters, 8–9
    sensing systems for biomolecular analysis, 5–7
    supply-chain security, 259
    threats assessment of, 226–229
    Trojan taxonomy, 229
Digital microfluidics (DMF), 4, 7, 209; *see also* Microfluidic(s)
    for gene-expression analysis, 39–43
    logic gates, 251–252
    technology, 2, 5
Direct-addressing method (DA method), 168–171
Directed-acyclic graph model (DAG model), 65, 120
Dirichlet-based mapping, 188
Dirichlet distribution, 188
Dirichlet regression, 203
Discrete-time Markov chain (DTMC), 113, 130–131, 133–135
Distilled water (DW), 182
Distributed-RC model, 171–173
Distributed-system
    architecture, 93
    interfacing and integration, 105–106

Distributed model, 159–160
DM, *see* Dynamic-variation model
DMF, *see* Digital microfluidics
DMFB, *see* Digital microfluidic biochip
DNA
    amplification, 36–37
    bands, 241
    ladder, 245
    processing, 17
    PSE of biocircuits in presence of DNA memory, 279
DNA-forgery attacks, 262
    attack models for gene-expression analysis, 241–242
    benchtop experiment of malicious DNA preparation, 242–246
    DNA preparation, amplification, and analysis, 240–241
    interpretation of gel-electrophoresis results, 246–248
DNA barcode/barcoding, 249
    as biochemical-level defense mechanism, 262–265
    methodology, 262–263
    security analysis, 263–265
    sequence information relating to BAR1 and BAR2, 267
    side-channel fingerprinting for piracy prevention, 280–282
    specifications, 266
    using traditional DNA-analysis flow, 266
dNTPs, *see* Deoxynucleotides
Double-stranded DNA (dsDNA), 263
Down-stream analysis, *see* Quantitative analysis
Downstream assay operations, 250
Droplet-aliquoting constraints, 193–195
Droplet barcoding, 115–118
Droplet-handling operations, 251–252
Droplet indexing, 111, 112
Droplet routing, 128–130
Droplet volume, 253
dsDNA, *see* Double-stranded DNA
DSS, *see* Decision-support systems
DTMC, *see* Discrete-time Markov chain
DW, *see* Distilled water
Dynamic-variation model (DM), 51, 55–56

E

EDTA, *see* Ethylenediaminetetraacetic
        acid
Electrode actuations, number of, 256,
        258
Electrodes, 259–260
Electrowetting-on-dielectric function
        (EWOD function), 2, 4–5,
        220, 251–252
Emerging cyber-pysical on-chip bioassay
        protocols, 261
Encrypted sequencing graph, 250–251
Encryption security analysis and
        simulation results, 257–261
    area overhead, 259–260
    bioassay execution-time overhead,
        260–261
    security analysis, 285–259
Epigenetic(s), 59
Epigenetic analysis, 59; *see also*
        Gene-expression analysis
        (GEA)
    experimental demonstration, 77–79
    miniaturization of
        epigenetic-regulation analysis,
        60–63
    simulation results, 73–77
    system model, 64–67
    task assignment and scheduling,
        68–73
Error recovery
    biochip layout, 214
    C^5 architecture to, 211–212
    dictionary-based, 216–219
    protocol design and, 213–214
Ethylenediaminetetraacetic acid
        (EDTA), 244
Evaporation rate, 255
EWOD function, *see* Electrowetting-
        on-dielectric function
Execution time
    of assay, 254–255
    overhead, 260–261

F

Fault-tolerant realization of
        biomolecular assays
    C^5 architecture to error recovery,
        211–212

dictionary-based error recovery,
        216–219
    experiment results and
        demonstration, 219–221
    physical defects and prior
        error-recovery solutions,
        209–211
    system design, 213–215
FCFS, *see* First-come-first-served
Field-programmable gate array
        (FPGA), 29, 218–219
Fingerprinting process, 282
Finite state machine (FSM), 211–212
Firmware, 102–103, 104–105
    for quantitative analysis, 49–51
First-come-first-served (FCFS), 70,
        74–75
Flow-based microuidic biochips, 12
Fluidic-processor set, 65
Fluidic-task set, 65
Fluidic multiplexers (FMUX), 251
    realization of, 251–253
    synthesis, 253–254
Fluidic processor, *see*
        Biochip—resources
Fluorescence detection, 5–6, 7
FMUX, *see* Fluidic multiplexers
Forward primer, 266
FPGA, *see* Field-programmable gate
        array
Fragment operation, 307
FSM, *see* Finite state machine
Fully connected routing crossbar,
        283–286
    modeling, 287–289

G

Gamification, 97–98
GAP, *see* Generalized assignment
        problem
GEA, *see* Gene-expression analysis
Gel-electrophoresis, 246–248
    analysis, 264
    apparatus, 269
    results, 270–272
Gene-expression analysis (GEA), 123;
        see also Epigenetic analysis
    attack models for, 241–242
    benchtop protocol for, 34–36

CFG representation of, 43
digital microfluidics for, 39–43
firmware for quantitative analysis, 49–51
protocol efficiency and quantitative analysis, 36–38
quantitative protocol for, 40
shared-resource allocation, 46–49
simulation results, 51–57
spatial reconfiguration, 43–46
Generalized assignment problem (GAP), 191–192
Generic bioassay, 250
Genomic association studies, 81, 82
biological pathway of, 84–86
design of microfluidics for, 100–105
distributed-system interfacing and integration, 105–106
integrative multi-omic investigation of breast cancer, 86–90
using microfluidic and computational methods, 83
miniaturized protocols for, 85
omic data, 84–86
GFP, *see* Green fluorescent protein
Glucose-test
case study, 232–237
experimental results, 237–238
incubation time and mixing time, 232
manipulation attacks on, 230
sample preparation, 230–232
Glucose assay, 233
Graph-theoretic routing, 127–128
Greedy microfluidic encryption, 255
Green fluorescent protein (GFP), 115
Guanine (G), 263

H

Hagen-Poiseuille equation, 157
Hardware Trojan attacks, protection against, 259
Hardware/wetware/software co-design methodology, 278–279
Heuristic algorithm, 69–71
High-level synthesis, 191–193
Human-interface system, 92
Hydraulic compliance, 157–158

I

Identification of biomolecules, *see* Up-stream analysis
ILP-based synthesis method, *see* Integer linear programming-based synthesis method
Imbalance factor, 167
Immunoprecipitation (IP), 131
In-vitro assays, 257–258
In-vitro glucose test, 233
Incremental routing, 128
Integer linear programming-based synthesis method (ILP-based synthesis method), 183–184
Integrative multi-omic investigation of breast cancer, 86
BioCyBig, 88–89
CanLib, 89–90
multi-omics of breast cancer, 86–88
Intellectual property (IP), 249, 280
Internet-of-Things (IoT), 81, 84, 276
Interpretation of gel-electrohoresis results, 270–272
IoT, *see* Internet-of-Things
IP, *see* Immunoprecipitation; Intellectual property

K

Knowledge-based reasoning systems (KBS), 99

L

Lab-on-a-chip technology, *see* Microflidic(s)—technology
Latin hypercubes, 185–187
Latin hypercube sampling (LHS), 186
Least-progressing bioassay, 70
Least-progression-first (LPF), 70, 74, 75
Lensless sensing, 7
Level of imbalance, 167
LHS, *see* Latin hypercube sampling
Lippmann-Young equation, 4–5
LPF, *see* Least-progression-first

M

MaaS, *see* Microfluidics as a service
Machine-readable barcode, 262
Machine Learning Library (MLlib), 106

Man-in-the-middle attacker (MITM attacker), 262n1
Manipulation attacks on glucose-test results, 230–238
Markov chains
    DTMCs, 130–131, 133–135
    probabilistic modeling approach, 131–133
    protocol modeling using, 130
Markov model analysis, 140–141
MBs, *see* Megabases
MCs, *see* Microelectrode cells
MC scheme, *see* Multiplexed-control scheme
MC unit, *see* Microelectrode unit
MEDA-enabled aliquoting, 201
MEDA, *see* Micro-electrode-dot-array
Megabases (MBs), 87n1
Memory module, 219
Messenger RNA (mRNA), 84
Micro-electrode-dot-array (MEDA), 2, 180, 182
    biochips for biomolecular mixture preparation, 9–10
Micro-electrodes, 2
Microbiology-driven synthesis, 3
Microelectrode cells (MCs), 9
Microelectrode unit (MC unit), 188
Microflidic(s), 82; *see also* Digital microfluidics (DMF)
    design automation and optimization of microfluidic biochips, 12–14
    design for genomic association studies, 100–105
    logic gates, 249
    microfluidics-driven biomolecular analysis, 3–4
    system, 1
    system model composition, 101–102
    technology, 1, 22
Microfluidic encryption, 250–255
    FMUX synthesis, 253–254
    methodology, 250–251
    realization of fluidic multiplexer, 251–253
    secure biochemistry synthesis, 255
    selection of multiplexer locations, 254–255

Microfluidics as a service (MaaS), 81
Miniaturization of epigenetic-regulation analysis, 60–63
MITM attacker, *see* Man-in-the-middle attacker
MLlib, *see* Machine Learning Library
Molecular biology, 3
Monolithic membrane valves, 10
mRNA, *see* Messenger RNA
Multi-omics of breast cancer, 86–88
Multi-sample protocol experimentation, 21
Multiplexed-control scheme (MC scheme), 292
Multiplexer-insertion optimization, 255
Multiplexer locations selection, 254–255

N

NAs, *see* Nucleic acids
NDoS attack, *see* Negative denial-of-service attack
Negative controls, *see* Negative mixes
Negative denial-of-service attack (NDoS attack), 242
Negative mixes, 35
No-template controls, *see* Negative mixes
Non-reconfigurable modules, 46
Non-reconfigurable scheme (NON scheme), 43–44, 52–53, 55
NON scheme, *see* Non-reconfigurable scheme
NR, *see* Unrestricted resource sharing
Nucleic acids (NAs), 34, 35

O

OA-LHS, *see* Orthogonal Array-based Latin Hypercube sampling
OD, *see* Optical detectors
Omic-Logging Utility, 91, 93–94
Omics-driven biochip synthesis, 102–103
Omics technologies, 81
On-chip heaters, 8–9
Operational fault, 210
Optical detectors (OD), 52
Orthogonal Array-based Latin Hypercube sampling (OA-LHS), 186

P

Parameter-space exploration (PSE), 180, 277, 279
 of biocircuits in presence of DNA memory, 279
 drawbacks, 182
 high-level synthesis, 191–193
 on MEDA biochips, 183–185
 3-parameter biocircuit, 181
 physical-level synthesis, 193–202
 prior sample-preparation techniques, 182–183
 sampling of CF space, 185–188
 simulation results, 202–207
 synthesis methodology, 188–191
Particle detection, magnetic-bead manipulation for, 7–8
Path-length model, 146
PCA, *see* Principal component analysis
PCR, *see* Polymerase chain reaction
PCTL, *see* Probabilistic Computation Tree Logic
PDMS, *see* Polydimethylsiloxane
PDoS attack, *see* Positive denial-of-service attack
Physical-level synthesis, 193–202
Physical, non-reconfigurable resources (PN), 64
Physical, reconfigurable resources (PR), 64
Physical space, 13
Pin-constrained design, 171
 of flow-based biochips, 145–146
Pin-sharing methods, 171
Plug-and-play control of microfluidic devices, 93–94
PMPs, *see* Programmable microfluidic platforms
PN, *see* Physical, non-reconfigurable resources
Point-of-care (PoC), 1
Polydimethylsiloxane (PDMS), 10
Polymerase chain reaction (PCR), 18, 266
 assays, 257–258
 master mix, 268–269
Positive denial-of-service attack (PDoS attack), 242

PR, *see* Physical, reconfigurable resources
Principal component analysis (PCA), 95
Probabilistic Computation Tree Logic (PCTL), 131
Probabilistic modeling approach, 131–133
Programmable microfluidic platforms (PMPs), 1–2
Protein assays, 257–258
Protocol
 miniaturization, 39–40
 model, 122–123
 protocol-condition space, 131
PSE, *see* Parameter-space exploration
Public-key cryptography, 264
Public key-encrypted message material, 264

Q

QMC techniques, *see* Quasi-Monte Carlo techniques
qPCR, *see* Quantitative polymerase chain reaction
Quantitative analysis, 3, 21, 36–38, 42
 applications, 275
 assays, 277
 cyber-physical adaptation for, 15–16
 firmware for, 49–51
 frameworks, 225, 249
Quantitative polymerase chain reaction (qPCR), 2, 6, 111–112
QuantiTect Reverse Transcription, 35
Quasi-Monte Carlo techniques (QMC techniques), 186–187

R

Ramp, *see* Reliable computation with amplification variations
Read-only memory (ROM), 219
Reagents, 189
Real-time multi-processor scheduling, 276
Realization of FMUX, 251–253
Reconfigurable flow-based microfluidic biochips (RFBs), 143
 for single-cell screening, 146–148
Reconfigurable modules, 46

Regression analysis, 206–207
Reinforcing security, 249
Relay board, 216–218
Reliable computation with amplification
    variations (Ramp), 51
    variants of probabilistic models in,
        55–57
Research methodology, 18
    glucose-calibration curve, 18
    qPCR standard curve, 19
    workflow-related design-
        automation challenges,
        21–24
    workflow-related security threats,
        25–26
    workflow of biomolecular
        quantitative analysis, 20–21
Resource-allocation
    scheme evaluation, 52–55
    sequence and constraints,
        47–49
Restricted resource sharing (RR),
    43–44, 52–53, 55
ReSyn, *see* Algorithmic baseline
Reverse primer, 266
Reverse Transcription (RT), 35
RFBs, *see* Reconfigurable flow-based
    microfluidic biochips
RMSE, *see* Root-mean-square error
ROM, *see* Read-only memory
Root-mean-square error (RMSE), 203
RR, *see* Restricted resource sharing
RT, *see* Reverse Transcription

S

SaaS, *see* Software-as-a-Service
Sample labeling and differentiation, 21
Sample-processing modules, 46
Sample switching, 242
Sampling of CF space, 185–188
Scalable protocol-DNA co-modeling,
    102–103
Scheduling-policy analysis, 71–73
Secure biochemistry synthesis, 255
Security analysis, 285–259
    of DNA barcoding, 263–265
Security assessment of biomolecular
    quantitative analysis,
    17–18

Security countermeasures of
        quantitative-analysis
        frameworks
    aging reinforces DMFB security,
        256–257
    Benchtop demonstration of DNA
        barcoding, 265–273
    DNA barcoding as
        biochemical-level defense
        mechanism, 262–265
    encryption security analysis and
        simulation results, 257–261
    microfluidic encryption, 250–255
    optimization of Amplicon copy
        numbers using serial dilution,
        272–273
Security vulnerabilities of
        quantitative-analysis
        frameworks, 225
    cyber-physical integration,
        238–239
    DNA-forgery attacks on DNA
        preparation, 240–248
    manipulation attacks on
        glucose-test results, 230–238
    threats assessment of DMFBs,
        226–229
Sensing systems
    for biomolecular analysis, 5–7
    custom-built fluorescence
        sensor, 6
Sensor spoofing attacks, 280
Sequencing-driven synthesis, *see*
        Staged-driven synthesis
Serial dilution, amplicon copy numbers
        using, 272–273
Shared-resource allocation, 46
    problem formulation, 46–47
    resource-allocation sequence and
        constraints, 47–49
Short-circuited electrodes, 210
Signal-processing module, 218–219
Single-cell analysis, 111
    co-synthesis methodology, 123–125
    hybrid platform and, 113–120
    integration of heterogeneous,
        111–112
    mapping to algorithmic models,
        120–123

protocol modeling using Markov
chains, 130–135
simulation results, 135–141
valve-based synthesizer, 126–130
Single-cell screening, 3, 20–23, 143, 144,
146
delay of pressure-driven fluid
transport, 157–159
distributed model, 159–160
example, 148–149
experimental results, 168–176
multiplexed control, 150–156
pin-constrained design of
flow-based biochips, 145–146
RFB for, 146–148
Sortex, 160–167
Single-nucleotide polymorphism (SNP),
84
Single-stranded DNA (ssDNA), 263
SM, *see* Static-variation model
SNP, *see* Single-nucleotide
polymorphism
Software-as-a-Service (SaaS), 94
Software-programmable microchannel
structure, 1–2
Sortex, 145, 160–167
Source
of imbalance, 167
mixtures, 189
Spatial reconfiguration, 43
degradation-aware resource
sharing, 44–46
resource-sharing schemes, 43–44
ssDNA, *see* Single-stranded DNA
STA, *see* Static-timing analysis
Staged-driven synthesis, 103–104
Standford Seaborn, 91
Static-timing analysis (STA), 157
Static-variation model (SM), 51,
55–56
Stochastic systems, 130–131
Stratified sampling, 185–187
Supernode, 42, 122
Synthetic biocircuit, 180
Synthetic biology, 278–279

T

Target mixtures, 188
t-distributed stochastic neighbor
embedding (t-SNE), 95
TDM, *see* Time-division multiplexing
TEM, *see* Transmission-electron
microscope
Thickness of dielectric layer, 256–257
Threshold voltage, 256
Thymine (T), 263
Time-division multiplexing (TDM), 150
TQ library, *see* Tracing and queuing
library
Tracing and queuing library (TQ
library), 49–50
Traditional DNA-analysis flow, 266
Transcription factors, 90
Transmission-electron microscope
(TEM), 60–61
Transposer, 1
Trojan, 228, 229
Two multiplexer-insertion algorithms,
257–258
Two-phase system, 12
260/280 ratio, 36
Type-driven cell analysis, 111
Type-driven single-cell protocol,
118–119

U

Unit-volume droplets, 253
Unmanned aerial vehicle (UAV), 280
Unrestricted resource sharing (NR), 44,
52–53, 55
Up-stream analysis, *see* Single-cell
screening

V

Valve-based crossbar modeling, 120–121
Valve-based routing schemes, 137–138
Valve-based synthesizer, 126–130
Virtual, reconfigurable resources (VR),
64
Visual analytics, 98–99

For Product Safety Concerns and Information please contact our EU
representative GPSR@taylorandfrancis.com
Taylor & Francis Verlag GmbH, Kaufingerstraße 24, 80331 München, Germany

www.ingramcontent.com/pod-product-compliance
Lightning Source LLC
Chambersburg PA
CBHW060802220326
41598CB00022B/2513

9 7 8 0 3 6 7 5 1 2 9 1 0